Environmental Chemical Analysis

Environmental Chemical Analysis

B.B. KEBBEKUS and S. MITRA

Department of Chemical Engineering, Chemistry and Environmental Science
New Jersey Institute of Technology
Newark, USA

BLACKIE ACADEMIC & PROFESSIONAL
An Imprint of Chapman & Hall

London · Weinheim · New York · Tokyo · Melbourne · Madras

**Published by Blackie Academic & Professional, an imprint of
Thomson Science, 2–6 Boundary Row, London SE1 8HN, UK**

Thomson Science, 2–6 Boundary Row, London SE1 8HN, UK

Thomson Science, 115 Fifth Avenue, New York, NY 10003, USA

Thomson Science, Suite 750, 400 Market Street, Philadelphia, PA 19106, USA

Thomson Science, Pappalallee 3, 69469 Weinheim, Germany

First edition 1998

© 1998 Thomson Science

Thomson Science is a division of International Thomson Publishing I(T)P*

Typeset in 10/12 Times by Blackpool Typesetting Services Limited, UK
Printed in Great Britain by St Edmundsbury Press Ltd, Bury St Edmunds, Suffolk

ISBN 0 7514 0456 X

A catalogue record for this book is available from the British Library

Library of Congress Catalog Card Number: 97-74427

To our children
and the environment they will inherit

Contents

3 Spectroscopic methods 54

4 Chromatographic methods 103

7   Methods for air analysis   209

8   Methods for water analysis   247

9 Methods for analysis of solid samples 279

10 Quality assurance and quality control 305

Preface

As our planet becomes more crowded and industrialization becomes more widespread, the deleterious effects on the environment have become obvious. This has led to significant efforts to study the effects of pollutants, and to regulate and reduce their discharge. In addition, methods are being developed and applied to clean up polluted areas and water resources. Understanding the sources and fates of environmental contaminants, regulating their discharge, or remediating polluted sites all require detection of the presence of the materials of interest and measurement of their concentrations. This is the task of the environmental analyst.

The environment is a massive sink where even large quantities of pollutants are diluted to very low concentrations. Even at these levels, pollutants can be detrimental to human health and the biosphere. Environmental measurements often involve 'finding the needle in the haystack'. In addition, the environment is made up of a huge number of different chemical species, so the matrix in which the components of interest must be measured is complex. The combination of low concentrations and complex matrices makes environmental measurements a challenging task.

Identification and quantitation of pollutants is necessary before decisions can be made or action taken. To study the depletion of the ozone layer in the stratosphere, one must measure the concentrations of chlorofluorocarbons in the upper atmosphere; to understand pesticide contamination, pesticide residues must be measured in food, soil, and water; to regulate industrial discharges, contaminant levels in effluents are monitored at regular intervals.

Environmental professionals will usually have to deal with a substantial amount of data generated in the analytical chemistry laboratory. While the persons developing impact statements or inspecting industrial facilities for compliance with environmental regulations may not have to perform the analyses themselves, they will have to interact with analytical chemists. They must have a grasp of the techniques which are most suited to a particular analytical problem, and must also have an understanding of the types of interference which can be expected under various conditions. The basics of sampling and quality assurance should also be part of their intellectual tools.

Modern, sophisticated analytical instruments have made it possible to identify and measure components in the environment at trace levels. Instruments such as chromatographs, mass spectrometers, and atomic spectrometers are used to measure organic and inorganic components present in concentrations as low as 1 part per billion (ppb). It is important to have an understanding of the principles behind these instruments in order to use them, or even to use the data produced, intelligently. A familiarity with other aspects of environmental measurements, sampling and sample preparation is also important.

This text was developed from our experience in teaching a class in environmental analysis which served both beginning graduate students majoring in environmental science and more advanced undergraduates in chemistry, chemical engineering, and environmental engineering. The text is intended for students interested in any aspect of environmental science and engineering. Environmental analysis is a field which has only recently been developed as a discipline in academe, and there are, no doubt, many opinions about the topics which are essential to it. We welcome comments from readers on areas which they feel should be covered in more detail in this text.

Based on our experience at NJIT, we expect that the academic backgrounds of a typical class will be quite heterogeneous, including students from chemistry, chemical engineering, environmental science and engineering, biological sciences, and others. Consequently, this text does not assume a background in analytical chemistry. The first chapters cover the important topics of sampling, sample preparation, basic statistics, and the operating principles and descriptions of major techniques such as spectroscopy, chromatography, and mass spectrometry. From the spectrum of available analytical instrumentation, only those widely used in environmental measurements have been presented. In the later chapters, the major environmental matrices, air, water, soils, and solids are explored in more detail. Techniques which are peculiar to specific determinations are found here. Within these chapters several 'example determinations' are presented. These follow particular determinations from sampling, through sample preparation and analysis. The intention is to integrate the information from several chapters into a practical example for the student. The information on sampling, sample preparation and the application of instrumental techniques is brought together here in the description of actual environmental measurements.

Students majoring in the environmental sciences need a grounding in measurement techniques. They need to study those techniques normally used in the environmental field. Important topics such as quality assurance plans, statistical data handling, and sampling, not always covered in detail in many chemistry courses, are given thorough coverage here.

The text is intended for a one semester course. We feel that it can take the place of an instrumental analysis course for those with an interest in the environment. We have attempted to make this text sufficiently complete, so that the reader will come away with enough information to judge the quality and interpret the information produced in the analytical laboratory.

Barbara B. Kebbekus
Somenath Mitra
Newark, NJ
February 1997

1 Introduction to Environmental Measurements

The entire world is the environmental scientist's laboratory. Environmental science takes for its own the study of the composition of the atmosphere, water, soil, and how materials are taken up and given off by plants and animals. It examines the transport of various chemical species, elements, and compounds, among the atmosphere, the hydrosphere, the geosphere, and the biosphere. Of great importance in the development of this branch of science has been the realization that human activities are having an increasing effect, frequently not for the better, on the natural environment.

Deterioration of the environment reached significant levels with the dramatic increase in human population, accompanied by industrialization, which took place over the past century or so. Major problems brought about by these changes are the pollution of air, water, and soil; growth in the amount of hazardous waste; depletion of arable land, energy, and other natural resources; increased exposure to toxic chemicals in food, water, and air; and exposure to radiation. It was not until the 1960s that an awareness of these problems grew all over the world. Since then, a plethora of environmental regulations have been promulgated to protect our environment.

1.1 Role of measurement in environmental studies

All environmental studies ultimately depend on the results of chemical analysis of samples of air, water, soil, and biological organisms. Policies on reduction of pollution cannot be designed if the extent and identity of the pollutants is not known. Emissions cannot be regulated if there is no accurate and reliable method of monitoring the emitted materials. Natural cycles cannot be understood if chemical compounds cannot be followed as they are passed from air to water to soil to the biosphere. The analytical chemist who specializes in the analysis of environmental samples is choosing to work on some of the most challenging problems in analysis, because the analytes are often present in extremely low quantities, and the matrix can be very complex. To add further to the problem, the material to be sampled is usually very non-homogeneous. The sampling of an area such as a landfill, or a constantly changing system such as a river or the ambient atmosphere, requires much thought about what is to be determined and where and when samples are to be taken.

The environmental scientist or engineer may not be the person who actually does the laboratory analysis, but he or she may well be the person who specifies what samples are taken and where, and which analyses should be performed. In addition, the decisions about how the laboratory data are to be interpreted and what conclusions can be legitimately drawn are also functions of the environmental specialist.

1.1.1 *Units of measurement*

When environmental samples are analyzed, the units in which the results are expressed vary with the type of sample. Concentrations are expressed in SI units, such as ml/m^3 or $\mu g/kg$. For concentrations of materials in air, data are sometimes expressed as parts-per-million (ppm) or parts-per-billion (ppb). For gases, ppm can be defined as moles of analyte per million (10^6) moles of air, or volume of analyte per million volumes of air. Moles of analyte per 10^9 moles of sample defines ppb. The equivalent SI units are ml/m^3 for ppm and $\mu l/m^3$ for ppb. Since at any given temperature and pressure, the number of moles of gas occupying a certain volume is the same for all gases, these two definitions are equivalent. Only in cases where the ideal gas law is not valid, that is at cryogenic temperatures and high pressures, would this not be a reasonable assumption. Since most environmental work involves moderate temperatures and pressures, the assumption is appropriate. For particulate matter in air, the results are usually expressed as milligrams or micrograms of analyte per cubic meter of air.

In water analysis, ppm usually implies grams of analyte per million grams of water. For water, this is equivalent to milligrams of analyte per liter of water. Remember that this equivalence between ppm and mg/liter is **valid only for water**, because it is derived from the fact that 1 liter of water weighs 1000 g. Table 1.1 lists some of the more common units of measurement.

In general, it is preferable to use the SI units when reporting concentrations. When ppm or ppb are used, ambiguity about the basis of the notation can be eliminated by using the specific notation ppm_w or ppb_v to indicate that the concentration is given in weight/weight or volume/volume terms. However, $\mu g/l$ or $\mu g/g$ are less prone to misinterpretation, and are the international standard.

Very low concentrations are given as $\mu g/kg$ or parts per trillion, and high concentrations may reach the percent range. The magnitude of the units should be chosen so that they are suitable to the amount being reported, although there

Table 1.1 Commonly used units of measurement

Unit	Meaning	Usual use
$\mu g/g$	Micrograms of analyte per gram of sample	Concentrations in solid samples
$\mu g/m^3$	Micrograms of analyte per 1000 l of sample	Concentrations of particulate matter in air
$\mu l/l$	Microliter of analyte per liter of sample	Concentration of liquid analyte in liquid sample
ppm_v	Volumes (or moles) of analyte per 10^6 Volumes (or moles) of sample	Concentration of gaseous analyte in air or other gases
ppm_w	Grams of analyte per 10^6 g of sample	Solid or liquid analyte in solid samples
ppb_w	Grams of analyte per 10^9 g of sample	Solid or liquid analyte in solid samples

may be a reason to report 7000 µg/kg instead of 7 µg/g. If one is comparing, for example, a series of samples which are all in the µg/kg range except for a few in the low µg/g range, it would be better to keep the same units for all. It is simpler to compare data if they are all reported in the same units. On the other hand, one should avoid trying to make data more eye-catching by reporting them in inappropriate units. One person's "enormous concentration of 5000 nl/m^3 of benzene in air" may be another's "hardly measurable trace at 0.005 ml/m^3" (or even "a half of a millionth of a percent"!). None of these is incorrect, but a good analyst would report the 5 µl/m^3 concentration, in the most appropriate units, and let the number speak for itself.

1.1.1.1 *Conversions between units.* Calculations usually have to be performed on measurement data to bring these to a useable form. If the units in which the quantities are expressed are handled carefully, the calculations are not difficult. In a simple case, the instrument signal is converted to a quantity of analyte in some quantity of sample, which is then converted to ppm in the original sample taken in the field. Dilutions of the sample must be taken into account in these calculations. The most systematic and error free way to go about unit conversion problems is by the factor label method. The basis of this method of setting up calculations is that multiplication of any quantity by unity does not change its value. If a number of grams is multiplied by the quantity 1000 mg/1 g, the original value does not change, since 1000 mg = 1 g and therefore the conversion factor is equal to unity. Only the units in which it was expressed have changed. This is always true if the conversion factor is made up of a numerator and denominator which are equivalent to each other. If the calculation is written down with all the factors carefully labeled, and if the factors cancel, leaving the answer in the correct units, there is little chance of error. One must be careful that the conversion factors are truly equivalent to unity, and that the labels are sufficiently detailed so that you do not cancel "grams of sample" with "grams of benzene" simply because both were labeled "g". Some useful conversion factors are found in Table 1.2.

A few examples will review the method.

Table 1.2 Some useful conversion factors

1 kilometer (km)	10^3 meters (m)
1 m	10^3 millimeter (mm)
1 m	10^6 micrometer (µm)
1 m	10^9 nanometer (nm)
1 m	100 centimeter (cm)
1 inch (in)	2.54 cm
1 mile (mi)	5280 ft
1 liter (l)	1.057 quart (qt)
1 kilogram (kg)	0.4536 pound (lb)
1 cubic meter	35.336 cubic feet

Example

Toluene is trapped from water in a cartridge, washed out and analyzed by liquid chromatography. A 500 ml portion of sample was used, and 10 ml of reagent used to wash out the trapped toluene. From the analysis, the 10 ml reagent solution was found to contain 5 µg toluene/ml. What was the concentration of toluene in the original sample?

$$\frac{5\ \mu\text{g toluene}}{1\ \text{ml reagent}} \times \frac{10\ \text{ml reagent}}{} = 50\ \mu\text{g toluene in the sample}$$

$$\frac{50\ \mu\text{g toluene}}{500\ \text{ml water}} \times \frac{1\ \text{ml water}}{1\ \text{g water}} = 0.1\ \mu\text{g/g in the original sample}$$

Example

Particulate matter is filtered from air. After the sample is collected for 8 h at a flow of 10 cfm (cubic feet per minute), the filter is extracted into 100 ml of solution. The solution is found to contain 23.3 ng/ml of lead. What is the concentration in the air in ng/m^3?

First calculate the number of cubic meters of air collected:

$$8\ \text{hours} \times \frac{60\ \text{min}}{1\ \text{h}} \times \frac{10\ \text{ft}^3}{\text{min}} \times \frac{1\ \text{m}^3}{35.3\ \text{ft}^3} = 135\ \text{m}^3$$

Then calculate the amount of lead in the sample:

$$\frac{23.3\ \text{ng Pb}}{\text{ml solution}} \times \frac{1\ \text{mg}}{10^6\ \text{ng}} \times 100\ \text{ml solution} = 2.33 \times 10^{-3}\ \text{mg Pb}$$

Finally, calculate the concentration in the original air sample:

$$\frac{2.33 \times 10^{-3}\ \text{mg Pb}}{135\ \text{m}^3} \times \frac{10^6\ \text{ng}}{1\ \text{mg}} = 17.3\ \frac{\text{ng}}{\text{m}^3}$$

1.1.2 Significant figures

The way a measurement is reported says something about the accuracy with which the measurement was made. If the mass of an object is reported as "1 g" it is supposed that the measurement was made roughly. However, if the reported mass is given as 1.0345 g, one immediately supposes that the object was weighed carefully on an analytical balance. Therefore, if values are calculated from measured quantities, it is important to keep the proper number of significant figures, so that the reader is not misled into thinking that a measurement was done more carefully than it was. For instance, one might prepare a solution by weighing out 1.0434 g of sodium chloride into a 100 ml volumetric flask and diluting it to the mark. This gives a solution containing 1.0434 g/100 ml, or 0.010434 g/ml or 10.434 mg/ml. Each of the measurements, the volume and the mass, was made to the same number of significant figures, five in both cases, so the final calculated concentration can be legitimately reported to five significant figures. If one-quarter of the sample is withdrawn using a calibrated pipette of similar accuracy, and this portion is diluted to 100 ml, the resulting solution is one-quarter the concentration of the original. A calculator will report that 10.434/4.0000 is 0.2608526. These are obviously more figures than are actually significant, and the number should be rounded to 0.26086 mg/ml, keeping the five significant figures of the original data.

The rules are:

- Zeros at the end of a number are usually significant. (Numbers such as 1000 are ambiguous and should be avoided. If four significant figures are intended, use 1.000×10^3. If only two are intended, use 1.0×10^3.)
- Zeros before a number are not significant. 0.00034 has only two significant figures.
- When numbers are multiplied or divided, the one with the fewest significant figures will govern the final answer. $1.234 \times 0.0032 = 0.0039$.
- When numbers are added or subtracted, the one with the smallest absolute accuracy will govern the final answer. Adding $123.44 + 2.9177 + 3243.1$ gives 3369.4577. This is rounded to one decimal place, 3369.5, as the third number had only one decimal place.
- Counted numbers, as opposed to measured numbers are assumed to have as many significant figures as needed in any calculation. This includes such quantities as 60 min in an hour, or 1000 mm in a meter.

1.2 Pollutants: sources and measurements

The atmosphere contains just a few major constituents: nitrogen (80%), oxygen (20%), argon (0.9%), carbon dioxide (290–500 $\mu l/l$), and water vapor, which varies widely. However, if we look at materials found in the low $\mu l/l$ and $\mu l/m^3$ concentration ranges, there are hundreds of compounds present in air. These are classified in several ways: gases and particulates, organics and inorganics, toxic and nontoxic materials. Water samples, likewise, have a wide variety of compounds present, both in dissolved and suspended forms. These vary even more widely than air samples, as natural waters can contain many substances at quite high concentrations. Think of spring water, sea water, and sulfur hot springs waters, all of which are natural, yet have very different compositions. In addition, polluted waters may have a vast number of anthropogenic materials in them.

Soils and biological materials are even more complex, since their basic matrix is not simple, as air and water are. Normal plant and animal tissues or even soil contain a large array of materials which are essential to their normal functioning. Added to this complex matrix may be compounds accumulated from the environment. These may be present at levels which are obviously harmful or they may be low concentrations of materials which have been shown to be harmful at much higher levels.

It is at this point where politics and science must fight their major battles. It is easy to say that any amount of a harmful compound in, for instance, human tissues is undesirable and that regulations should be written to prevent this contamination. This is overly simplistic, however, since modern analytical methods are often so sensitive that some compounds may be detected at concentrations below 1 µg/kg. Unfortunately, it is at this very point that it is most difficult to make real scientific judgments about the degree of harm any particular exposure will do. When an exposure causes a verifiable problem for one in a hundred thousand people, it will be exceedingly difficult to prove.

Results of animal testing are often extrapolated to determine which substances are harmful and to decide what exposure levels are acceptable. This is inherently risky, because it has been found that, even between similar species, the susceptibility to different compounds is widely different. Therefore, extrapolations from mice to humans cannot be taken as having a strong scientific grounding. However, it is also not morally acceptable to experiment on controlled groups of humans, so regulations do not always have the kind of scientific grounding we would like to see. Mostly, regulations are based on some combination of scientific evidence of harm in animal experiments, political interests, epidemiological data, guess work, and informed judgment.

1.2.1 *Classes of environmental contaminants*

The term **contaminant** refers to a material which is present in a part of the environment where it is not expected. When the material begins to cause a problem, having a negative effect on health of humans or other organisms in the area, or causing widespread changes in the local environment, it is termed a **pollutant**.

1.2.1.1 *Products of combustion.*   Combustion processes are one of the major sources of air contamination. There is a continual interchange of material between all the components of the environment, so that a substance in the air will eventually also contaminate both water and soil. Particulates, visible as smoke, chiefly contain carbon, but may also carry substantial amounts of organic compounds such as polynuclear aromatics, as well as metal salts. Unburned and partially burned fuels are emitted into the air as vapors, containing hydrocarbons and other volatile organic compounds (VOC). When combustion occurs at high temperatures and pressures, as in an internal combustion engine, the nitrogen from the fuel and the air also react, forming various oxides of nitrogen, NO_x, which have detrimental effects on the atmosphere, as they are involved in atmospheric smog-forming reactions. Partially oxidized carbon from fuel burned under oxygen-poor conditions will form toxic carbon monoxide.

Sulfur-containing fuels form sulfur oxides, (SO_x). Indirect pollutants from products of combustion are ozone and compounds formed under sunlight from ozone, NO_x, and hydrocarbons. Acids form in air from reactions between NO_x, SO_x, and water vapor. Nitric, nitrous, sulfuric, and sulfurous acid are present in cloud and fog water droplets, and wash down as acid rain.

1.2.1.2 *Industrial emissions.* Most industrial processes emit some materials to the environment. These may take the form of gaseous emissions through a stack, aqueous wastes dumped into a river or sewer line, or solid materials, which are often landfilled. Much regulation is directed at these types of emissions and major improvements in environmental quality have resulted from such regulation over the past few decades. Industries such as mining and ore refining are responsible for a range of metal ions, in air, water, or soil, as well as acidic mine drainage water. Petroleum refining may emit hydrocarbon gases and liquids, while the burning of petroleum products produced is the source of carbon monoxide, sulfur oxides, nitrogen oxides and a wide variety of organic compounds. The power generation industry may emit sulfur oxide gases or sulfate particulate from coal burning plants, radioactive elements from nuclear plants, and carbon dioxide, a greenhouse gas, from any fossil fueled plant. The chemical manufacturing industry may emit many different compounds, both organic and inorganic, depending on the particular products being made. In addition, there are many examples of compounds which were very well designed for a particular use, but ended by being a major environmental problem, because these substances could not be readily destroyed. Chlorofluorocarbon synthetic refrigerants became an environmental threat when they began to destroy the earth's ozone shield, while polychlorinated biphenyls, very stable insulating liquids for electrical transformers, were found to bioaccumulate in the aquatic biota, and cause serious health effects.

Agriculture also contributes an assortment of chemical contaminants to the environment. Many tons of insecticides, herbicides, fungicides, as well as fertilizers are applied to crops and soil each year. These compounds have widely varying lifetimes in the environment and determining these, both in their original form and their breakdown products or metabolites, is an analytical challenge.

1.2.1.3 *Other sources of environmental contamination.* Another major contributor to environmental degradation is domestic activity, especially in areas of concentrated population. The solid waste collected from residences, as well as domestic sewage and air pollution caused by use of internal combustion engines for transportation, each have a major environmental impact. There are changes which are being carried out to mitigate this impact. Recycling bulky materials such as paper reduces the volume of waste sent to landfills. Removing household batteries from the garbage stream allows incineration with lowered emissions of toxic metals to the air. There have been substantial reductions in air emissions from automobiles because of improvements in fuel efficiency and mandated pollution controls.

One should also recognize that there are natural sources of some of the compounds we look upon as contaminants. Nitrogen oxides are generated in the air by lightning, as well as by automobiles. Sulfur oxides and hydrogen sulfide, as well as particulate matter, are emitted by volcanoes as well as by power plants. Radioactive radon is emitted from certain rock formations, and the radioactive elements we are exposed to arise much more from the natural environment than from anthropogenic sources. The sources of some contaminants are listed in Table 1.3.

Table 1.3 Sources of some environmental pollutants

Contaminant	Sources
Gases	
Radon	Soil, volcanic action
Ozone	Lightning, photochemical reactions, oxygen
NH_3	Microbial biodegradation, fertilizers, tobacco smoke
NO_x	Auto exhaust, fossil fuel combustion, lightning
SO_2	Coal, fuel oil combustion, volcanic action
H_2S	Petroleum and ore refining, anaerobic biodegradation, pulp and paper manufacturing
CO	Incomplete fuel combustion, auto exhaust
CH_4	Natural gas, microbial biodegradation
Metals and metal salts	
Pb	Paints, leaded gasoline, batteries
Cd	Batteries, mining and metal refining
Cr	Electroplating, printing
Hg	Batteries, laboratory equipment, biocides
Other metals	Mining, metal refining and processing, paint, steel manufacturing, coal combustion
Anions	
Sulfate	Power plants, coal burning, industrial uses of sulfuric acid
Phosphates	Fertilizers, detergents
Nitrates	Fertilizers, explosives
Organic compounds	
Alkane, alkene, cycloalkane hydrocarbons	Petroleum refining and use, automotive exhaust, solvents from manufacturing processes, tobacco smoke, pharmaceutical production, natural gas
Polycyclic aromatic hydrocarbons (PAH)	Combustion of fuels, waste combustion, tobacco smoke, wood stoves, auto exhaust, forest fires
Aromatic compounds	Solvent production and use, auto exhaust, pharmaceutical production
Dioxins	Incineration, pulp and paper manufacturing, other manufacturing
Halogenated organics	Refrigerants, chlorinated solvents
Mercaptans (thiols)	Pulp and paper manufacture
Insecticides, herbicides, fungicides	Agriculture, landscape maintenance, wood preservation
Polychlorinated biphenyls (PCB)	Electrical equipment coolant, heat transfer fluid, plasticizer, de-inking agent

1.2.2 *Regulating the environment*

When the degradation of the environment finally became a topic of concern for people all over the world, governments responded by setting up agencies and issuing regulations. For example, in 1969 the National Environmental Policy Act (NEPA) was enacted in the United States. This act established the council on Environmental Quality to develop environmental policy and advise the President. The following year, the Environmental Protection Agency (EPA) was established. EPA's mandate was to manage environmental protection laws issued by Congress. EPA advised Congress in the establishment of the Federal Water Pollution Control Act and the Clean Air Act. The agency performs research on many environmental and health related issues and develops and enforces standards for pollutant emissions. Individual states have also drafted legislation to address their internal environmental problems. Other Federal agencies which directly or indirectly participate in study and regulation of the environment in the USA are:

- Occupational Safety and Health Administration (OSHA): sets and enforces safety and health related issues pertaining to the work place.
- National Institute for Occupational Safety and Health (NIOSH): identifies substances which pose potential health problems; recommends exposure levels to OSHA.
- Food and Drug Administration (FDA): protects consumers from harmful food, drugs and cosmetics.
- National Institute of Health (NIH): studies a variety of health related issues, including environmentally related diseases such as cancer.
- National Oceanic and Atmospheric Administration (NOAA): conducts research and monitoring of oceans and atmospheres.

Government agencies similar to these exist in many countries. Environmental issues are quite complex and the jurisdiction of agencies can overlap. Often, different aspects of the same environmental problem are addressed by different agencies.

Measurement techniques play an important role. Measurements are necessary to assess the level of contamination as well as to ensure compliance with regulations. A particular substance cannot be easily regulated if a suitable measurement method is not available to determine its concentration. Consequently, regulating agencies expend a significant amount of effort in establishing measurement techniques for different determinations. They also provide guidance on the acceptability of newly developed measurement techniques. Among the agencies mentioned above, EPA and OSHA, in particular, publish standard analytical methods for measurement of pollutants. These cover wide ranges of substances which may be present in air, water, soil, and hazardous waste. As additional regulations are promulgated and the list of regulated pollutants increases, additional methods will be developed to address these regulatory needs.

1.3 Design of environmental studies

It is difficult not to be somewhat vague about what goes into a good study design. This is because each study has different purposes, different financial constraints, different problems to be studied, and is staffed by people with different amounts of expertise and different preferences. At least, we can point out the final results desired – that the study yield data which is useful in the solution or delineation of a certain problem – and list some of the topics which should be examined and dealt with before the study starts.

The most important step in the design of an environmental study is the determination of the **purpose of the study**. It is much easier to design a study with the purpose firmly in mind. Once the targets of the analyses to be done are decided, then the criteria for site selection should be laid out. The **available methods** for doing the analysis have to be reviewed with an eye to the accuracy required. If differences among sites are to be examined, the methods used must have an accuracy which exceeds the differences expected. Methods also have to be examined for the amount of **labor and laboratory** time required to perform them, and the time and money available for the study. It is not advisable to design the study so tightly that any equipment breakdown or absence of a key worker will cause irreparable damage to the overall structure. Murphy's law, often stated as "If something can go wrong, it probably will" seems to be operative in all fields of endeavor. Most studies are ultimately limited by the **amount of financing**. One can always take more samples, subject them to more tests, have more redundant equipment in case of malfunctions, run more calibrations and blanks, etc. The art of planning is to design the best study which can be done with the resources available, and which can obtain the maximum amount of information.

What must not be neglected in project design, however, is **sufficient statistical information** so that the confidence levels of the data can be accurately determined. Data are useless if their inherent level of error is not known. Environmental studies are especially prone to such problems, since they are often measuring concentrations near the limits of detection of the methods, where the uncertainties are the highest. In addition, the error due to area inhomogeneity and interference from the complexity of the samples may not be easy to determine. Statistical treatment of data is essential before conclusions about differences between samples and background levels can be made. Systems must be observed carefully and experiments designed to distinguish between variance caused by real differences between samples or sample portions, and that caused by variance in the analytical methods.

1.3.1 *Sampling and analysis*

An environmental analysis is comprised of several steps. **Sampling** is the process by which we collect enough of a material to adequately represent the system which we wish to characterize. The sample should represent both the overall concentration of whatever substance is being measured, and the distribution of the substance in the system. Samples must be properly preserved, so that their compositions do not change before analysis.

After the samples are taken, they must be **prepared for analysis**. Solids or liquids may be dissolved in a suitable solvent. Sample preparation may require removing the analyte from the matrix in which it is held. Extraction or column clean-up can be used for this purpose. Volatile materials in gaseous samples may be chilled to condense them out of the gas phase, or the analytes may be adsorbed on a suitable solid trapping material. Sometimes the samples contain substances which may seriously interfere with the proposed analysis. These must be removed or changed in some way to remove the interference. Also, a proper blank must be specified for the sample, to insure that the results are due to the substance sought and not to some artifact.

Finally, records must be kept so that the analyst can readily prove that significant errors were not introduced, and that proper steps were taken to make sure the analysis was accurate. **Quality assurance and control** is a systematic handling of the sampling, analysis, and record keeping processes to ensure that each step, from sampling to final reporting, was under proper control. If this is done, error levels will be known and one will be able to show that unusual errors were not introduced. The handling of each sample can be reconstructed and examined if necessary. Quality control involves such processes as analyzing control samples which are replicas of typical samples containing a known amount of the analyte, as well as the selection and analysis of blanks. One of the major reasons for taking and analyzing environmental samples is to check for compliance with regulations. Therefore, records are not only for scientific use, but also may become of interest to regulatory agencies and sometimes to the courts of law.

1.4 Basic statistical data handling

The objective of environmental measurements can be qualitative or quantitative. For example, the presence of lead in household paint is a topic of concern. The question may be "Are there toxic metals present in the paint in a certain home?" An analysis designed to address this question is a **qualitative analysis**, where the analyst screens for the presence of a certain pollutant or class of substances. The next obvious question, of course, is "How much lead is in the paint?" This type of analysis is called **quantitative analysis** and it not only addresses the question of the presence of lead in the paint, but also its concentration.

Let us say that the analyst uses a technique that can measure as low as 1 µg/kg of lead in paint. For a particular sample he says that no lead was detected. This means that the concentration of lead in that sample is less than 1 µg/kg. It is not proper to report the results of this analysis as "zero lead" since it may well be present at some lower concentration, which our measuring device cannot detect. The report should read "lead not detected above 1 µg/kg" or "lead not detected; detection limit 1 µg/kg". Quantitative analyses of environmental samples are particularly challenging because the concentrations of pollutants in environmental samples are often very low, and the matrix is usually complex.

1.4.1 Errors in quantitative analysis

All numerical results obtained from experimentation are accompanied by a certain amount of error and an estimate of the magnitude of this error is necessary to validate the results. **Error** is a statistical term, and should not be confused with the common meaning of the word, which implies a blunder and blame to be laid on someone! **All** measurements have some error associated with them. One cannot eliminate it, but with careful work, the magnitude of the error can be characterized and, sometimes, the magnitude can be reduced with improved techniques.

In general, the types of error can be classified as **random** or **systematic**. If the same experiment is repeated several times, the individual measurements will fall around an average value. The differences are due to unknown or uncontrollable factors, and are termed **random error**. Random errors, if they are truly random, should cluster around the true value, and have equal probability of being above or below it. Chances are that the measurements will be slightly high or slightly low more often than they will be very high or very low. The measure of the amount of random error present in a set of data is the **precision** or **reproducibility**.

On the other hand, **systematic error** tends to deviate or **bias** all the measurements in one direction. So **accuracy**, which is a measure of deviation from the true value, is affected by systematic error. **Accuracy** is defined as the deviation of the mean from the true value.

$$\text{Accuracy} = (\text{mean} - \text{true value})/\text{true value} \qquad (1.1)$$

Often the true value is not known. For the purpose of comparison, measurement by an established method or by an accepted institution is sometimes accepted as the true value. In colloquial English, accuracy and precision are often used interchangeably, but in statistics they mean quite different things.

The difference between accuracy and precision is further illustrated by the following example. A sample containing 10 µg/kg of lead is analyzed by acid digestion and atomic absorption spectrometry. Five repeat analyses are performed by four different laboratories, resulting in the data presented in Table 1.4 and Figure 1.1. The analyses performed by laboratory A show less variability but are not close to the true value. So, this laboratory's work has high precision but low accuracy. This is a strong sign that this laboratory is subject to some identifiable source of bias. Its calibrations, standards, and methods should be examined carefully. Laboratory D's data on the other hand show little variability

Table 1.4 Example data for an interlaboratory comparison of a 10 µg/kg lead sample

Lab	Analysis 1	Analysis 2	Analysis 3	Analysis 4	Analysis 5	Average	s
A	8.99	8.67	9.03	8.92	8.77	8.87	0.152
B	9.80	7.99	8.01	7.55	10.07	8.68	1.16
C	8.43	9.88	11.21	10.77	9.66	9.99	1.08
D	9.87	9.79	10.04	10.12	9.99	9.96	0.132

and good accuracy. Laboratory B and C both have data with poorer precision, but one is accurate while the other is not. Actually, for a small number of replicates, when good accuracy is obtained with poor precision, it is usually accidental!

1.4.2 *Statistics of repeated measurements: precision*

1.4.2.1 *Precision and standard deviation.*   Since every measurement contains a certain amount of error, the result from a single measurement cannot be accepted alone as a true value. An estimate of the error is necessary to predict within what range the true value may lie. The random error in a measurement is estimated by repeating the measurement several times, which provides two valuable pieces of information: the **average value** and the **variability of the measurement**. The most widely used measure of average value is the arithmetic mean, $\bar{x}$

$$\bar{x} = \frac{\sum x_i}{n} \tag{1.2}$$

where $\sum x_i$ is the sum of the replicate measurements and n is the total number of measurements.

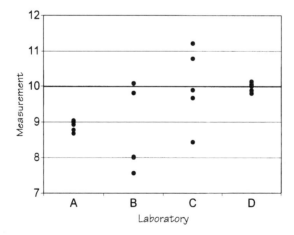

Figure 1.1 Sets of replicates on a performance evaluation sample done by four laboratories. The accepted value for the sample was 10. Labs A and D show good precision, while C and D have good accuracy.

The most useful measure of variability (or precision) is the standard deviation, σ. This is calculated as:

$$\sigma = \sqrt{\frac{\Sigma(x_i - \bar{x})^2}{N}} \tag{1.3}$$

When only a few pieces of data are available, the calculated standard deviation may be underestimated, since the mean is used as the true value, and the mean was calculated from the same small data set. To obtain an unbiased estimate of σ, which is designated s, $N - 1$ is used in the denominator.

$$s = \sqrt{\frac{\Sigma(x_i - \bar{x})^2}{N - 1}} \tag{1.4}$$

As the number of data points becomes larger, the value of s approaches that of σ. When N becomes as large as 20, the equation for σ may be used. Another term commonly used to measure the variability is called the **coefficient of variation** (CV) or the **relative standard deviation** (RSD). It may also be expressed as a percentage

$$\text{RSD} = (s/\bar{x}) \quad \text{or} \quad \%\text{RSD} = (s/\bar{x}) \times 100 \tag{1.5}$$

Relative standard deviation is the parameter of choice for comparing the precision of data of different units and magnitudes and is used extensively in analytical sciences.

1.4.3 Distribution of error

In the absence of systematic error and if a large number of measurements are made, the results fall into a normal or Gaussian distribution. Normal distribution is defined by the equation

$$y = \frac{1}{\sigma\sqrt{2\pi}} e^{-(X-\mu)^2/2\sigma^2} \tag{1.6}$$

where y is the frequency of occurrence and μ is the arithmetic mean. The normal distribution has a bell-like shape. In a set of data which is normally distributed, 68% of the measurements will lie within $\pm 1\sigma$ of the mean, 95% within $\pm 1.96\sigma$, and 99.7% within $\pm 2.97\sigma$. It is worth pointing out that it is not always possible to prove that repeated measurements will fall into a normal distribution, but in general it provides a good approximation. That is especially true when the number of measurements are not very large.

In environmental measurements, it is often the case that the analytical method is being used near its limit of detection. This may result in skewed data sets which do not show a normal distribution. In cases where the distribution is not normal, the mean may not be the best indicator of the "center" of such a set. These data sets will usually contain some high outliers, but low outliers are simply reported as "not detected". The effect is to cut off the normal distribution on the low side. When high errors are more likely than low ones, the log normal distribution often fits the data better.

Figure 1.2 shows the normal and log normal distribution curves. There are several ways of treating log normal data. As the name "log normal" hints, the logarithms of the data are normally distributed. This means that one can take the log of each data point and then do the statistical calculations. At the end, the antilog of the final value is taken. For instance, the geometric mean, calculated as the antilog of the arithmetical average of the logs of n data points, may give a better estimate of the center of the data, if a log normal distribution is present.

$$\log \bar{x} = \frac{\log x_1 + \log x_2 + \cdots + \log x_i}{n} \tag{1.7}$$

Sometimes, instead of setting "not detected" sample values to zero, they are set to one-half of the method detection limit, but this method is not universally accepted, and its statistical validity is not proven.

Normal Distribution

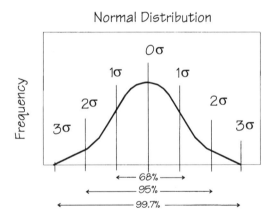

Log normal Distribution

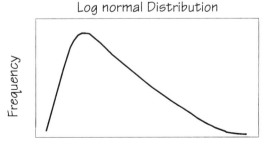

Figure 1.2 Normal distribution and log-normal distribution. Normally distributed data have equal chance of showing positive or negative deviations from the mean, while in the log-normal case, there is a greater chance of deviations on the high side.

In dealing with non-normally distributed data, the **median**, the central value when all the points are arranged by magnitude or the **mode**, the most frequently occurring value, in some cases better describe the central tendency of the data.

1.4.4 Confidence interval and the t-distribution

If there is no systematic error, then the distribution of random error can be computed from a series of measurements. Thus, it is possible to predict, with a certain amount of "confidence", a range within which the true value should lie. This range is called the **confidence interval** and the end values of this range are called **confidence limits**. The larger the confidence interval is, the higher the probability that the true value lies within that confidence interval. If we assume a normal distribution of error, then 95% of the measurements will lie between

$$\bar{x} - 1.96\left(\frac{\sigma}{n}\right) \quad \text{and} \quad \bar{x} + 1.96\left(\frac{\sigma}{n}\right).$$

This is the 95% confidence interval, i.e. there is a 95% probability that the true value lies within this range. Similarly, the 99.7% confidence interval falls between

$$\bar{x} - 2.97\left(\frac{\sigma}{n}\right) \quad \text{and} \quad \bar{x} + 2.97\left(\frac{\sigma}{n}\right).$$

If the number of measurements is not large, the estimated experimental standard deviation (s) is not equal to σ. So in this case, instead of the normal distribution, the confidence interval is calculated using Student's t-distribution. The t-distribution table is presented in Appendix 1. In this case, the confidence interval is computed as:

$$x = \bar{x} \pm \frac{ts}{\sqrt{n}} \tag{1.8}$$

The value of t is determined from the t-distribution table and its value depends upon the number of degrees of freedom, which is one less than the number of measurements ($n - 1$), and the desired confidence level.

1.4.4.1 Estimation of mean from several sets of measurements.

Sometimes several data sets are combined and a grand mean is calculated from the pooled data. For example, several carbon monoxide measurements are made in a city every day, and at the end of the month a grand average is to be obtained. If the precision of the various data sets is not significantly different, the overall average can be computed as:

$$\bar{\bar{X}} = \frac{\bar{X}_1 W_1 + \bar{X}_2 W_2 + \cdots + \bar{X}_k W_k}{W_1 + W_2 + \cdots + W_k} \tag{1.9}$$

and the corresponding standard deviation is:

$$s_{\bar{x}} = \sqrt{[\sum s_{\bar{x}}^2 / W_1 + W_2 + \cdots + W_k]} \tag{1.10}$$

where the weights, W_k, are the corresponding number of measurements in each subaverage.

If the precision of the different sets varies significantly, the values are weighted inversely according to their variance (s^2):

$$W_i = n_{x_i} / s_{x_i}^2 \tag{1.11}$$

then Equation 1.9 is applied.

1.4.4.2 *Estimation of standard deviation from several sets of measurements.* Just as several means may be combined to obtain a grand average, standard deviations also can be combined to obtain a single estimate. If there are k sets of measurements whose standard deviations are not significantly different, then the pooled standard deviation, s_p, can be calculated as:

$$s_p = \sqrt{\frac{s_1^2(n_1 - 1) + s_2^2(n_2 - 1) + \cdots + s_k^2(n_k - 1)}{(n_1 - 1) + (n_2 - 1) + \cdots + (n_k - 1)}} \tag{1.12}$$

where s_1, s_2 and s_i are the standard deviations of i separate sets of measurements. The number of measurements in these sets are n_1, n_2, and n_i respectively.

1.5 Significance tests

As mentioned before, because of random error, the measured mean value is seldom exactly equal to the true value. Significance tests are used to decide if the difference between a known value and the measured value or between two measured values can be due to random error alone. The quantities being compared could be means and standard deviations from two different sets of measurements, to see if the means are really different, or to see if a certain measurement is a statistically improbable outlier.

1.5.1 *Comparison between a measured and a known value*

It is often important to know whether the mean of several measurements is significantly different from a known value. The known value could be a specified standard value or regulatory threshold for environmental compliance. The significance of the difference between the experimental and the known or target value must be determined, to see if the difference is due to random error only or if there is truly a difference, a **statistically significant difference**, between the two numbers. To do this, a confidence interval for the measured mean is calculated. If the target value falls within the same range, then it can be said that the two values are not different at that level of confidence.

1.5.2 Comparison of the mean of two samples

This type of comparison is necessary, for instance, when the two sets of measurements are made at different laboratories or using different analytical techniques. It can also be useful when one is comparing two sets of samples from different populations or areas, to determine if the measured variable, (concentration, temperature, pH or whatever), is truly different between the two populations, or if the difference can be ascribed to random error alone.

Let x_a and x_b be the means from the two sets of measurements whose standard deviations are s_a and s_b respectively. Then we must determine if the difference between these means, $(x_a - x_b)$, is significantly different from zero, within a chosen confidence interval.

Case 1. When the standard deviations of the two measurement sets are not thought to be significantly different from each other, for instance, if water samples were taken from two well-mixed sources, and were all done by the same analyst, using the same method. If there is a doubt as to whether the standard deviations are different or not, an F test, described below, will give that information. The variances of both means are calculated as:

$$V_A = s_p^2/n_A \quad \text{and} \quad V_B = s_p^2/n_B \tag{1.13}$$

using the pooled value of s. Then the desired confidence level is chosen. The uncertainty of the difference between the two means is calculated as:

$$U_\Delta = t\sqrt{V_A + V_B} \tag{1.14}$$

using a value for t, from the t-distribution table, corresponding to the probability level chosen and the number of degrees of freedom, $(n_A + n_B - 2)$. If the difference between the two means being examined is not greater than the uncertainty in the difference, U_Δ, the means are not considered to be different.

Case 2. If it is thought that s_1 and s_2 might be significantly different from each other, then the variances of the means are:

$$V_A = s_A^2/n_A \quad \text{and} \quad V_B = s_B^2/n_B \tag{1.15}$$

The effective number of degrees of freedom, f, is:

$$f = \frac{(V_A + V_B)^2}{\dfrac{V_A^2}{n_A - 1} + \dfrac{V_B^2}{n_B - 1}} \tag{1.16}$$

The result from the above equation is rounded to the nearest whole number. Finally

$$U_\Delta = t^*\sqrt{V_A + V_B} \tag{1.17}$$

using the f value calculated above and the desired confidence level to determine the t^* value to be selected from the t-table. If the difference between the two means being examined is not greater than the uncertainty in the difference, U_Δ, the means are not significantly different.

Example

The following data were obtained for PCB in fish tissues from two different rivers. Are the two populations of fish contaminated to a significantly different degree at the 95% confidence level?

River A	River B
2.34 ng/g	1.55
2.66	1.82
1.99	1.34
1.91	1.88

First the means and standard deviations of each group must be calculated. The mean for A is 2.225 and s is 0.3449. For B the mean is 1.648 and s is 0.2502.

The variances of each of the means are:

$$V_A = \frac{0.3449^2}{4} \qquad V_B = \frac{0.2502^2}{4}$$

The effective number of degrees of freedom are:

$$f = \frac{(0.0297 + 0.0157)^2}{\dfrac{0.0297^2}{4-1} + \dfrac{0.0157^2}{4-1}} = 5.4 \cong 5$$

The t-value for 95% CL and 5 degrees of freedom is 2.571.

$$U_\Delta = 2.571\sqrt{0.0297 + 0.0157} = 0.548$$

The difference between the two means is $2.225 - 1.648 = 0.577$. The uncertainty in the difference is 0.548. The uncertainty is less than the difference, so there is a significant difference in these two batches of fish, at the 95% CL.

1.5.3 Comparison of standard deviations using the F-test

The F-test is used to compare the precision of two sets of analytical measurements. The measurements can be from different methods, laboratories, or instruments. They can also be a series of different samples from different populations, as in the example above. If s_1 and s_2 are the standard deviations from the two measurements, the F value is calculated as:

$$F = \frac{s_{larger}^2}{s_{smaller}^2} \qquad (1.18)$$

The value of F is compared to a critical value, F_c, from standard tables. The value of F_c depends upon the degrees of freedom of the two measurements, and the chosen level of confidence. If the calculated F does not exceed the F_c from the table, then it can be concluded that the standard deviations do not differ.

Example
If we want to compare the deviations in the PCB content of the fish in the two rivers, we can apply an F test to the data given in the previous example.

$$F = \frac{(0.3449)^2}{(0.2502)^2} = 1.9$$

F from the table (Appendix 2) for 3 degrees of freedom and 95% CL is 15.4. The calculated F is less than that from the table so the variation within each population can be considered to be the same.

1.5.4 Outliers

When several measurements are taken, certain values may be unusually different from others in the data set. These points are called **outliers**. It is not always easy to tell from statistics if an outlier is due to inherent random error present in the measurement or is due to some identifiable error. It is always best to search carefully for a reason before a measurement is discarded as an outlier. When a system or population is not well known, or when only a few measurements have been made on it, it is very difficult to determine with a high degree of confidence, that an outlier is truly due to an error, not just an extreme sample in the normal distribution.

For example, ten repeat measurements (in μg/g) for lead in a soil sample are as follows: 1.0, 1.1, 1.2, 0.9, 1.0, 0.8, 1.0, 1.1, 0.9, 2.8. The last number looks like an outlier and may have been caused by instrument malfunction, human error, or contamination. One reason why the last measurement in the above data set is immediately considered to be an outlier is because it does not fit a normal distribution.

1.5.4.1 *Rule of the huge error.* A simple and rapid method of determining if an outlier may be rejected is to divide the difference between the questioned value and the mean by the standard deviation.

$$\frac{|X_q - \bar{X}|}{s}$$

If this ratio is less than 4, the value should be retained, when s is well defined. If s has been estimated from only a few measurements, a value of 6 should be reached before it can be discarded at a confidence level of about 98%. For the data set above, $\bar{X}$ is 1.18 and $s = 0.58$. The ratio is only 2.1, so the last figure is not an outlier by this test.

1.5.4.2 *Dixon test for rejection of outliers.* The Dixon test is based on the normal distribution of error. In this test the data is arranged in the order of increasing numerical value: $x_1 < x_2 < x_3 < \cdots < x_n$

The suspect value may be either x_n or x_1. The ratio τ is calculated as follows depending upon the number of measurements.

For $n =$	If x_n is suspect	If x_1 is suspect
3–7	$\tau_{10} = (x_n - x_{n-1})/(x_n - x_1)$	$(x_2 - x_1)/(x_n - x_1)$
8–10	$\tau_{11} = (x_n - x_{n-1})/(x_n - x_2)$	$(x_2 - x_1)/(x_{n-1} - x_1)$
11–13	$\tau_{21} = (x_n - x_{n-2})/(x_n - x_2)$	$(x_3 - x_1)/(x_{n-1} - x_1)$

The ratio is compared to critical values from Table 1.5 at a predetermined level of risk of false rejection. If the computed ratio is greater than the tabulated value, the value may be considered an outlier at that risk level.

1.5.5 *Reporting data*

From the above discussion, you should have a sense that the simple reporting of an average value is not much use to the reader. At minimum, one needs to report the mean, the standard deviation and the number of degrees of freedom, or the number of measurements. Otherwise, the user of the data has no way of applying statistical tests to the data. The confidence the user can place in the reported data is essentially described in the information given in these three parameters.

1.5.6 *Hypothesis testing*

The determination of the significance or non-significance of measurements is important because an environmental study can usually be proposed as a hypothesis. The data obtained after the measurements are performed are tested statistically to see if the hypothesis is proven or not. What kinds of hypotheses can be

Table 1.5 Critical values for the Dixon test

| | n | \multicolumn{4}{c}{Risk of false rejection} | | | |

	n	0.5%	1%	5%	10%
τ_{10}	3	0.994	0.988	0.941	0.886
	4	0.926	0.889	0.765	0.679
	5	0.821	0.780	0.642	0.557
	6	0.740	0.698	0.560	0.482
	7	0.680	0.637	0.507	0.434
τ_{11}	8	0.727	0.683	0.554	0.479
	9	0.677	0.635	0.512	0.441
	10	0.639	0.597	0.477	0.409
τ_{21}	11	0.713	0.679	0.576	0.517
	12	0.675	0.642	0.546	0.490
	13	0.649	0.615	0.521	0.467

tested? If the question is "Is this river polluted by PCB?" the hypothesis will be stated as "The river contains a significantly higher concentration of PCB than another river which we consider to be clean (i.e., our background site)". Another study might look at tests of an emission control system: "Is the emission of SO_2 from this stack too high?" The hypothesis can be stated as: "The concentration of SO_2 in the stack effluent is not significantly higher than the concentration stated in the company's release requirement." The statistical tests discussed above can be used to test these hypotheses, as long as all the required data have been collected – the means and standard deviations of both the area of interest and the background or standard to which it is being compared.

1.6 Standards and calibration

Almost all analytical systems require calibration against chemical standards. Mass measurements can be made using a known mass to calibrate a balance, but determination of the relationship between a signal from a gas chromatograph and the amount of analyte which caused the signal requires that one run a known quantity of analyte through the detector for comparison. This requires standards. Standards may be prepared in the laboratory, by weighing an amount of a pure chemical, or measuring out an accurate volume, and dilution to an appropriate concentration range.

There are also many commercial sources for prepared standards. One important consideration in choice of standards is the matrix. For some types of analysis, for instance, X-ray fluorescence of solid samples, the matrix is very important, while for others it is of less importance. For final validation of an analytical method, a **reference material** can be very useful. These are available prepared and certified by several non commercial sources such as the National Institute of Standards and Technology in the USA (NIST), the Community Bureau of Reference (BCR) of the Commission of the European Communities, and the Laboratory of the Government Chemist (LGC) in the UK. There are also commercial sources for reference materials. These materials are usually real samples of common substances which have been carefully homogenized and analyzed, with the determination of each component being verified by several different analytical techniques.

1.6.1 Calibration methods

The most common method of calibration is to prepare surrogate samples or **standards** of known concentrations, covering the range of concentrations expected in the samples. The standard concentrations should bracket the sample concentrations, and should be as close to the samples as possible. These standards are prepared so that they match the samples as much as possible. If the sample is extracted into an organic solvent, for instance, the standard would be prepared by dissolving a weighed portion of pure analyte in a measured volume of the same solvent, and diluting the resulting stock solution to several different concentrations. The solutions are analyzed on the instrument to be calibrated, and the response is plotted against the known concentrations. The plot, or its electronic equivalent, is used to determine the amount of analyte in the unknown samples. If the response/concentration line is a straight line or is a curve with a readily determined equation, the equation of the line is often used to calculate the concentration in a sample, in place of the physically plotted data.

1.6.2 Standard addition method

The matrix of the sample should be similar to that of the prepared sample. In some cases, this is difficult to do. One way of attempting to match the calibration standards to the samples is to use the **standard addition** method. In this method, the sample is analyzed. Then a measured portion of standard is added to the sample, and the determination is repeated. The standard should be added as a small quantity of a fairly concentrated solution, so that the matrix is not significantly diluted. Several sequential additions may be made. The results are plotted, and the response of the instrument per unit of analyte is determined from the slope of the line. The concentration of analyte in the original sample can then be determined.

When a measurement is made, the signal S is proportional to the total concentration. $S = kC_x$, where k is the proportionality constant. Then V_s ml of standard is added to V_x ml of sample and another measurement is made. The signal is due to the total concentration of analyte, from both the sample and the standard.

$$S_{sample+standard} = kC_x\left(\frac{V_x}{V_x + V_s}\right) + kC_s\left(\frac{V_s}{V_x + V_s}\right)$$

This can be rearranged to solve for C_x, the unknown concentration.

$$C_x = \frac{C_s V_s}{R(V_s + V_x) - V_x} \qquad \text{where } R = \frac{S_{sample+standard}}{S_{sample}}$$

A graphical method may also be used, when all the samples, the original and those with standard added, are brought to the same volume before analysis. This is shown in Figure 1.3.

Example
A water sample is analyzed for cadmium by atomic absorption. The initial sample gave a reading of 0.051. Then 10 ml of the sample was mixed with 0.5 ml of a standard containing 1 µg/l. The instrumental reading for this

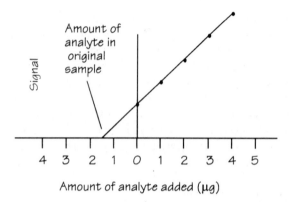

Figure 1.3 Graphical calculation of the concentration of a sample using standard addition. The signal is plotted against quantity of analyte added. Each sample is brought to the same volume. When the line is extrapolated, the amount of analyte in the sample is found. The amount in this case is 1.5 µg.

mixture is 0.092. What is the concentration of Cd in the original sample?

$$R = \frac{0.092}{0.051} = 1.80$$

$$C_x = \frac{1 \times 0.5}{1.80(10 + 0.5) - 10} = 0.056 \ \mu g \ Cd/l$$

The units in which the answer is obtained are the same as those of the standard.

This method ensures that the sample and standards are affected by the same background interferences, but it is also considerably more labor intensive, as each sample must be run two or more times. In comparison, a single set of calibration standards can be run and many samples analyzed using the curve produced.

1.7 Performance of analytical methods: figures of merit

The criteria for evaluating an analytical methodology are called figures of merit. Based on these characteristics, one can predict whether a method meets the needs for an application. These figures of merit are listed in Table 1.6. If two or more methods satisfy the figures of merit requirements, the selection of a method can be based on additional desirable characteristics listed in the table. Precision and accuracy have already been discussed. Other important characteristics are sensitivity, detection limits, and the range of quantitation.

Table 1.6 Figures of merit for instruments or methods

Parameter	Definition
Accuracy	Ability to produce the correct data
Precision	Reproducibility of replicate measurements
Sensitivity	Magnitude of signal generated by small change in the sample
Detection limit	Smallest amount which can be detected above the noise level
Linear dynamic range	Range over which the signal is proportional to the amount or concentration of analyte
Selectivity	Ability to distinguish the analyte from other materials
Speed of analysis	Rapidity of measurement, combined with time needed for sample preparation
Ruggedness	Durability of instrument, and ease with which a method produces good results
Portability	Ability to use instrument on site. For method, ease with which it can be moved from one laboratory to another with equally good results
Throughput	Numbers of samples which can be run in a given time period
'Greenness'	Method can be used without producing large amounts of toxic or otherwise undesirable waste, or using an inordinate amount of supplies
Cost	Initial investment, cost of consumable supplies, and cost of training and labor all must be considered

1.7.1 *Sensitivity*

The sensitivity of an instrument (or a method) is a measure of its ability to distinguish between small differences in concentration of analyte at a desired confidence level. The simplest measure of sensitivity is the slope of the calibration curve at the concentration of interest. This is also known as calibration sensitivity. Usually, calibration curves are linear of the form,

$$S = mc + s_{bl} \qquad (1.19)$$

where S is the signal at concentration c and s_{bl} is the signal in absence of any analyte, i.e., the blank. Then m is the slope of the calibration curve and hence the sensitivity. Since the precision of measurements decreases at low concentrations, the ability to distinguish between small concentration differences also decreases. Therefore, sensitivity is a function of the precision. A sensitivity measurement which includes this factor is known as **analytical sensitivity**, a, and is expressed as:

$$a = m/s_s \qquad (1.20)$$

where s_s is the standard deviation of the signal. While calibration sensitivity does not depend upon concentration, analytical sensitivity is often concentration sensitive, as s_s varies with concentration.

1.7.2 *Detection limit*

A measurement is acceptable only when the signal measured is larger than the uncertainty associated with the measurement. The detection limit is defined as the lowest concentration or the weight of analyte which can be measured at a specific confidence level. So, as the detection limit approaches, the signal generated by the instrument approaches that of the blank. Therefore, the smallest distinguishable signal, S_m, is

$$S_m = X_{bl} + ks_{bl} \qquad (1.21)$$

where X_{bl} and s_{bl} are the average blank signal, and its standard deviation. The constant k depends upon the confidence level, and the accepted value is usually 3 at a confidence level of 89%. The detection limit can be determined experimentally by running several blank samples to establish the mean and standard deviations of the blank. Substitution of Equation 1.21 into Equation 1.19 and rearranging gives:

$$C_m = (S_m - S_{bl})/m$$

where C_m is the the minimum detectable concentration and S_m is the signal obtained at that concentration.

Example
Samples are analyzed for mercury using a cold vapor atomic absorption spectrometer. Calibration against standards yields the following equation:

$$S = 1.15\,C_{Hg} + 0.025$$

where S is the instrument response and C_{Hg} is the concentration of Hg in µg/ml. A blank sample was analyzed 10 times, giving a mean value of 0.28 and a standard deviation of 0.008. Five replicate measurements at 1.0 µg/ml showed a standard deviation of 0.01.

• Determine the calibration sensitivity

By definition, the calibration sensitivity is the slope of the calibration curve m, which, in this case, is 1.15.

• What is the analytical sensitivity at 1.0 µg/ml?

$$a = m/s_s$$

$$a = 1.15/0.01 = 115$$

• What is the detection limit?

The minimum measurable signal is $S_m = S_{bl} + 3\sigma_{bl}$

$$S_m = 0.28 + 3(0.008) = 0.304$$

$$C_m = (0.304–0.28)/1.15 = 0.021 \text{ µg/ml}$$

1.7.3 Linear dynamic range

The lowest concentration level at which a measurement is quantitatively meaningful is called the limit of quantitation (LOQ). The LOQ is arbitrarily defined as ten times the standard deviation of the blank ($10 \times s_{bl}$). For all practical purposes, the upper limit of quantitation is the point where the calibration curve tends to become non-linear. This point is called the limit of linearity (LOL). A typical calibration curve is shown in Figure 1.4. Analytical techniques are expected to have a linear quantitation range of at least two orders of magnitude.

Sensitivity, limits of detection, and limits of quantitation can be applied to both instruments and to entire methods. The definitions are the same, and the methods of determining these limits are also similar. For an instrument, blanks and standards are run and the limits are calculated from these. For a method, blanks are taken through all the steps, and the standards are replaced by portions of a matrix similar to that of the samples, which have been spiked with the analyte at various concentrations, covering the range of concentrations expected. These are again taken through the entire procedure.

1.7.4 *Validation of new methods*

When existing, documented methods are not suitable because new sample types or analytes are to be studied, or because new instrumentation has become available, a new method must be developed. A new method or instrument may be an improvement to an existing method, or it may be completely novel. In either case, the method must be validated. During validation, the various figures of merit for the entire analytical process are determined quantitatively. The random and systematic error is measured in terms of precision and bias. The

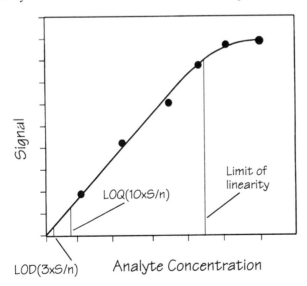

Figure 1.4 A typical calibration curve. The lower concentration points in this case are linear, but the curve deviates from linearity at the high end. It is preferable to use this system in the linear range.

detection limit is established for each of the analytes. Precision and bias are determined at concentrations covering the range for which the method is to be used. The linear dynamic range is established and the calibration sensitivity is measured. The studies involved in a good method validation provide a comprehensive picture of the capabilities and limitations of the new method or instrument.

The validation process involves the following steps:

- Determination of the single operator precision, bias, and figures of merit. Parameters such as minimum detection limits, bias, and precision are studied using standards at several concentrations including some near the detection limit. The linear dynamic range is established.
- Analysis of unknown samples: this step involves the analysis of samples whose concentrations are unknown to the analyst. Both qualitative and quantitative measurements should be performed. Reliable unknown samples are obtained as certified reference materials from commercial sources, or governmental agencies. The accuracy and precision obtained are reported.
- Equivalency testing: once the method has been developed, it is often a good idea to see if the results obtained are equivalent to those from an existing comparable method, if such a method exists. Statistical tests are used to determine if the new and established methods give equivalent results or not. The statistical tests to determine the significance of the difference in two means can be used to determine whether the results are comparable.
- Collaborative testing: once the method has been validated in one laboratory, it may be subjected to collaborative testing. Here, identical test samples and the procedure are distributed to several laboratories. Analysts from each laboratory analyze these samples, following the developed procedure. The results are studied statistically to determine bias and interlaboratory variability. This step is necessary to determine if a method is rugged enough to be passed on to the general analytical community, and if it will yield reliable results when followed by others.

The analytical chemist has always been looked upon as a practical problem solver. The environmental analyst is required to use techniques which were initially designed to monitor raw and manufactured materials of industry, for the day-to-day characterization of an enormously complex system, the environment in which we live. However, in the effort to save the environment of our planet from further degradation, the environmental analyst must be in the forefront of the battle.

Suggested reading

Caulcutt, R. and Boddy, R. (1983) *Statistics for Analytical Chemists*, Chapman & Hall, London, New York.
DeLevie, R. (1997) *Principles of Quantitative Analysis*, McGraw Hill, New York.

Keith, L.H., Crummett, W., Deegan, J., Libby, R.A., Taylor, J.K. and Wentler, G. (1983). Principles of environmental analysis. *Anal. Chem.* **55**, 2210–2218.

Manahan, S. (1994) *Environmental Chemistry*, 6th edition, Lewis Publishers, Boca Raton.

Miller, J.C. and Miller, J.N. (1994) *Statistics for Analytical Chemistry*, 3rd edition, Ellis Horwood, New York.

O'Neill, P. (1993) *Environmental Chemistry*, 2nd edition, Chapman & Hall, London.

Prichard, E., MacKay, G.M. and Points, J. (1996) *Trace Analysis, A Structured Approach to Obtaining Reliable Results*, Royal Society of Chemistry, Cambridge, UK.

Study questions

1. Benzene is present in air at 10 ppm_v. Express this in $\mu g/l$, in mg/m^3.
2. Ca^{2+} is found in a water sample at a level of 10 $\mu g/l$. How much calcium, expressed as calcium carbonate, is present in 1 m^3 of the water?
3. What are the major steps in the process of performing an environmental analysis?
4. Distinguish between random and systematic errors. Which of these can be treated statistically?
5. Explain how it is possible to have high precision without high accuracy.
6. For the following set of analyses of an urban air sample for carbon monoxide determine the mean, standard deviation, median, and coefficient of variation. 325, 320, 334, 331, 280, 331, 338 $\mu g/m^3$.
7. In the above data set, is there any value which can be discarded as an outlier?
8. Samples of bird eggs were analyzed for DDT residues. The samples were collected from two different habitats. The question is: are the two habitats different from each other in the amount of DDT to which these birds are exposed? The data reported are:

Sample collection area	Number of samples	Mean conc. DDT (ppb)	Standard deviation (s)
Area 1	4	1.2	0.33
Area 2	6	1.8	0.12

   At the 95% CL, are the two sets significantly different or not?

9. Soil samples were collected at different areas surrounding an abandoned mine and analyzed for lead. At each area several samples were taken. The soil was extracted with acid, and the extract analyzed using flame atomic absorption spectrometry. The following data were obtained:

Area	Number of samples	Pb concentration in ppm
A	4	1.2, 1.0, 0.9, 1.4
B	5	0.7, 1.0, 0.5, 0.6, 0.4
C	3	2.0, 2.2, 2.5
D	5	1.4, 1.1, 0.9, 1.7, 1.5
E	4	1.9, 2.3, 2.5, 2.5

Calculate:

   a. The overall mean of all the measurements.
   b. The pooled estimate of the standard deviation of the method.
   c. The 95% and 80% confidence interval for the measurement at each spot.
   d. Which of the above areas show mean Pb concentrations which are not significantly different from each other at a 90% CL, indicating a similar pattern of contamination?

10. The analysis of waste water for benzene using purge and trap GC/MS yielded a pooled standard deviation of 0.5 $\mu g/l$. A waste water sample from a oil refinery showed a benzene concentration of 5.5 $\mu g/l$. Calculate the 75, 85, 95 and 99% confidence interval, if the concentration reported was based upon:

  a. a single analysis
  b. the mean of 5 analyses
  c. the mean of 15 analyses
11. Distinguish between the following terms:
  a. Sensitivity and detection limit
  b. Limit of quantitation and limit of linearity
  c. Quantitation by calibration curve and the method of standard addition
  d. Confidence interval and confidence limits
12. The following data were obtained in calibrating a halogen specific GC detector. A linear relationship between detector response in mV and concentration of dichloroethane (DCE) is expected. Analysis of calibration standards yield the following results:

DCE concentration, ng/ml	Detector output, mV
1.0	−54.0
2.1	−28.2
3.05	+2.8
4.0	+32.2
5.05	+65.8

  a. Plot the calibration data and draw a line through the points by eye.
  b. Determine the best straight line by a least squares fit.
  c. Calculate the concentration of an unknown for which the detector output was 7.8 mV.
  d. What is the calibration sensitivity?

13. The following calibration data were obtained for a total organic carbon analyzer (TOC) measuring TOC in water:

Concentration, µg/l	Number of replicates	Mean signal	Standard deviation
0.0	20	0.03	0.008
6.0	10	0.45	0.0084
10.0	7	0.71	0.0072
19.0	5	1.28	0.015

  a. Calculate the calibration sensitivity.
  b. What is the detection limit for this method?
  c. Calculate the relative standard deviation for each of the replicate sets.

2 Environmental sampling

We might imagine a satellite which could scan the earth's surface and provide a complete analysis of every part of the environment. This is, of course, in the realm of science fiction. Instead, we must collect representative samples of a small part of the environment in which we are interested, and analyze these to provide information about the composition of the area. For example, it is obviously impossible to analyze all the water in a lake, so portions of the water must be collected and analyzed to determine the true concentrations of materials in the lake. Similarly, to study contamination around a leaking underground gasoline tank, numerous soil samples are needed to map the extent of the pollution.

We must keep in mind that only a small amount of sample (a few grams or milliliters) is collected from a vast heterogeneous area. It is imperative that the samples collected represent the environment as accurately as possible. Major decisions are based on the results of the analyses. The steps involved in environmental sampling are:

- Development of a sampling plan, including where and when samples will be collected and the number of samples required.
- Collection of the samples.
- Preservation of samples during transportation and storage.

2.1 The sampling plan

The importance of good sampling cannot be overstressed. The sample is the source of information about the environment. If it is not collected properly, if it does not represent the system we are trying to analyze, then all our careful laboratory work is useless. Care must be taken to avoid the introduction of bias or error.

Sampling is done for monitoring purposes, as well as for research. Data may be collected to monitor air and water effluents, or to characterize pollutant levels in environmental media (air, water, soil, biota). The objectives may be to comply with regulatory requirements, to identify long and short term trends, to detect accidental releases, or to develop a data base or inventory of pollutant levels. Research may involve studying the fate and transport of pollutants or identifying pollutant exposures for humans and animals. It is important to design these studies scientifically so that they are cost effective and generate statistically significant information.

Intuitively, we see that the number of increments which go to make up a sample, the size of a sample, depends on the inhomogeneity of the system itself. If the system to be analyzed is a bag of marbles, with some red and some blue, an adequate sample size would depend on how homogeneous the sample is. If there are nearly equal numbers of marbles of each color, and they are well mixed, a handful would probably give a good idea of the overall composition. However, if there are only two or three red ones per 100 blue ones, it takes a much larger sample to be sure that the sample is representative. While there are mathematical formulas to determine how much sample is needed, depending on the variation in composition of the particles and the size of these particles, the parameters needed to use these equations are not often available when environmental samples are considered. The number of aspirin tablets coming off an assembly line which must be taken for testing may be calculated when the standard deviation of the usual aspirin tablet and the tolerable variation are known. It is not as easy to determine the number of portions of soil required to characterize a landfill. The actual inhomogeneity of the landfill is unknown, and it would probably take more analyses to determine it than anyone would be willing to do.

2.1.1 Spatial and temporal variability

If an environmental domain was completely homogeneous, a single sample would adequately represent it. However, we seldom come across such a situation, as the environment is highly heterogeneous. One must also distinguish between a **static** and a **dynamic** system. A static system is one which does not change much with time. It must be sampled so that the sample reflects all the inhomogeneity of the system. If a field is to be tested for a long-lived pesticide in the soil, that could be considered to be a relatively static system. Sample increments could be taken over a grid pattern on the field or randomly selected spots. Remember that the random spots should be chosen in a truly random fashion, and that strolling around with a shovel is not likely to generate a random sample. A calculator with a random number generator can be used in imaginative ways to generate such a sample. For instance, a field might be sampled by making equally spaced traverses across it, generating a new random number every 10 paces, and taking a spadeful of soil each time the random number generator gave a preselected digit.

A dynamic system is one whose content changes with time. Most regions which we wish to characterize by taking samples are dynamic to some extent, and show both spatial and temporal variation. When a river or a waste effluent stream is to be characterized, its concentration will probably change over a period of minutes, days, or hours. This system must be sampled at many different times to collect a representative sample. This may be done by collecting a small constant volume of sample and compositing it for a day or a week, or it may be done by collecting a given volume at random times or on a regular schedule. The rule of random sampling, that any portion has the same chance of being selected, applies here. For instance, if samples are taken at random, these random times should include all periods of time including weekends and nights, as well as business hours.

In air sampling, the concentration of VOCs will vary from neighborhood to neighborhood within a town. It will also change with the time of day. Concentrations of compounds from automobile exhausts are generally higher during peak traffic hours in urban areas. Consequently, to gain a good understanding of the air quality in an area, samples have to be taken or measurements made at different locations and at different times of the day. Even further variation must be considered, since there are changes due to seasonal and weather factors.

When the sample is collected from a large environmental domain, it can be conceptualized as a point in time and space. Space units, S_1, S_2...denote sites, cities, even countries. Specific sampling locations are located within each space unit, described as a three dimensional space, using x, y, and z coordinates. So, measurements may be taken at each location at different points in time, and at different locations at the same time. For example, we might be monitoring ozone levels in an urban area, where S_1 and S_2 denote two cities on opposite sides of a river. Within city S_1, several locations (L_{111}, L_{112}, ...) are chosen and measurements are taken at two different vertical distances from the ground (V_{11} and V_{12}). This is illustrated in Figure 2.1. Time periods T_1, T_2 might represent different seasons of the year. Within each season, t_{11}, t_{12}... would represent daily or weekly averages.

2.1.2 Development of the plan

To do a successful environmental study it is necessary to have a "plan of action", a sampling plan. If the content of heavy metals in a river is being studied, for example, the purpose might be to examine the effect of these metals on fish, or it might be to monitor the content because the river is a drinking water source. The sampling plan will be different for each of these purposes. The first step is to define clearly the problem being studied and identify the environmental "population" of interest. Some of the major steps involved in the development of a successful study are as follows:

- Clearly outline the goal of the study. Decide what hypothesis is to be tested and what data should be generated to obtain statistically significant information.
- Identify the environmental population or area of interest.
- Obtain information about the physical environment. Weather patterns, for instance, are important if air samples are to be taken.
- Research the site history.
- Carry out a literature search and examine data from similar studies previously carried out. This can provide information about trends and variability in the

Each sample has a time dimension, and further
contains subsamples and repeat measurements

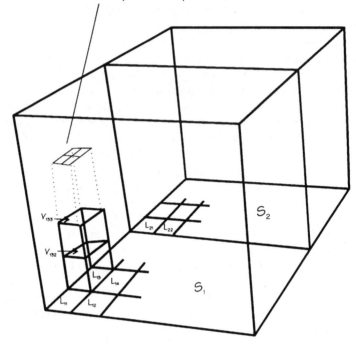

Figure 2.1 Sampling in space and time. Samples are taken over wide areas (S_1, S_2), then in three-dimensional space. Time is an additional dimension. Within each sample block, samples are taken at randomly chosen or planned times.

data. In the absence of previous data, a pilot study may be necessary to generate preliminary information on which to base a more detailed study.

- Identify the measurement procedures to be used, because these affect the way samples are collected and handled.
- Develop an appropriate field sampling design. Decide how many samples are to be collected and delimit the time and area to be covered by the study.
- Determine the frequency of samples to be taken, both in time and space, depending upon the project objectives. Decide if, for example, 24-h integrated samples will be collected or individual samples will be taken every few hours.
- Develop a plan to insure and document the quality of each of the processes involved in the study: sampling, laboratory analysis, contamination control, etc.
- Once the sampling and analysis are complete, assess the uncertainty of the measurements.
- Perform statistical analysis on the data. Determine mean concentrations, variability, and trends with time and location.

- Evaluate whether study objectives have been achieved. If not, additional work may be necessary to provide the needed information.

2.1.3 *Sampling strategies*

When it comes to sampling, the essential questions are: where to collect the samples, when to collect them, and how many samples to collect. Most environmental measurement domains are large and it is not easy to answer these questions. Some of the factors to be considered in determining a sampling strategy are:

- The study objectives: different objectives require different sampling strategies. For example, if the objective is to measure the total release of heavy metals into a river by an industry, a 24-h integrated sample may be taken. However, if the goal is to monitor for accidental releases, then sampling and analysis may have to be done almost continuously.
- The pattern and variability of environmental contamination: the number of samples to be collected in space and time depends upon the variability in the concentrations to be measured. For example, pollutant levels in air can vary significantly depending upon meteorological conditions, or traffic patterns. In general, if the spatial or temporal variability is high, a larger number of samples needs to be analyzed.
- Cost of the study: if more samples are analyzed, the information obtained will have higher precision and accuracy. However, more samples also require more money, time, and resources. So, it is necessary to design an effective sampling plan within the available resources.

Other factors such as convenience, site accessibility, limitation of sampling equipment and regulatory requirements often play important roles in developing a sampling plan, as well. A well designed strategy is needed to obtain the maximum amount of information from the number of samples. The strategy may be a statistical or a non-statistical one.

There are several approaches to sampling: systematic, random, judgmental (non-statistical), stratified, and haphazard. More than one of these may be applied at the same time. Very often, not much is known about the environmental area to be studied. A statistical approach is taken to increase the accuracy and decrease bias. An industrial discharge into a lake is shown in Figure 2.2. It would be expected that the concentration of the pollutants present in the wastewater outfall are at maximum near the discharge point. A systematic sampling plan would divide the water surface into a grid, and take samples in a regular pattern. Sampling a few of the grid blocks chosen in a genuinely random way constitutes random sampling. Judgmental sampling would concentrate on the area around the outfall. Taking a few samples at locations chosen by the person doing the sampling would be termed haphazard sampling. Finally, a **continuous monitor** may eliminate the time factor by giving real-time measurements all the time. This is still a sampling process, however, as the location of

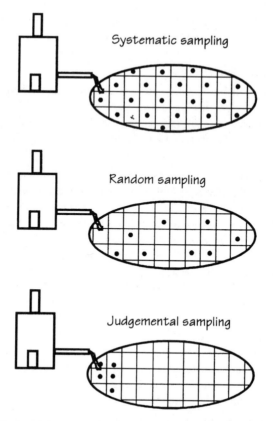

Figure 2.2 Sampling strategies. Here sampling is being done after the area is divided into equal areas. Sampling points are chosen using systematic, random, or judgmental sampling schemes.

the sensor must serve as a typical location to give information about a larger area.

2.1.3.1 *Systematic sampling.* Measurements are taken at locations and/or times according to a predetermined pattern. For example, the area to be analyzed may divided by a grid, and a sample taken at each point of the grid. For air pollution studies, an air sample might be taken at fixed intervals of time, say every 3 h. This approach does not require the prior knowledge of pollutant distribution, is easy to implement, and should produce unbiased samples. However, systematic sampling may require more samples to be taken than some of the other methods.

2.1.3.2 *Random sampling.* The basis of random sampling is that each population unit has equal probability of being selected. Random methods are good if the population does not have any obvious trends or patterns. When we think of random surveys of public opinion, for instance, we can readily see that a survey might come to very wrong conclusions if it relied on a door-to-door canvass taken on weekday mornings. All people who held 9-to-5 jobs would be essentially eliminated from the sample, probably skewing the results. Likewise, it would be foolish to rely on the opinions expressed by sampling a single street, when most of the people who live there are likely to be of the same class or background. If a system varies with time, as a stream might, we must sample at a variety of times, so that any time has an equal chance of being chosen. If the system varies with location within it, as a landfill would, we have to sample across the surface and down into it, so that any point in the three dimensional space of the landfill has an equal chance of being chosen.

Typically, the area to be sampled is divided into triangular or rectangular areas with a grid. Three dimensional grids are used if the variation in depth (or height) also needs to be studied. The grid blocks are given numbers. A random number generator or a random number table is then used to select the grid points at which samples should be collected. If a waste site contains numerous containers of unknown wastes and it is not possible to analyze every container, a fraction of the containers are selected at random for analysis.

2.1.3.3 *Judgmental sampling.* This is a non-statistical sampling procedure. Here, the prior knowledge of spatial and temporal variation of the pollutants is used to determine the location or time for sampling. In the lake example, samples might be collected just around the outfall point. This type of judgmental sampling introduces a certain degree of bias into the measurement. For example, it would be wrong to conclude that the average concentration at these clustered sampling points is a measure of the concentration of the entire lake. However, it is the point which best characterizes the content of the waste stream. In many instances, this may be the method of choice, especially when purpose of the analysis is simply to identify the pollutants present. Judgmental sampling usually requires fewer samples than statistical methods, but the analyst needs to be aware of the limitations of the samples collected by this method.

2.1.3.4 *Stratified sampling.* When a system contains several distinctly different areas, these may be sampled separately, in a stratified sampling scheme. The target population is divided into different regions or strata. The strata are selected so that they do not overlap each other. Random sampling is done within each stratum. For example, in a pond or a lagoon where oily waste floats over water and sediment settles to the bottom, the strata can be selected as a function of depth, and random sampling can be done within each stratum.

The strata in a stratified scheme do not necessarily have to be obviously different. The area may be divided into arbitrary subareas. Then a set of these is selected randomly. Each of these units is then sampled randomly. For example, a hazardous waste site can be divided into different regions or units. Then, the soil samples are collected at random within each region or within randomly selected regions. Stratification can reduce the number of samples required to characterize an environmental system, in comparison to fully random sampling.

2.1.3.5 *Haphazard sampling.* A sampling location or sampling time is chosen arbitrarily. This type of sampling is reasonable for a homogeneous system. Since most environmental systems have significant spatial or temporal variability, haphazard sampling often leads to biased results. However, this approach may be used as a preliminary screening technique to identify a possible problem before a full scale sampling is done.

2.1.3.6 *Continuous monitoring.* An ideal approach for some environmental measurements is the installation of instrumentation to monitor levels of pollutants continuously. These real-time measurements provide the most detailed information about temporal variability. If an industrial waste water discharge is monitored continuously, an accidental discharge will be identified immediately and corrective actions can be implemented while it is still possible to minimize the damage. A grab sample would have provided information about the accidental release only if a sample happened to be taken at the same time as the release was taking place, and that might well not have been when the problem began. A sample composited frequently enough could have identified the accidental release, but the time for preventive action would likely have passed.

Continuous monitoring is often applied to industrial stack emissions. Combustion sources, such as incinerators, often have CO monitors installed. A high CO concentration implies a problem in the combustion process, with incomplete combustion and high emissions. Corrective action can be triggered immediately. Continuous monitoring devices are often used in workplaces to give early warnings of toxic vapor releases. Such monitors can be lifesaving, if they prevent or minimize chemical accidents such as the one which occurred in Bhopal, India.

At present, a limited number of continuous monitoring devices are available. Monitors are available for gases such as CO, NO_2, and SO_2 in stack gases, and for monitoring some metals and total organic carbon in water. These automated methods are often less expensive than laboratory-analyzed samples, because they require minimal operator attention. However, most of them do not have the sensitivity required for trace level determinations.

2.2 Types of samples

Grab sample: a grab sample is a discrete sample which is collected at a specific location at a certain point in time. If the environmental medium varies spatially or temporally, then a single grab sample is not representative and more samples need to be collected.

Composite sample: a composite sample is made by thoroughly mixing several grab samples. The whole composite may be measured or random samples from the composites may be withdrawn and measured.

A composite sample may be made up of samples taken at different locations, or at different points in time. Composite samples represent an average of several measurements and no information about the variability among the original samples is obtained. A composite of samples which all contain about the same concentration of analyte can give a result which is not different from that obtained with a composite made up of samples containing both much higher and much lower concentrations. During compositing, information about the variability, patterns, and trends is lost. When these factors are not critical, compositing can be quite effective. When the sampling medium is very heterogeneous, a composite sample is more representative than a single grab sample. For example, in a study of the exposure to tobacco smoke in an indoor environment, a several hour composite sample will provide more reliable information than several grab samples.

Composite samples may be used to reduce the analytical cost by reducing the number of samples. A composite of several separate samples may be analyzed and if the pollutant of interest is detected, then the individual samples may be analyzed. This approach can be useful for screening many samples. A common practice, for example, in clinical laboratories screening samples for drug abuse among athletes is to analyze a composite of about 10 samples. If the composite produces a positive result, then the individual samples are tested.

A typical compositing scheme is shown in Figure 2.3. Here a field sample is taken at a random time point once within each hour. These 24 field samples per day are mixed to form two composites. From each composite two subsamples are taken and each subsample could also include two repeat samples.

2.3 Sampling and analysis

Even a perfect analytical procedure cannot rectify the problems created by faulty sample collection. A good sampling plan will ensure that the samples obtained will, on average, closely represent the bulk composition of the environment being measured. In addition, the sample must be collected and handled in such a way that its chemical composition does not change by the time it is analyzed. Finally, the sampling must be done with the requirements of the analytical method in mind.

Proper steps should be taken so that the pollutants are not lost or chemically altered during sample collection, preservation, and transport. Organic materials in water or soil samples, for instance, can be readily attacked and digested by bacteria present in the sample. A preservative to prevent bacterial action may be

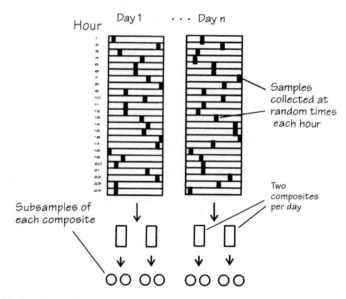

Figure 2.3 Sampling a flowing system, with compositing. One liter samples are collected at a random time each hour. These are composited into two samples which are then subdivided for analysis. This scheme assumes that the flow is constant.

added as samples are collected, or the samples may be frozen or chilled to reduce these losses. Of course, the preservative must be carefully selected so that it does not interfere with the analyses to be done.

The sample size must be adequate. If a 1-ml water sample containing 1 ng/l of a pesticide is collected, the sample would contain such a small quantity of pesticide that it could not be detected by conventional analysis. Therefore, a larger volume of water must be collected, from which the pesticide can be concentrated.

The most common environmental samples are air, water, soil, biological materials, and wastes (liquids, solids, or sludges). Each matrix is sampled using different techniques, but the underlying concepts are the same in each case. It is always good to know as much about the sampling site as possible, especially about the sources of the pollutants being investigated, and the mechanisms for their removal. Before choosing a site for air sampling, pollution sources in the vicinity, such as industries and traffic should be surveyed. Knowledge of previous activities at a hazardous waste site may be helpful in finding the location of maximum contamination. Another important consideration is the physical environment. To predict the migration and distribution of pollutants in a contaminated site, for example, one should know about factors such as soil type, ground and surface water flows. Similarly, for air sampling it is important to take into account factors such as wind direction, temperature, and relative humidity.

2.3.1 *Samples in the laboratory*

When the sample, composited or not, reaches the laboratory, it may have to be reduced in size. A **reduced sample** is prepared by taking a representative portion of the original sample, usually by a mixing and dividing process. These processes depend strongly on the form of the sample and the analytes being sought. A loose solid, such as a soil sample, may be screened, ground, dumped into a pile and quartered, with opposite quarters being selected, and the other two quarters discarded. This process can be repeated several times to reduce a large sample to a reasonably sized reduced sample.

Subsamples are portions of this sample, and after a reduced sample or a subsample is subjected to the laboratory processes needed to prepare it for analysis (grinding, dividing, mixing), it is referred to as a **test sample**. From the test sample, **test portions** are removed for the analysis. These must be of the proper size and concentration to be readily run on the instrument or to be analyzed by the chosen method. Often this test portion is dissolved, digested, or extracted to obtain a **test solution**, and this is sometimes further treated with chemicals to derivatize or react some of its components. In that case, it becomes a **treated solution**. Sometimes the test solution is subdivided into equal portions, often to allow replication of the analytical method. These portions are termed **aliquots**, and this term almost always refers to a portion of a liquid. When a solution is made up in a 100-ml volumetric flask, and a 25-ml portion is taken out by pipette, that portion is a one quarter aliquot of the original solution.

2.4 Statistical aspects of sampling

Uncertainty in environmental measurements can come from both the sampling and the analytical measurement. The total variance is the sum of the two factors:

$$\sigma_T^2 = \sigma_s^2 + \sigma_a^2 \tag{2.1}$$

where the subscript T stands for total variance; subscripts s and a stand for the sampling and analysis processes, respectively. The variance in the sampling process is more difficult to account for. The total variance can be estimated by collecting and analyzing several samples which are expected to produce identical results. The variance of the analytical process can then be subtracted from the total variance to obtain the sampling variance. The variance in the sample not only comes from variation in the sample population, but also from variability during sampling:

$$\sigma_s^2 = \sigma_p^2 + \sigma_{sa}^2 \tag{2.2}$$

where p and sa represent population and sampling procedure, respectively. The variation in the population may be due to stratification or to temporal variability.

The most important question is the minimum number of samples needed for meeting the measurement objective. In environmental sampling, the situations vary case by case, and it is not easy to form a sampling strategy based on classical statistical prediction. The discussion here offers a simplified approach to predicting the minimum number of samples required to estimate the average pollutant concentration in a certain population. The total uncertainty, E, at a specific level of confidence, is selected. The value of E, and the confidence limit can be used to estimate the quality of the measurement:

$$E = z\sigma/n \qquad (2.3)$$

where σ is the standard deviation of the measurement, z is the percentile of standard normal distribution depending upon the level of confidence and n is the number of measurements. In environmental measurements, the E, σ and n can be assigned to the sources from which the variations arise.

If the variance due to sampling, σ_s^2, is negligible and the major source of uncertainty is in the analysis, the minimum number of analyses per sample is given by:

$$n_a = [z\sigma_a/E_a]^2 \qquad (2.4)$$

The number of analyses can be reduced by choosing an analytical method which has higher precision, i.e., a lower σ_a or by using a lower value of z, which means accepting a higher uncertainty.

If the measurement uncertainty is negligible ($\sigma_a \to 0$), the minimum number of samples, n_s is given by:

$$n_s = [z\sigma_s/E_s]^2 \qquad (2.5)$$

Again the number of samples can be reduced by accepting a higher uncertainty or by reducing σ_s. The sample variance can be reduced by using a larger number of samples or by taking composite samples.

When σ_a and σ_s are both significant, the total error E_T is given by:

$$E_T = z\left[\frac{\sigma_s^2}{n_s} + \frac{\sigma_a^2}{n_a n_s}\right]^{1/2} \qquad (2.6)$$

This equation does not have an unique solution. The same value of error, E_T, can be obtained by using different combinations of n_s and n_a. Combinations of n_s and n_a should be chosen based on scientific judgment and the cost involved in sampling and analysis. In the usual environmental case, the parameters of the equation are probably not known accurately or even well estimated. Therefore, while statistical equations for determining the number of samples and the number of replicates can be derived, these are seldom of practical use, except as general guidelines.

A simple approach to calculating the number of samples is to collect and analyze a few samples to estimate an overall standard deviation, s. Using Student's t-distribution, the number of samples required to achieve a given confidence level is calculated as:

$$n = (ts/e)^2 \qquad (2.7)$$

where t is the t-statistic value selected for a given confidence level, and e is the acceptable level of error or uncertainty. The degrees of freedom which determine t can be first chosen arbitrarily and then modified by successive iterations. If an experimental value of s is not available, an estimate may be done from previous similar studies.

Example

Preliminary analysis of a few samples from a contaminated site showed Cr(VI) concentrations between 5 and 20 µg/g, and a standard deviation of 3.25. Calculate the number of samples required so that the sample mean would be within ± 1.5 µg/g of the population mean at the 95% confidence level. Let us assume 10 degrees of freedom. Using Equation 2.7 and the t-table:

$$t = 2.23, \qquad s = 3.25, \qquad \text{and} \qquad e = 1.5$$
$$n = (2.23 \times 3.25/1.5)^2 = 23$$

Since 23 is significantly larger than 10, an iteration must be done with a new value of t corresponding to 23 degrees of freedom:

$$n = (2.069 \times 3.25/1.5)^2 = 20$$

Therefore, 20 samples should be tested. To reduce the number of samples a higher level of error or a lower confidence level may be accepted.

2.5 Water sampling

Water samples can come from many sources: ground water, precipitation (rain or snow), surface water (lakes, river, runoff, etc.), ice or glacial melt, saline water, estuarian water and brines, waste water (domestic, landfill leachates, mine runoff, etc.), industrial process water and drinking water. Pollutants are distributed in the aqueous phase and in the particles suspended in the water. Solids and liquids with densities less than water (such as oils and grease) tend to float on the surface, while those with higher density sink to the bottom. The composition of stagnant water varies with the seasons and also with ambient temperatures. In rivers, lakes, and oceans the concentration of pollutants varies with depth and may also depend on the distance from the shore.

Precipitation water changes with meteorological conditions and atmospheric concentrations of the species of interest. The concentration of rainwater components may be higher when precipitation begins, and drop as the pollutants are washed out of the atmosphere. Concentration of water soluble gases such as H_2S, SO_2, NO_x are also higher in the early part of a precipitation event. Ground water shows seasonal variation and is especially affected by rain or snow. Many of these sources exhibit spatial and temporal variation and sampling devices should be chosen with these variations in mind.

Many different types of manual and automatic samplers are commercially available. They are designed to collect grab samples or composite samples. Particular attention is given to the material of construction of the sampler. Stainless steel or Teflon are preferred because of their inert nature.

2.5.1 *Surface water sampling*

Sampling surface water sources such as lakes, ponds, lagoons, flowing rivers and streams, sewers, and leachate streams can be quite challenging. Shallow depths can be sampled as easily as dipping a container and collecting water. However, sampling at depth in stratified sources can offer unique challenges. Prior to sampling, surface water drainage around the sampling site should be characterized. In a flowing water stream, sampling should be carried out downstream before sampling upstream, because the disturbance caused by sampling may affect sample quality. Similarly, if water and sediment samples are to be collected at the same point, the water sample should be collected before the sediment is stirred up.

The simplest sampling device is a dipper (or a container) made of stainless steel or Teflon. The device is filled by slowly submerging the sampler into the water with minimum disturbance and the water is transferred to the sample bottle. This type of device is not good for volatile pollutants, which can be lost during sample transfer or by sticking to the surface of the dipper. An example of a surface sampler is shown in Figure 2.4. For sampling from a pond or lagoon, a telescopic pole is attached to the dipper so that the sample can be collected at a distance.

Several different devices are commercially available for collecting samples at different depths. Most of them work on the general principle that a weighted

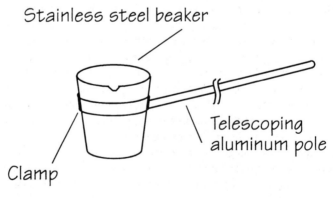

Figure 2.4 Pond sampler. A simple beaker on an extending pole can be used to take grab samples from a body of water.

bottle is lowered to the specified depth. At this point, a stopper or a cap is opened and the bottle is allowed to fill. Then the stopper is closed to prevent any water from flowing in or out and the bottle is pulled out. An example of such a device is shown in Figure 2.5.

2.5.2 *Ground water well sampling*

To obtain ground water samples, monitoring wells, from which water samples can be collected, are drilled into the ground. Care should be taken so that the water does not get contaminated during the drilling process or contaminants do not enter the water from the surface through the well. To ensure that the sample represents the water in the well and contaminants from drilling are not present in the sample, some water is removed from the well before a sample is collected. The amount of water purged depends upon the diameter, depth and the refill rate of the well. The purge amount is usually three to 10 times the well volume. In some cases, the pH, conductance, or temperature are monitored until a constant value is reached. Then a sample is collected.

Various bailers and pumps are used in ground water sampling. Bailers are made of stainless steel or Teflon with a check valve at the bottom. The check valve opens to fill the sample, but closes when the sample is brought up. A sketch of the bailer is shown in Figure 2.6. They can be used to obtain samples

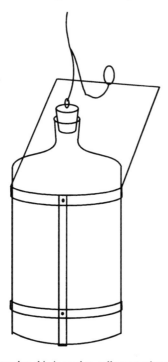

Figure 2.5 Weighted bottle sampler: this is used to collect samples at a predetermined depth.

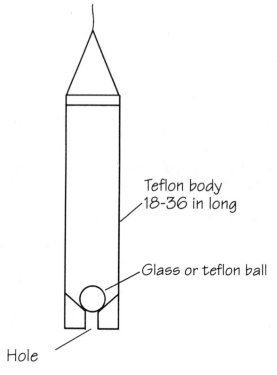

Teflon body
18-36 in long

Glass or teflon ball

Hole

Figure 2.6 Groundwater bailer: sampler fills automatically, and the check valve in the bottom keeps the sample from flowing out as the bailer is retrieved from the well.

with minimum disturbance and are useful for samples containing volatile pollutants or those which may degrade with contact with oxygen.

Peristaltic pumps, among others, are used in ground water sampling. Peristaltic pumps are common because the water does not come in contact with any pump parts. Teflon tubing is inserted into the well and is attached to a flexible tubing which is wrapped around the pump rotor and connected to the discharge tube. As the rotor turns, the tubing is squeezed and released and water is forced through the tubing with a peristaltic action.

2.6 Biological tissue sampling

Contaminants in water or soil often find their way into the food chain and bioaccumulate in plant or animal tissues. The sampling and analysis of various specimens from the biota may be a good way to establish the extent of contamination. Often more pertinent information on the extent of damage done by a particular contaminant can be found in this way, rather than by analyses of water and sediment. A contaminant which is present in a very stable, insoluble form in sediment may remain inert for long periods of time, with no effect on the ecosystem. Yet, the substance would be found in a routine analysis of the sediment. As an example, the extent of water contamination is often determined by a study of fish or other aquatic organisms. This is not only useful for checking on water quality, but also provides a guide to the acceptability of the fish for consumption. Fish tissues are often analyzed for metals and for organic pollutants such as PCBs, and pesticides.

The fish may be caught by hook and line, in nets or traps, or by stunning them with an electrical shock. Because of the uncertainty involved with individual fish, replicates are important. Fish should be chilled immediately by placing them on wet ice in an insulated container. Data on the weight of the fish as well as the time and location of the catch should be recorded. Fish can be stored on ice for a maximum of 24 h, but should be frozen if longer storage is necessary. In preparing samples for analysis, the fillets should not be skinned. Many fat soluble pollutants such as chlorinated pesticides tend to be concentrated in the layer of fat underneath the skin. Many pollutants tend to accumulate in the liver, so this should be dissected and analyzed. All sampling equipment, as well as dissecting tools should be carefully cleaned and decontaminated by washing with detergent, isopropyl alcohol and finally with analyte-free water, to avoid cross contamination of samples.

2.7 Soil sampling

Soil is quite heterogeneous, containing rocks, trapped gases, and liquids. It varies across the surface, and with depth. This variation is caused by contact with the atmosphere and the biosphere, as well as by the flow of ground water. Soil sampling devices must be made of tough material which can be forced into the soil. These are usually brass, steel, or plastic, sometimes Teflon coated to prevent contamination of the samples by the metals used in construction of the sampler. Stainless steel sampling devices are most popular. Chrome and nickel plated devices should be avoided, since scratches and flaking can contaminate samples with trace elements. When the sampling device is forced into the soil, there is much friction between the tool and the soil sample. Since most of the possible contamination will occur on the surface of the sample which comes in contact with the tool, contamination can be reduced by collecting samples with high volume to surface ratio.

Soil samples collected from the uppermost foot of soil can be obtained using a sample scoop. The soil can be loosened with a shovel or a spade and a scoop can be used to collect the sample. A device for collecting an undisturbed sample is a thin walled tube 3–8 cm i.d. and 30–60 cm long. This tube sampler is pressed or hammered into the soil and then is pulled out, bringing up a core sample which preserves differences in the soil composition with depth.

For obtaining samples from a greater depth, a device that can drill into the ground has to be used. Samplers of many different designs are available for doing this. A thin walled tube sampler is usually used with an auger bit to drill a hole to the desired depth in the soil. The auger bit is then replaced with a tube corer, which is lowered down to the bottom of the hole and is pushed into the soil to the desired depth. The tube is then withdrawn and the sample is collected. This device is quite versatile. Samples can be collected at the surface by using only the tube corer. Devices like this may be used to collect samples down as far as 6 m under ideal conditions. However, because rocks may be encountered, or the bore hole may collapse in softer soils, the normal sampling depth is usually less than 2 m. Different types of cutting tips are available for coring dry, moist, sandy, or hard rocky soil. This sampler is depicted in Figure 2.7.

The Veihmeyer sampler consists of a sampling tip, sampling tube, a drive head, and a drop hammer. The sampling tube is constructed of chromium molybdenum steel, and its length can be anywhere from 3 to 16 ft. The tube is calibrated every 12 in. The drive head is attached to the top of the tube to prevent the hammer from deforming the tube when it is driven into the ground. The sampling tip is removable, and different tips are available for different types of soils. The drop hammer is made of cast iron, weighs about 15 lb, and is used to drive the sampling tube into the ground. A puller jack may be used to pull the tube sampler out when sampling is done at a greater depth, or when the soil is hard.

2.8 Sampling stratified levels in containers

Sampling the contents of containers of non-homogeneous materials, for example, barrels of hazardous waste, can offer additional challenges especially when the liquid is stratified within the container. An oily liquid will float over an aqueous waste, while solids may settle to the bottom. Several different types of samplers have been developed for taking samples from containers of this type.

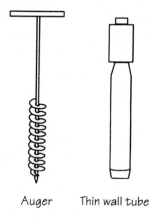

Auger   Thin wall tube

Figure 2.7 Auger and tube sampler. The auger is used to form a hole down to the desired depth. The tube is then driven into the soil to bring up a core sample.

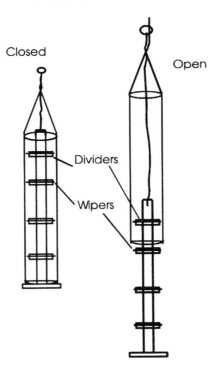

Figure 2.8 Stratified sample thief: the sampler can bring up an example of the layered structure of a liquid sample. The sampler is dropped into the sample, and the cover retracted. After the sample has flowed around the dividers, the cover is slid down, trapping samples of the individual layers.

An example is the stratified sample thief, shown in Figure 2.8. This consists of an outer sheath and an inner rod. Along the rod round, flexible wipers are positioned to hold the sample in place, and prevent the different layers of liquid from mixing with one another. First, the outer sheath is raised, exposing the center rod. The rod is gently lowered into the container from which the sample is to be collected. Then the outer sheath is moved down, trapping the liquid in place. When the sampler is withdrawn, samples of the liquid at each level can be recovered.

2.9 Preservation of samples

The sample must be representative of the environment. Both physical and chemical processes may be involved in changing the composition of a sample after it is collected. Common physical processes which may degrade a sample are volatilization, diffusion, and adsorption. Possible chemical changes include photochemical reaction, oxidation, and microbial degradation.

The collected sample is invariably exposed to conditions very different from the original source. For example, if a ground water sample is exposed to sunlight after it is collected, photochemical reactions may degrade some of the analytes of interest. Samples often must be preserved in some way, to keep them stable until the analysis is completed. Table 2.1 shows some sample types and the appropriate preservation methods. These methods must not only keep the sample stable, but must not interfere with the analyses to be done.

The steps commonly taken to minimize sample degradation are the use of proper containers, temperature control, addition of chemical preservatives, and observance of the recommended maximum sample holding time. The holding time depends upon the analyte of interest and the matrix under consideration. For example, some metals in water can be held for months, while Cr(VI) analysis should be performed within 24 h. If the holding time is unknown, one can make up a spiked sample, or store an analyzed actual sample, and analyze it at fixed intervals to determine the optimum holding time.

Table 2.1 Sample preservation techniques

Sample	Container type	Preservation method	Holding time
Metals			
Dissolved	plastic or glass	filter on site, acidify to pH 2 with HNO_2	6 months
Total	plastic or glass	acidify to pH 2 with HNO_2	6 months
Cr(VI)	plastic or glass	Cool to 4°C	24 h
Hg	plastic or glass	acidify to pH 2 with HNO_2	28 days
Inorganic anions			
Bromide, chloride, fluoride	plastic or glass	none	28 days
Chlorine	plastic or glass	none	analyze immediately
Iodide	plastic or glass	Cool to 4°C	24 h
Nitrate, nitrite	plastic or glass	Cool to 4°C	48 h
Sulfide	plastic or glass	Cool to 4°C, add zinc acetate and NaOH to pH 9	7 days
Organics			
Organic carbon	plastic or brown glass	Cool to 4°C, add H_2SO_4 to pH 2	28 days
Purgeable halocarbons	glass with teflon septum cap	Cool to 4°C, add 0.008% $Na_2S_2O_3$	14 days
Purgeable aromatics	glass with teflon septum cap	Cool to 4°C, add 0.008% $Na_2S_2O_3$ and HCl to pH 2	14 days
PCBs	glass or teflon	Cool to 4°C	7 days to extraction, 40 days after
Organics in soil	Glass or teflon	Cool to 4°C	as soon as possible
Fish tissues	Wrap in aluminium foil	Freeze	as soon as possible
pH			immediately, on site
Temperature			immediately, on site
Biochemical oxygen demand	plastic or glass	Cool to 4°C	48 h
Chemical oxygen demand	plastic or glass	Cool to 4°C	28 days

2.9.1 *Volatilization*

Analytes with high vapor pressures, such as volatile organic compounds and dissolved gases, such as HCN, SO_2, will readily escape from the sample by evaporation. Filling sample containers to the brim, so that they contain no head space is the most common practice to minimize volatilization. The volatiles cannot equilibrate between the water and the vapor phase above, if no air space is present at the top of the container. The samples are usually held at 4°C, on ice, to lower the vapor pressure. Agitation during sample handling should also be avoided, to minimize air–sample interaction.

2.9.2 *Choice of proper containers*

The surface of the sample container may interact with the analyte. For example, metals can adsorb irreversibly on glass surfaces, so plastic containers are often chosen for water samples to be analyzed for their metal content. These samples are also acidified with HNO_3 to help keep the metal ions in solution.

Organic molecules may also diffuse in or out of the sample if the proper container is not used. Plasticizers such as phthalate esters can diffuse from plastic containers into the samples. For organic analytes, it is best to collect samples in glass containers. Bottle caps should have Teflon liners to preclude contamination from the plastic caps.

Oily materials in water samples will adsorb strongly on plastic surfaces, and samples to be analyzed for such materials are usually collected in glass bottles. Oil which remains on the bottle walls should be removed by rinsing with a solvent and returned to the sample. Sometimes, oily samples are emulsified with a sonic probe to form a uniform suspension of the oil and then removed for analysis.

2.9.3 *Absorption of gases from the atmosphere*

Water samples can dissolve gases from the atmosphere as they are being poured into containers. Such components as O_2, CO_2 as well as volatile organic compounds may dissolve in the samples. Oxygen may oxidize species such as sulfite or sulfide to sulfate. Absorption of CO_2 may change conductance or pH measurements. This is one reason why pH measurements are always done in the field. Dissolution of organic compounds may lead to the detection of compounds that were actually absent. Field blanks should show if the samples have been contaminated with organic compounds which have been absorbed during sampling or transport.

2.9.4 *Chemical changes*

A wide range of chemical changes in the sample is possible. For inorganic samples, controlling the pH can be useful in prevention of chemical reactions.

For example, metal ions may react with oxygen to form insoluble oxides or hydroxides. The sample is usually acidified with HNO_3 to a pH less than 2, as most nitrates are soluble and excess nitrate ions will prevent precipitation. Other ions such as sulfide, or cyanide, are also preserved by pH control. Samples collected for NH_3 are acidified with sulfuric acid to stabilize the NH_3 as NH_4SO_4.

Organic species can also undergo changes due to chemical reactions. Photo-oxidation of polynuclear aromatic hydrocarbons, for example, is prevented by storing the sample in amber glass bottles. Organics can also react with free chlorine to form chlorinated organics. This type of problem is common for samples collected in treatment plants after the water has been chlorinated. Sodium thiosulfate, added to the sample, will remove chlorine.

Samples may also contain microorganisms which may biologically degrade the sample. High or low pH conditions, and chilling can minimize microbial degradation. The microbes can also be killed by addition of mercuric chloride or pentachlorophenol, if these preservatives will not interfere with the planned analyses.

2.9.5 Sample preservation for soil, sludges, and hazardous wastes

Handling of water samples is better understood than solid and sludge samples, as these can be more varied in composition, but similar methods are used. Commonly encountered problems are biodegradation, oxidation–reduction and volatilization. Storing the sample at low temperature is always recommended to reduce volatilization, chemical reaction, and biodegradation.

A preservation temperature of 4°C is most commonly used, because ice storage is convenient, and because a lower temperature may freeze the water, and separate the organic phase from the aqueous. Minimizing head space is also important for reducing volatilization losses. This also eliminates oxygen so that aerobic biodegradation or chemical oxidation are reduced. Samples to be analyzed for volatile organic compounds are sometimes collected directly into a known quantity of a solvent. In the laboratory, the analytes are either extracted or purged from the solvent. Methanol and polyethylene glycol have been used for this purpose.

Suggested reading

Csuros, M. (1994) *Environmental Sampling and Analysis for Technicians*, Lewis Publishers, Boca Raton.

Gilbert, R.O. (1987) *Statistical Methods for Environmental Pollution Monitoring*, Van Nostrand Reinhold, New York.

Grieco, P. and Trattner, R. (1990) *Sampling for Environmental Data Generation*, SciTech Publishers, Matawan, NJ.

Keith, L.D. (1988) *Principles of Environmental Sampling*, American Chemical Society, Washington, DC.

Study questions

1. What must be considered in taking a reduced sample from a bulk sample? How is this done on a practical basis?
2. What must be considered in determining the number of samples to be taken in a particular study?
3. What is the purpose of discarding the initial quantity of water from a well, before a sample is taken?
4. Describe some sampling situations in which a sampler which takes a stratified sample would be necessary.
5. Fish have been dying in a river downstream from a company which makes batteries. The public is outraged, and a local news reporter is publishing articles blaming the industry, saying that dissolved mercury from their discharge pipe must be causing the fish to die. If you were sent in as an impartial investigator, what questions would you want to find answers to? What samples would you take (how, when, where, how frequently?) and what analyses would you do on these samples? Which samples would be essential to do and which would be nice to do if the money is available? Give your reasons for your answers.
6. What are some considerations which must be taken into account in selecting suitable sampling containers?
7. A field is to be surveyed for pesticide residues. Other fields treated in similar ways have been shown to have residues of 40–200 µg/g, with a standard deviation of about 5 µg/g. If an error level of ±2 µg/g is acceptable, how many samples are needed to be 95% confident that the requirement is met?
8. What kind of sampling strategy would you use in the following situations? Take into consideration the spatial and temporal variability you expect to encounter. Which type of sampling would be preferred: grab sample, composite sample or continuous monitoring?
   a. To study the contamination of fish in a river where a chemical company has a waste water discharge outlet.
   b. To identify accident release of chemicals by the industry mentioned above.
   c. To implement strategies to reduce smog formation in your city.
   d. To study the effect of auto exhaust on the air quality in your city.
   e. To identify ground water contamination around an abandoned chemical factory.
   f. To determine if apples from a sprayed orchard are contaminated with pesticides.

3   Spectroscopic methods

The interaction of electromagnetic radiation, the energy inherent in light, with matter is useful in many ways to determine both the identity of compounds and their concentration in mixtures. The electromagnetic spectrum, shown in Figure 3.1, ranges from high energy γ-rays to very low energy radio waves. Many regions of the spectrum are used for obtaining information about material samples. Because of the wide range of energies involved, the methods used in the various spectral areas seem quite different, but they all are based on similar principles.

3.1 Spectroscopic methods for environmental analysis

Environmental analysts have used visible and ultraviolet spectroscopic methods for years. Common colorimetric tests for properties of water, such as acidity, have been reduced to simple kit forms, using visual color matching or hand-held portable colorimeters. Atomic absorption and emission spectroscopy in the ultraviolet and visible regions are used to determine metals in samples derived from air, water or solids. These usually require that the analyte be put into solution before analysis. However, some solid or semisolid samples can be analyzed directly when electrothermal atomization is used in atomic absorption spectrometry.

Infrared spectroscopy is also finding a place in the environmental analysts' arsenal of weapons, with the development of long range IR sensors. These operate by beaming an IR signal to a reflector mounted several hundred yards away, and analyzing the returned beam to determine the concentration of certain compounds in the intervening air mass. Ultraviolet long path methods are also being used, although not as commonly as IR.

X-ray methods, primarily X-ray fluorescence, are used to determine the atomic composition of solid materials, and have the advantage of operating on

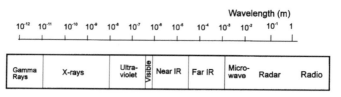

Figure 3.1 The electromagnetic spectrum.

solids without prior dissolution. X-ray fluorescence (XRF) has found applications in determining metals in particulate matter filtered out of the air, and in soil samples. Hand-held XRF units are used in the field for quick screening, especially for lead in painted surfaces or dust. Magnetic resonance spectroscopy and microwave region spectroscopy have been applied in some environmental research, but have not yet found a niche in everyday environmental analysis methods.

3.1.1 *Properties of electromagnetic radiation*

Electromagnetic radiation has both a wave and a particle character. It is often envisioned as a wave being propagated through space, which can be characterized by its frequency. **Frequency** is defined as the number of oscillations passing a point in a specified time interval. The **wavelength**, the distance measured from the maximum of one wave to the next, is related to the frequency by the speed of the wave. Both wavelength and frequency are related to the energy borne by each oscillation. The relationships are expressed as:

$$E = h\upsilon = \frac{hc}{\lambda} \quad \text{and} \quad \upsilon = \frac{c}{\lambda} \tag{3.1}$$

where E is the energy, h is Planck's constant, 6.62×10^{-34} J s, λ is the wavelength, υ is the frequency, and c is the speed of light in a vacuum, 3.00×10^8 m/s. Since the speed of the wave propagation depends on the matter through which the wave passes, the frequency and the energy are the only truly inherent characteristics of a wave. The wavelength will change when the medium changes. The relationship between the speed of light in any material and that in a vacuum is described by the **refractive index**, η, of the material.

$$\eta = \frac{\text{speed of light in vacuum}}{\text{speed of light in material}} \tag{3.2}$$

It should be noted that the speed of light in a material, and therefore the refractive index of the material, changes with wavelength, which has some important consequences in spectroscopy.

The important factors in the characterization of a beam of radiation are its frequency and its amplitude. Radiation detectors in instruments cannot usually measure the amplitude of radiation. The radiant power, P, often referred to as the intensity, is measured instead. P is related to the square of the amplitude of the wave. It is expressed in terms of the energy and the photon flux, the number of photons per unit time, ϕ.

$$P = E\phi = h\upsilon\phi \tag{3.3}$$

Electromagnetic radiation can be produced as a **monochromatic** beam, which consists of radiation composed of a very small range of wavelengths, ideally a single wavelength. A beam which contains a wide distribution of wavelengths is called **polychromatic**. Most spectrometric instruments select a discrete band of wavelengths to be used for measurements. The range of wavelengths included in this band is called the **bandpass** of the instrument. The narrower the bandpass, the closer to monochromatic the measuring beam will be. However, the selection of a very narrow beam will always reduce the power of the light being used.

The wave properties of light lead to the phenomena of dispersion and interference. These can be used in selecting certain bands of radiation and separation of radiation into discrete wavelengths. When a beam of polychromatic radiation is passed through a glass or quartz prism, the light beam is deflected as it passes from air into glass and back again. The deflection is due to the difference between the refractive index of the two media. However, the refractive indices differ with the wavelength of the light. Therefore, red radiation is bent through a different angle than green radiation. The different wavelengths are focused at different points in space, and can be observed sequentially, or **scanned**, by moving either the prism or the detector. Figure 3.2 shows beams of different wavelengths as they are diffracted through a prism.

Interference is the effect which occurs when light rays of the same wavelength are brought together. If they are **in phase**, with their waves synchronized, the resulting beam will be constructively interfered with and the beam will be reinforced. If the beams are **out of phase** they will destructively interfere with each other and disappear. Light rays passed through closely spaced slits or reflected from narrow, closely spaced, angled surfaces will undergo interferences so that different wavelengths will be reflected at different angles. This is the principle under which diffraction gratings operate.

3.1.2 *The electromagnetic spectrum*

While the entire spectrum of radiation obeys the same laws, travels at the same speed, and has the same basic nature, different spectral regions are used in different ways. Short wavelength, high energy radiation in the X-ray region has sufficient energy to cause changes in the inner electron structures of atoms. The

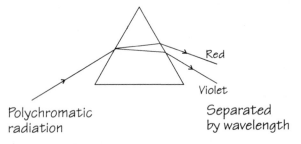

Figure 3.2 Dispersion of light through a prism. As white light passes through the prisms each different wavelength ray is diffracted at a different angle, because the refractive index and wavelength are related.

chemical state of an atom has little effect on the inner shell electrons. Therefore, X-rays can be used to probe the elemental composition of a sample, without regard to the chemical state of the atoms present. Ultraviolet radiation interacts with the outer electronic levels, promoting electrons to higher energy levels, so this region of the spectrum will yield information on the bonding of the atoms into molecules, and their oxidation state. At even lower energy, infrared radiation affects the rotational and vibrational energy levels in molecules, giving spectra that indicate the identity of functional groups and are rich in molecular structure information. In the radio frequency area, the energies are so low that only reorientation of nuclear spins occurs in the molecules. The absorption of energy in this area is informative about structure and conformation of molecules. In Table 3.1, the various types of spectroscopy, their applications, and the wavelength range used are listed.

3.1.3 *Radiation and matter*

The interaction of radiation and matter can be put to use in either the emission or absorption modes. In emission spectroscopy, molecules or atoms are stimulated in some way to raise them to a higher energy level. When they return to the ground state, energy is emitted as radiation. The radiation is detected and its frequency and intensity give information about the identity as well as the amount of the radiating species. In absorption, radiation is passed through the sample. The reduction in the intensity of the radiation emerging from the sample indicates the concentration of absorbing species, while the wavelengths absorbed provide information about the identity of the absorbing species.

Table 3.1 Spectroscopic methods and applications

Type of spectroscopy	Application	Wavelengths used
X-ray fluorescence	Short λ X-rays are absorbed by sample atoms, and longer λ X-rays, characteristic of the sample are emitted and detected. Identifies elemental composition	0.3–3.0 Å
UV-visible absorption spectroscopy	Visible or ultraviolet radiation at selected λ is passed through the sample and the amount absorbed is measured. Used to measure molecular or ionic species in solution	2.5 µm to 2400 Å
Infrared absorption	Identity of molecules or functional groups in liquid, solid or gaseous samples, and their composition are determined by measuring IR absorbance	1 mm to 2.5 µm
Atomic absorption	Samples (liquids or solids) are decomposed to free atoms which absorb light from a line source in the UV and visible ranges	2.5 µm to 2400 Å
UV fluorescence	Molecular sample components are stimulated with short λ radiation and emitted light at longer λ is detected	
Chemiluminescence	Radiation in the visible or ultraviolet is generated during a reaction. The quantity of light is a measure of the amount of reactant	

Fluorescence occurs when radiation is absorbed, and the excited species formed loses part of its excess energy by a non-radiative means. Then the remaining energy is emitted as radiation. This radiation is of a longer wavelength than that which caused the excitation. The instruments used to measure absorption, emission and fluorescence are depicted in Figure 3.3.

According to quantum theory, atoms and molecules may only exist in certain energy states. Their lowest energy level is called the ground state. They can be promoted to a higher energy level or an excited state by irradiation with an electromagnetic wave of suitable wavelength:

$$S + h\nu \rightarrow S^* \tag{3.4}$$

where S is a low energy species, and S^* is its excited state. The amount of energy absorbed is exactly equal to the energy difference between the lower

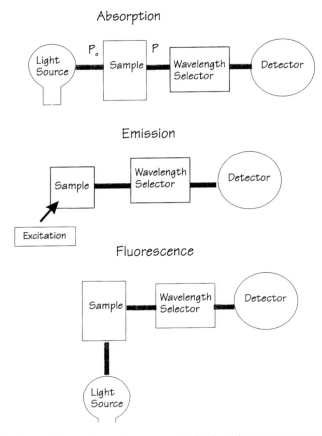

Figure 3.3 Instruments for spectroscopic measurements. In absorption spectroscopy, radiation from the source, P_0, is partly absorbed by the sample and that remaining, P, is detected. In emission, the sample is excited to a higher energy state at a high temperature, and the emitted radiation is detected. In fluorescence, emission is stimulated by an external light source, and the emitted radiation, of a longer wavelength, is detected.

energy state and the excited state. This energy can only be supplied from radiation of the specific wavelength which has the same energy as that needed for the transition. The excited species can then lose its energy by a process in which no radiation is emitted, or through a radiative process that involves emission of radiation. These phenomena are shown in the energy diagram presented in Figure 3.4.

Both atoms and molecules exist in discrete energy states. These energy levels can be attributed to the electronic states, as well as rotational and vibrational levels. In an electronic transition, an electron is promoted to a higher energy level. In rotational and vibrational transitions, the molecule absorbs or emits energy to undergo rotational or vibrational changes. The total energy is expressed as E_t:

$$E_t = E_e + E_r + E_v \tag{3.5}$$

where E_e, E_r and E_v denote the energy associated with electronic, rotational, and vibrational states. Figure 3.5 shows some molecular motions which lead to vibrational and rotational absorbances. Single atoms have no rotational or vibrational levels. In absorption or emission spectroscopy with atomic vapors, only electronic transitions are possible. Therefore, the spectra of atoms consist of a series of narrow lines (0.02–0.05 Å wide), each corresponding to a discrete

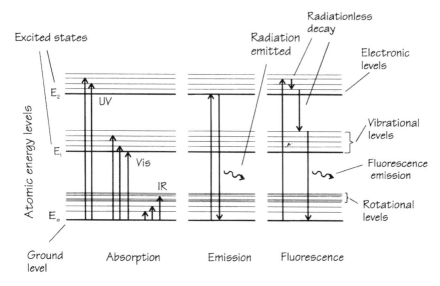

Figure 3.4 Absorption and emission of electromagnetic radiation. Absorption of radiation moves the atom to a higher energy level. It may return to the ground state by emission, or it may lose some of its energy as thermal energy, then return to the ground state by emitting a new, longer wavelength, fluorescent radiation. Transitions among the vibration and rotational states give rise to absorption at IR wavelengths, while those between electronic levels involve more energetic visible or UV radiation.

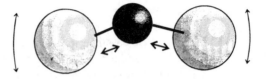

Vibrational motions

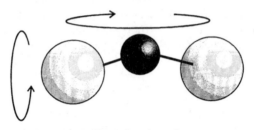

Rotational motions

Figure 3.5 Molecular vibration and rotation motions, which give rise to IR absorbances.

electronic transition. This is called a **line spectrum**. Since each element produces a unique set of spectral lines, they can be used to identify the element. Atomic spectroscopy is, therefore, an excellent tool for qualitative analysis even in a complex sample.

In a molecule, each electronic state is associated with rotational and vibrational sublevels, in addition to the electronic levels. Consequently, for a molecule, there are numerous possible transitions which are quite close in energy. As a result, a **continuous spectrum** is produced, which contains broad absorbance or emission bands, rather than discrete lines. Figure 3.6 shows the difference between a typical molecular absorption spectrum and an atomic absorption spectrum. Since the molecular spectrum does not have many distinctive features,

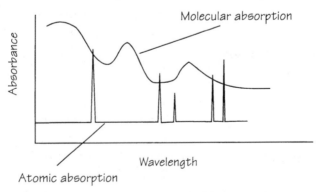

Figure 3.6 Comparison of molecular and atomic spectra. The atomic spectrum shows very narrow absorption or emission lines, while the molecule shows broad bands.

identification of molecules in complex environmental samples is much more difficult, unless the component of interest is first separated from the rest of the matrix. However, molecular absorbance spectrometry can provide excellent quantitative results.

3.2 Absorption spectroscopy

In absorption spectroscopy, a sample is irradiated with electromagnetic radiation, and the amount which passes through is monitored. An absorption spectrum is a plot of amount of radiation absorbed as a function of wavelength. Based on the wavelengths which are absorbed, the absorbing atoms and molecules may be identified. Similarly, from the amount of energy absorbed, the amount or concentration of the analyte can be determined. Consequently, spectroscopy provides both qualitative and quantitative information.

Absorption spectroscopy dealing with absorption of radiation by atomic vapors is called **atomic absorption**. This technique is mainly used for determination of metals. Absorption of radiation in the UV-visible region by molecules in solution or vapor form is called **UV–visible molecular spectroscopy**, and absorption of molecules in the infrared region is called **IR spectroscopy**. These techniques are used for qualitative and quantitative determination of molecular species and are widely used in environmental analysis.

Absorption spectroscopy requires a light source to generate the incident radiation, a method of selecting the wavelength at which the measurement is to be made, and a detector to measure the unabsorbed light which passes through the sample. An absorption spectrum is generated by making absorbance measurements at different wavelengths by changing the monochromator setting.

3.2.1 Beer's law

The absorption of incident radiation is governed by Beer's law. This law can be readily derived by considering an amount of absorbing material diluted in a nonabsorbing solvent. Monochromatic light is passed through the solution, and the power of the incident beam, P^o, as well as the power of the emerging beam, P, are measured. If the path of the beam is divided into many infinitesimally thin segments, the power of the light will be diminished by the same fraction as it passes through each segment. This can be assumed because the number of particles encountered by the beam in each segment will be the same. Expressing this mathematically, the fractional decrease in power, $-dP/P^o$, depends on the number of absorbing particles in the segment, dn times a proportionality constant, k.

$$-\frac{dP}{P^o} = \text{k } dn \qquad (3.6)$$

To find the total amount of absorption over the whole path, $-dP/P^o$ is integrated from P^o to P and k dn is integrated between 0 and n, giving

$$\ln \frac{P}{P^o} = -kn \qquad (3.7)$$

This is not a useful equation as it stands, but the number of absorbing particles can be calculated from the concentration. The total number of particles, N, in 1 cc of solution is expressed as:

$$N = cN_A bS \qquad (3.8)$$

where c is the concentration in moles/cc, N_A is Avogadro's number, b is the length of the path in cm, and S is the cross sectional area of the beam. A parameter called **absorbance** (A) is defined as log P^o/P, and is substituted into the above equation. For convenience, the logarithmic term is converted from natural to common logarithm, reducing the relationship between absorbance and concentration to a simple law:

$$A = \varepsilon bC \qquad (3.9)$$

known as Beer's Law. The constant ε, the molar absorptivity, is a function of the compound that is absorbing the radiation and the wavelength of the radiation. The units for concentration are moles per liter, while the path length is given in centimeters. In addition, since each molecule absorbs independently, Beer's law is additive. If there are several absorbing species with differing molar absorptivity (ε_1, ε_2, ...etc.) and different concentrations (C_1, C_2, ...etc.) the overall absorbance can be expressed as:

$$A = \varepsilon_1 bC_1 + \varepsilon_2 bC_2 + \cdots + \varepsilon_n bC_n \qquad (3.10)$$

Example
Chromate and permanganate can be determined in the same sample by absorbance spectroscopy. Chromate absorbs strongly at 440 nm while permanganate absorbs better at 525 nm. A standard containing 0.01 M $KMnO_4$ and one containing 0.02 M $K_2Cr_2O_7$ are prepared and their absorbances are measured at both wavelengths. The sample containing the two analytes is also read. The following data are obtained. What is the concentration of permanganate and dichromate in the sample?

Solution	A_{440}	A_{525}
(A) 0.01 M $KMnO_4$	0.0094	0.442
(B) 0.02 M $K_2Cr_2O_7$	0.320	0.0039
Sample	0.177	0.380

Since the pathlength is the same in all these measurements, it can be combined with the molar absorbances into a constant k. For each standard calculate k at each wavelength:

$$A = k(C)$$

$$0.0094 = k_{440, A} (0.01) \qquad k_{440, A} = 0.94$$

$$0.442 = k_{525, A} (0.01) \qquad k_{525, A} = 44.2$$

$$0.320 = k_{440, B} (0.02) \qquad k_{440, B} = 16$$

$$0.0039 = k_{525, B} (0.02) \qquad k_{525, B} = 0.195$$

This generates two simultaneous equations:

$$A_{440} = k_{440, A} (C_A) + k_{440, B} (C_B) \quad \text{and} \quad A_{525} = k_{525, A} (C_A) + k_{525, B} (C_B)$$

Substituting and solving:

$$0.177 = 0.94 (C_A) + 16 (C_B) \quad \text{and} \quad 0.380 = 44.2 (C_A) + 0.197 (C_B)$$

$$C_A = [KMnO_4] = 0.0087$$

$$C_B = [K_2Cr_2O_7] = 0.0066$$

Beer's law predicts a linear relationship between absorbance and concentration. However, in practice, nonlinearity at higher concentrations is often found when concentration is plotted versus concentration. There are several reasons why nonlinearity may occur.

- The refractive index of the solution changes as the solution becomes more concentrated, which changes the absorbance.
- The radiation being absorbed is not monochromatic. Beer's law is valid for monochromatic light.
- The use of a measuring wavelength in an area of the absorbance spectrum where the value of ε is changing rapidly may lead to a nonlinear relationship. In fact, if the portion of the absorption spectrum within the bandpass has a fairly constant slope, Beer's law will appear to be obeyed, as long as the instrument allows accurate resetting of the wavelength. The problem of nonlinearity is exacerbated when the absorbance peak being used for measurement is narrow, and the bandpass of the instrument is wide.
- Chemical deviations may occur if the species being measured is taking part in an equilibrium. When a species ionizes, the ionization equilibrium shifts as the samples are diluted. Changes in pH may also cause shifts in ionization, so samples may need to be buffered to hold the pH constant.
- Stray radiation also contributes to nonlinearity. This radiation (P_s) reaches the detector without passing through the sample, so the apparent absorbance can be expressed as:

$$\varepsilon bC = \log \left(\frac{P^o + P_s}{P + P_s} \right) \qquad (3.11)$$

which no longer describes a straight line. The ratio of stray radiation to incident radiation is an important limiting factor for a spectrometer. The maximum value of A which can be read depends upon the value of P_s.

3.3 Emission spectroscopy

In emission spectroscopy a species, S, is raised to a higher energy state, S^*. As it returns to a lower state, it emits some of the absorbed energy in the form of radiation, which is detected and measured.

$$S^* \rightarrow S + hv \qquad (3.12)$$

The excitation energy can be supplied by raising the sample to a high temperature, by irradiating it with electromagnetic radiation, or by exposing it to an electrical arc or spark. The energy emitted corresponds to the energy difference between the initial and final states. Radiation of a specific wavelength (Equation 3.1) is generated. An emission spectrum is a plot of the intensity of the emitted radiation as a function of wavelength. The wavelength of this radiation contains information about the type of atom or molecule undergoing the energy transition, and so provides qualitative identification. The intensity of emission is proportional to the number of atoms and molecules undergoing the transition and provides quantitative information.

Emission by excited atoms in the vapor state is measured in **atomic emission spectroscopy**, used most often for the determination of metals. In **molecular fluorescence**, molecules are excited by UV radiation and emit at longer wavelengths. In **chemiluminescence** measurements, emission is stimulated by a chemical reaction.

3.3.1 Fluorescence

Sample molecules may be excited by absorbed radiation. These may undergo a radiationless transfer to a lower energy state, before emitting the remaining excess energy as radiation and dropping back to the ground state. This is called **molecular fluorescence**. The radiation emitted, the fluorescence, is of lower energy than the stimulating radiation, and must be at a longer wavelength. Fluorescence can be easily seen when materials are radiated in the ultraviolet, which is invisible to the eye, and the re-emitted radiation occurs in the visible. The molecule absorbs at characteristic wavelengths and emits a spectrum which is also characteristic of the compound. Both the emission and absorption spectra are examined to select the best wavelengths for excitation and for monitoring the emitted fluorescent radiation. Figure 3.7 shows the fluorescence spectrum for the three-ring polynuclear aromatic compound, fluoranthene, at a fixed excitation wavelength.

Molecular fluorescence spectroscopy is not widely used in environmental applications since not every compound fluoresces. The fluorescence detector in

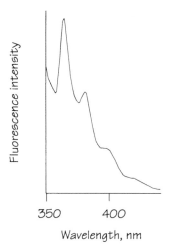

Figure 3.7 The fluorescence spectrum of fluoranthene. Radiation is emitted when fluoranthene is excited by radiation at 240 nm. Note that the emission is at longer wavelengths than the excitation radiation.

HPLC is probably the most common application of this type of spectrometric measurement in environmental analysis. For example, many polycyclic hydrocarbons and their derivatives fluoresce strongly. The fluorescence detector is used for HPLC determination of these compounds. In general, fluorescence is a very sensitive technique. When it can be used, detection limits are typically an order of magnitude better in fluorescence than in UV absorption.

Atomic fluorescence is used for the determination of metals. Here metals are first transferred to an atomic state in a flame. Radiation is used to excite the ground state metal atoms to an excited state. When they return to the ground state, they give off part of this energy as fluorescence.

3.3.2 *Atomic emission*

Atomic emission spectroscopy uses the radiation emitted from excited atoms in a vapor state. Excitation is accomplished by exposing the sample to high temperature in a flame or plasma, by an electric arc, or by a high voltage spark. These excitation methods are not used for molecular species because most molecules will decompose under such highly energetic conditions.

The intensity of emission is determined by the concentration of atoms in the elevated energy state and by the probability of these excited atoms dropping to a new level, emitting the wavelength being monitored. When a population of atoms is excited by heating, the fraction of atoms reaching a certain energy level can be expressed by the Boltzman equation. This equation is:

$$\frac{N_1}{N_2} = e^{-\Delta E/RT} \tag{3.13}$$

where N_1 is the number of excited atoms and N_2 is the number of atoms in the ground state. ΔE is the energy difference between the excited and ground states, and can be calculated from the wavelength of the emitted radiation. R is the gas constant and T the absolute temperature.

Even at temperatures of 2000–3000°C, the excited atoms are a very small fraction of the total number. Not only is the fraction small, but the dependence on temperature is exponential. Therefore, relatively small differences in temperature will have a large effect on the number of emitting atoms. However, temperature has little effect on the number of ground state atoms upon which atomic absorption depends.

Excitation by arc and spark are not widely used in environmental analysis, although they are often used for analysis of ores and geological samples. The arc and spark generate a short-lived burst of radiation, which must be separated and recorded before it disappears. The emitted light is passed through a monochromator and recorded on a film or by phototubes situated at the point where the particular lines of radiation of interest will be brought to focus. This makes it impossible to scan the spectrum, bringing each emitted line onto the detector in sequence. The precision of these methods is poor for sophisticated quantitative work. Therefore, arc and spark emission methods have fallen into disuse, except for some rather specialized applications, and most current instruments use either flames or inductively coupled argon plasma torches for excitation of atomic emissions. Both of these sources provide a steady flow of sample into the flame or plasma, so that the emission lines can be scanned by a single detector.

Emission instruments use most of the same components as absorption systems. A source of light is not usually required, except in the case of fluorescence spectroscopy. Fluorescence is stimulated in a sample by a beam of radiation which is usually aimed perpendicular to the line of the detector.

3.4 Spectroscopic apparatus

All spectrometers, although they use very different spectral regions and produce different types of information, use certain common components. These fall into the categories of light sources, monochromators, and radiation detectors. In absorbance spectroscopy, a source of radiation must be provided in order to measure the amount absorbed. In fluorescence spectroscopy, a source is needed to excite the fluorescence that will be measured. After light is transmitted through or emitted by the sample, it is necessary to measure its intensity at one or more wavelengths. A detector is a device to convert the energy of the radiation into a current or voltage in the measuring circuitry. Often the electrical signal is very small and requires amplification before it can be analyzed. The type of detector needed to determine the intensity of emitted or transmitted light depends on the wavelength. For all detectors, the desired properties are linearity, sensitivity, stability and a wide linear dynamic range. Of course, all these properties are not always available, but the quality and usefulness of a detector can be measured by comparing these qualities among different choices. Finally, it is often necessary to select a band of wavelengths for use in the measurement. This is done by filtering out unwanted wavelengths or by dispersing the radiation from the sample or from the source into its component wavelengths, thus separating them in space.

3.4.1 *Light sources*

Radiation sources are usually classified as broad band sources or line sources. Broad band sources emit a continuous spectrum, as their name would indicate, over a wide range of wavelengths. Line sources, on the other hand, are those in which the emitted radiation arises from specific transitions within atoms of a certain element which are stimulated to emit by being excited electrically or thermally. In the visible region, the common tungsten filament light bulb is an example of a broad band source. Its tungsten filament is heated to about 3000 K. Line sources are exemplified by the hollow cathode lamps used in atomic absorption systems. A particular element is incorporated in the cathode of a lamp specifically designed to be used for the determination of that element. The lamp's emitted light is made up of the spectral lines due to atomic transitions in the vaporized element.

3.4.2 *Wavelength selection*

Wavelength bands are selected for measurement. Certain spectral regions may be selected using filters, when a fairly broad band of radiation is permissible, and when scanning is not required. Such instruments are termed **filter photometers**. **Spectrophotometers** usually contain a **monochromator**, a device to disperse the various wavelengths, so that they are focused at different points in space. Spectrophotometers are often designed to scan, allowing the absorption or emission of the sample across a range of wavelengths to be recorded. Filter based photometers are relatively uncomplicated instruments, which may be more rugged and simpler than instruments containing a monochromator. They often have larger light throughputs, and so require less amplification and simpler electronics. These are often chosen for field portable instruments in situations where their inherent simplicity is more important than their lack of scanning ability and limited wavelength bands. Another apparatus for separating wavelengths, in time rather than in space, is the **interferometer**. This is used for longer wavelength applications, in the IR region of the spectrum.

3.4.2.1 *Filters.* Absorption filters are made of a variety of materials such as gelatin, plastic, or glass. These are designed to scatter or absorb light of wavelengths higher or lower than their cutoff wavelength. Two of these, one to

remove short wavelengths and one to remove higher wavelengths, can serve to allow only a narrow band of wavelengths to pass. The range of wavelengths passed, the bandpass, is characterized by the width of the passed peak at half height, the **bandwidth**. Figure 3.8 shows the light transmission characteristics of two cutoff filters and that of a composite filter composed of the two. Because the overlap of these filters increases as the bandpass between them becomes narrower, these filters cannot be used for narrow bandpasses, without losing much of the illumination.

Interference filters work on the principle that light beams will interfere either constructively or destructively, if they are in phase or out of phase with each other. These are constructed from two half-silvered mirrored surfaces held apart by a transparent spacer that is an integer number (1, 2, or 3) of half wavelengths in thickness. As a beam penetrates the front surface, it is reflected from the second surface and back from the first surface. It returns to the original beam having traveled a path that is just one wavelength longer and is still in phase. Therefore, the beam is reinforced. When light of a different wavelength passes

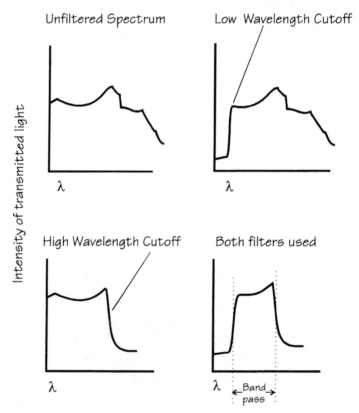

Figure 3.8 Radiation spectra passed through different cutoff filters. When both low wavelength and high wavelength filters are used, a reasonably narrow band of radiation can be selected.

this filter, the incident and reflected beams will interfere with each other and the beam will be destroyed. Therefore, only wavelengths that equal twice the spacing will be passed efficiently, giving a filter that has a fairly narrow bandpass of 10–15 nm. The wavelength which will be most strongly reinforced, the center and peak of the bandpass, is given by

$$f = \frac{2\eta b}{m} \tag{3.14}$$

where η is the refractive index of the transparent spacer, b is the thickness of the spacer and m is the order of the filter, taking an integer value.

3.4.2.2 *Monochromators.*   When a filter is not adequate, for instance when it is desired to make measurements at discrete wavelengths over a span of wavelengths, one must use a monochromator. This is a device that takes the incident polychromatic radiation and spreads it out in space. Different wavelengths of light are thus directed to different points. A prism, or a grating designed to disperse the middle area of the spectrum, is usually used. Of course, this method is not suitable for very short wavelength radiation, which would interact with the material of the prism or grating, or for very long wavelengths, in the radio and microwave regions. Through the ultraviolet, visible, and infrared, the grating is the most frequently used dispersing element. It consists of a base material which is either reflective or transparent. This is ruled with closely spaced grooves that serve to diffract the light. Gratings are often made of aluminum or of plastic coated with aluminum. These can be formed from a master grating in a pressing process, which makes them quite inexpensive.

Light reflected from the successive closely spaced surfaces of a grating is reinforced when the angle between the incident and reflected rays (Θ) obeys the equation

$$b \sin \Theta = m\lambda \tag{3.15}$$

where b is the spacing of the grooves and m is the order of the spectrum. When a grating produces a spectrum, the brightest is the first order spectrum, but successively dimmer ones at higher orders are also produced. Filters may be used to remove the unwanted higher order spectra. The angle at which the grooves are made and the spacing are designed to give the best dispersion in the wavelength range at which the grating will be used. The efficiency is greatest at the angle where the angle of diffraction is equal to the angle of specular reflection from the face of the groove. This is the **blaze angle** and the wavelength that it reflects with maximum efficiency is called the **blaze wavelength**. Figure 3.9 shows how the rays being diffracted from a grating surface interfere with each other so that each wavelength is reinforced at a certain angle and destroyed at other angles.

Gratings will cause some stray radiation, which arises from imperfections in the smoothness of the reflective surfaces and from irregularities in the rulings.

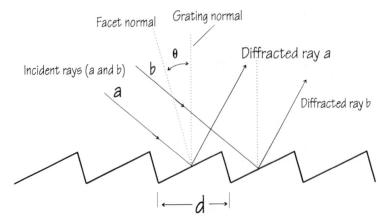

Figure 3.9 Diffraction of light by a grating. The diffracted rays *a* and *b* are in phase when ray *b* has traveled exactly an integral λ further than ray *a*. Otherwise, they will be out of phase and vanish.

The quality of the grating determines the amount of such stray radiation. Since most gratings are duplicated from master gratings, and are not individually ruled, they are much less expensive than they were in the past. This gives even high quality grating monochromators a cost advantage over prism monochromators.

Gratings have other major advantages over prisms and have largely supplanted them in modern instruments. The radiation does not have to pass through the grating but can be reflected from it, so the material of the grating itself is not important. With prisms, on the other hand, the radiation must pass through the prism. It must be transparent at the wavelengths to be studied. This means that prisms for infrared work were constructed of delicate materials such as sodium chloride or potassium bromide, while prisms for ultraviolet radiation were made from expensive quartz. Gratings, on the other hand can be made of aluminum or aluminized plastic surfaces, which are more readily produced and require much less care in use.

The monochromator consists of:

- the dispersing device, usually a grating for the range from the ultraviolet through the infrared,
- a means of adjusting the angles between the incident beams, the dispersing device, and the detector,
- optics as focusing lenses, collimators, and mirrors which keep the light traveling in organized beams through the monochromator.

3.4.3 Detectors

A detector is necessary to measure either radiation produced by the sample in emission or fluorescence or the radiation transmitted through the sample in absorbance spectroscopic methods. Qualities to be sought in a detector are usually a sensitive response to radiation over a reasonably wide wavelength range, as well as stability, and low noise. Gain and response time will also have a significant effect on the use of the detector. Detectors often show a **dark current**, a signal that is produced even when the detector is not being exposed to radiation. The ratio between the dark current and the signal from the incident radiation is important in determining the sensitivity.

The signals arising from these devices are often very small, and may require substantial amplification. In some detectors, several sensing elements may be joined to provide a larger signal. All detectors produce current or voltage signals which may be subject to some noise or drift. To improve the signal to noise ratio of these devices, the signal is usually modulated into an AC signal. This can be done by chopping the incident radiation with a rotating sector disk, which produces a signal consisting of dark current alternating with the desired signal. The difference between these two levels is the actual signal, and drifts in the baseline are compensated. Random noise is also reduced, because this noise will exist in both the signal and dark current, and so will be canceled.

3.5 Ultraviolet and visible absorption spectroscopy

Molecules in solution have continuous absorption spectra in the UV and visible region. Absorbance arises from transitions in the valence electrons of molecules and is affected by such things as solvation and molecular interactions. Figure 3.10 shows an example of a spectrum of an organic molecule, fluoranthene. Changes in oxidation states of species have large effects on absorption spectra. Because the absorption peaks are relatively broad, UV/Vis spectrometry is most frequently used for quantitative analysis. It is also widely used in liquid

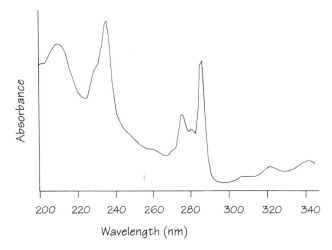

Figure 3.10 The UV spectrum of fluoranthene, a three ring polynuclear aromatic compound.

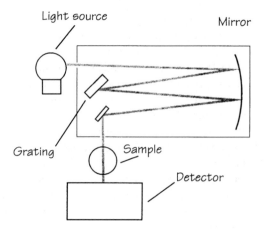

Figure 3.11 A UV–visible spectrometer.

chromatography detectors. A schematic diagram of a UV–Vis spectro-photometer is shown in Figure 3.11.

Quantitative analysis is performed by determining the absorptivity coefficient, ε, using solutions of known concentrations. A wavelength best suited for analysis is chosen. This is usually selected at a place in the spectrum where the absorbance is not changing rapidly, and where interfering substances have low absorbance. While it is not necessary to select a point at which the absorbance is at a maximum, selecting the highest absorbance point will give the highest sensitivity. If the samples to be analyzed are not at trace levels, a wavelength at which lower absorbance takes place may be preferable. The best accuracy in absorbance spectroscopy is achieved at fairly low absorbances. The point of best accuracy is at an absorbance of 0.38. At higher concentrations, because of the log term in Beer's law, a large change in concentration causes only a small change in the transmitted light. Therefore it is better to use a wavelength at which absorbance is less, if it avoids measuring solutions with absorbances above 1, or diluting the sample.

Several analytes can be determined simultaneously in the same solution, because of the additivity of Beer's law. Wavelengths are selected where the differences in the extinction coefficients are as large as possible, to minimize error. For instance, if one is determining a mixture of A and B, the absorption spectra of the pure compounds should be examined to find a point at which A absorbs strongly and B minimally, and another point where B is absorbing strongly, and the absorbance of A is low. The extinction coefficients for each component at each selected wavelength are determined from standards. The cell path length, b, is constant. Absorbance measurements are then made at different wavelengths, producing a set of simultaneous equations with only the concentra-tions as unknowns. One wavelength measurement is required for each compo-nent, to give one equation for each unknown concentration. In practice, the method is limited to two or three components, as the errors of measurement rapidly increase when components are added.

3.5.1 UV and visible instrumentation

3.5.1.1 *Light sources.* Tungsten filament lamps are common broad band sources used in the visible region of the spectrum. They have an operating temperature of about 3000 K. At this temperature, the peak of radiation is actually in the near infrared, and the emission in the visible range is only a small fraction of the total energy emitted from the lamp. While running the lamp filament at a much higher temperature will give more radiant energy in the visible, it will also shorten the lamp's useful lifetime. The use of bromine or iodine vapor in the lamp fill gas, combined with a fused silica envelope, allows a longer lifetime at elevated temperatures. So these quartz–halogen lamps are widely used. Xenon arc lamps require higher voltages but are often used as excitation sources for fluorescence measurements, because they produce a wide continuum, which extends into the ultraviolet region. In the ultraviolet, hydrogen or deuterium electrical discharge lamps are used. These lamps are filled with a low pressure of hydrogen or deuterium, and a DC voltage of about 40 V is applied. The envelope is made of quartz or fused silica. The low wavelength cutoff of these lamps depends on the transmission of the window material, and is usually about 180–200 nm.

3.5.1.2 *UV–Vis detectors.* Photoemissive vacuum tubes are useful in the range from 120 to about 1000 nm. The tubes are composed of a photocathode which will emit electrons when light falls upon it. The electrons are collected on an anode, producing an electrical current. Various combinations of photocathode materials and window materials make different tubes suitable for different ranges of wavelengths. Because of the low currents generated by low radiation levels, amplification may be necessary. These tubes have a **dark current** that flows when there is no incident radiation, so the lowest measurable currents must exceed the dark current by a significant amount. Amplification cannot extract signals which are not significant with respect to the dark current because both the signal and the dark currents are amplified to the same extent.

Photomultiplier tubes are considerably more sensitive because these, as their name implies, multiply the effect of each photon that falls on the detector. They are composed of a photoemissive cathode and a series of dynodes at successively higher voltages. The incident photon ejects an electron from the cathode. This falls on the first dynode, which emits several electrons in response. These are focused onto the next dynode, where the same process is repeated. Thus, the response to each photon is multiplied by the number of electrons emitted for each incident electron (f) and by the number of stages (n). The overall multiplication of signal, or **gain** (G), of the photomultiplier is $G = f^n$. The value of f depends not only on the material of the dynodes but also on the potential

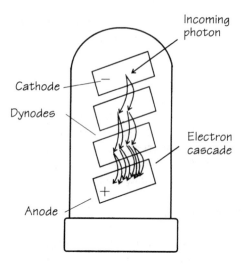

Figure 3.12 The photomultiplier tube. Each photon striking the photocathode ejects several electrons, which cascade down to successive dynodes, multiplying the number of electrons at each stage. The electrons are collected at the anode, forming the signal current.

difference imposed on these dynodes. Therefore, the sensitivity can be controlled over a wide range by adjusting the potential on the multiplier tube. Because of their sensitivity, these are used only for detection of low light levels. A photomultiplier, shown in Figure 3.12, should be protected from light as much as possible when it is being handled or mounted into equipment, even when it is not powered.

Photodiodes are solid state devices. In these silicon-based devices, electrons are promoted to the conduction band when they are irradiated with light in the visible or near infrared areas of the spectrum. These give a response that is larger than that of photoemissive tubes, but very much less than that of photomultipliers. Since they are solid state devices, they can be made very tiny. This allows a row of them to be placed so that each reads a narrow wavelength band, at the same time. From the output of such a **diode array detector**, an absorption or emission spectrum can be obtained instantly, without the need for scanning the spectrum across a single detector. This is both more rapid and less expensive to implement, since scanning requires expensive precision moving parts.

3.5.1.3 *Ultraviolet–visible spectroscopy samples.* Samples used in ultraviolet and visible spectroscopy are usually in solution form, although gas phase measurements can be made using long path gas-tight cells. The sample cells are made of quartz for UV analysis and of glass or plastic for use in the visible region. Round test tubes are often used for low precision work. It is difficult to get a reproducible path length in a round tube, because of slight differences in the placement of the tube in the light beam. It is common practice to mark the tube so that it can be replaced in the spectrometer in the same orientation each time. The tubes used to hold samples and standards can be matched by filling them with an absorbing solution, and selecting the tubes which give the same absorption readings. For more accurate work, cuvets which have flat parallel sides are used. These are also available in various materials, and are considerably more expensive than test tubes. Cuvets are made in various path lengths, with a common size being 10 mm.

The samples and standards must be free of particles and the cuvets should be carefully wiped to avoid fingerprints on the surface. These impurities can scatter light and reduce the transmission of light through the sample. The calibration standards should be held in matching cuvets and a solvent blank should also be prepared for zeroing the instrument.

Gas phase UV measurements are useful for stack monitoring purposes. A non-dispersive technique is most commonly used for monitoring such gases as SO_2, NO_2 and CO_2. A filter system is used, rather than a monochromator. An optimal wavelength for the analyte is selected, as well as a background wavelength. For SO_2, these are 280 nm for the analyte and 578 nm as the background. The filters are mechanically rotated through the light path so that the absorbance is measured at the analytical wavelength and the background wavelength alternately. The sample is drawn through a long path gas-tight cell, where its absorbance is measured continuously. This apparatus is shown in Figure 3.13.

3.5.2 Colorimetry

To determine individual compounds in samples, a reagent which reacts specifically with the component of interest is added to the sample, forming a colored species. The intensity of color produced is proportional to the concentration of analyte in the sample and can be measured using the UV–visible spectrophotometer.

$$\text{Analyte} + \text{Colorimetric reagent} \rightarrow \text{Colored complex} \qquad (3.16)$$

The selectivity is provided by the colorimetric reaction and the absorbance in the visible or UV region is used only for quantitation.

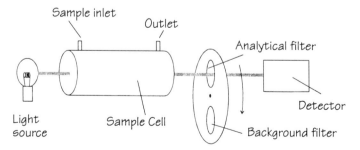

Figure 3.13 Nondispersive UV absorption apparatus for monitoring gases in stack effluents.

Table 3.2 Some colorimetric reagents

Analyte	Color system	Measurement wavelength (nm)
Metals		
Cr(VI)	1,5-Diphenylcarbazide	540
Pb	Dicyclohexyl-18-crown-6-dithizone	512
Fe(III)	Thiocyanate	460
Fe(II)	Pyrocatecol violet	570
Cd	Iodide/Malachite green	685
Hg	2-Pyridylketone 2-quinolylhydrazone	
Organics		
Phenol	1-nitroso-2-naphthol/Ce(IV)	
Inorganic ions		
NO_2^-	$TiCl_3$/sulfanilamide	530
SO_4^{2-}	Fe(III)/$HClO_4$	355
CN^-	Isonicotinic acid, 3-methyl-1-phenyl-2-pyrazoline-5-one	548
Gases		
O_3	KI	352
NH_3	Glutamate dehydrogenase	340

Many colorimetric reagents are available for specific metal ions as well as for organic pollutants. Some of these are listed in Table 3.2. A major advantage of this type of measurement is its simplicity. For example, to measure the amount of Cr(VI) in a water sample, the pH of the sample is adjusted to around 2 by adding H_2SO_4. A few milliliters of a 1,5-diphenyl carbazide solution is added to the sample. In 5–10 min, a pink color is formed, whose intensity can be measured at 540 nm using a UV–Vis spectrophotometer. Similarly, Cr(III) can be determined by first oxidizing it to Cr(VI) by boiling with $KMnO_4$ and continuing as above.

Another example of a colorimetric test is the measurement of NO_3^- in soil. Nitrates are extracted from a soil sample with a 0.01 M $CuSO_4$ solution containing Ag_2SO_4. The latter prevents interference from chloride. The extract is treated with phenoldisulfonic acid. The colorimetric reaction depends upon nitration of position 6 of 2,4-phenoldisulfonic acid:

$$C_6H_3OH(HSO_3)_2 + HNO_3 \rightarrow C_6H_2OH(HSO_3)_2NO_2 + H_2O \qquad (3.17)$$

In alkaline solution, the product is yellow in color.

The selectivity and sensitivity of these methods are often not as good as other methods. For example, atomic absorption usually provides interference-free measurement of metals at lower detection limits than the colorimetric methods. The main advantage of colorimetry is that the analysis can be done using a simple, inexpensive spectrophotometer. The availability of battery powered spectrophotometers makes this technique easily adaptable to field use. While colorimetric tests are usually designed to be specific for a selected analyte, they are prone to interferences. Detection limits depend upon the analyte/colorimetric system, but measurements down to ng/g levels are possible.

Colorimetry can be used for air, water, soil, and biological samples. In the case of air analysis, the air is usually drawn through a chemical reagent to trap the target pollutant. Then, a colorimetric reagent is added to produce the color. In water analysis, the colorimetric reagent may be added directly to the water. In soil (or solid) samples, the soil is first extracted and the colorimetric reagent is added to the extract.

3.6 Infrared spectroscopy

Infrared spectroscopy has found its major use in environmental analysis in the field of long range sensing of air pollutants. To obtain the necessary sensitivity, the path length of the sample is lengthened. Long path cells can be used. Open path measurements where the analyzing beam is reflected from a remote mirror to telescopic detection apparatus are used in the field. The air through which the beam travels is the sample.

The infrared absorption spectrum of a compound is governed by the absorbance of energy corresponding to the energy transitions in the vibrational and rotational modes of the molecule. Each bond in a molecule has a vibrational frequency which depends on the mass of the two atoms connected by the bond and by the strength of the bond. Therefore, certain frequencies are indicative of certain bonds. Table 3.3 shows the characteristic absorption wavelengths for some functional groups. For instance, a peak found at 2960–2870 cm^{-1} is usually due to the absorbance by the methyl group. Other frequencies indicate certain functional groups such as aromatic rings, OH groups, NH groups, etc. Absorbance peaks correlated with the presence of certain functional groups can be found in any reference work on IR. Figure 3.14 shows the spectra of benzene and trichlorobenzene. The difference in the absorbance in the C—H stretching band near 3100 cm^{-1} shows that the chlorinated benzene has fewer C—H bonds, while the appearance of peaks in the region of 650–750 indicates the presence of C—Cl bonds. The environmental analyst usually uses spectra to identify molecules by matching their spectra with library spectra, or to confirm the identity of

Table 3.3 Some characteristic absorbance frequencies

Compound type	Bond	Frequency (cm^{-1})	Intensity
Alkane	C—H	2850–2970	Strong
Alkene	C=H	3010–3095	Medium
		675–995	Strong
Aromatic ring	C—H	3010–3100	Medium
		690–900	Strong
Alcohols	O—H	3590–3650	Varies
Amines, amides	N—H	3300–3500	Medium
Alcohols, ethers, carboxylic acids	C—O	1050–1300	Strong
Aldehydes, ketones, carboxylic acids	C=O	1690–1760	Strong
Nitro compounds	O—N—O	1300–1370	Strong
Chlorinated compounds		600–776	Strong

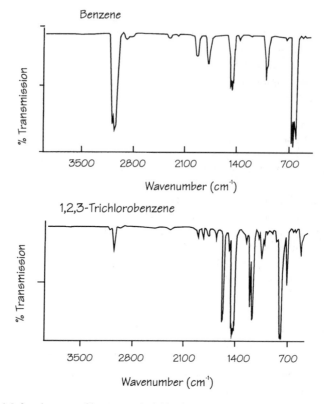

Figure 3.14 Infrared spectra of benzene and trichlorobenzene.

a molecule which is thought to be present. The IR spectrum is not often used in environmental work to deduce the structure of compounds which do not yet appear in spectral libraries, as is frequently done in synthetic organic chemistry. Therefore, we will not discuss the interpretation of spectra in detail.

IR absorption obeys Beer's law and can be used to determine concentration of the absorbing species. IR spectra are usually presented as a plot of percent transmission versus wavenumber. More modern instruments with computerized data handling can also display an absorbance spectrum. However the data are presented, the difference between the baseline and the peak absorbance (not transmission) must be measured before a concentration/absorbance calibration may be made.

3.6.1 *Scanning infrared instrumentation*

3.6.1.1 *IR sources.* In the infrared region, the common sources are electrically heated elements made of ceramic or alloys. The Nernst glower is composed of rare earth oxides, operates up to about 1800 K, and has a negative coefficient of electrical resistance. This means that the resistance becomes lower as the source is heated, and it may require preheating before a current can be passed at all. The globar is a silicon carbide rod, which operates at a lower temperature, about 1600 K, and gives more radiation in the region below 1500 cm^{-1} than does the Nernst glower. A characteristic of all the IR sources is their generally low output of radiation. This means that IR spectroscopy is generally energy limited, and requires sensitive detection.

3.6.1.2 *Infrared monochromators.* Since infrared radiation will not pass through glass or quartz optics, the monochromators are constructed using reflective gratings and front surface curved mirrors to diffract and focus the radiation. The range of wavelengths covered is too large to be diffracted efficiently by a single grating, so the instrument usually contains several gratings supported on a rotating post. The grating is rotated slowly when the sample is being scanned, then the scan is halted, and another grating is turned into place to scan the next region of the spectrum.

3.6.2 *Fourier transform infrared spectrometry*

The inherent sensitivity of IR spectroscopy is low, due to the limitation of the energy available from the source and the low sensitivity of the IR detectors. Therefore, a design using a Michaelson interferometer instead of a mono-chromator is often used. This is the basis of Fourier transform infrared (FTIR). To understand the functioning of an FTIR, one has to understand time domain spectroscopy.

Conventional spectroscopy involves measurements in the frequency domain, i.e., radiant power (expressed as absorbance or transmittance), is measured as a function of frequency. In time domain spectroscopy, radiant power is measured as a function of time. However, both these measurements contain the same information. Conversion between time and frequency domain can be achieved using a mathematical technique called Fourier transform. The conversion between time and frequency domain is shown in Figure 3.15. If signal intensity is plotted as a function of time, a time domain spectrum is produced. The frequency of this signal, when plotted, yields a single line, indicating that only one frequency was present in this signal. This is the frequency domain signal. When multiple frequencies are present, the spectrum becomes more complex. This is shown in Figure 3.16, where the time domain spectra contains two frequencies. The A plot shows the two waves, and B shows their sum. The frequency domain plot shows that only two frequencies were present.

In FTIR, the measurements are done in the time domain rather than in frequency domain. A Michelson interferometer is used for this purpose. The interferometer is shown in Figure 3.17. There are two mirrors, one fixed, the other movable. The moveable mirror travels at a constant velocity. Radiation from the source is passed through a beam splitter, so that half of the beam reaches the movable mirror while the other half is reflected from the fixed

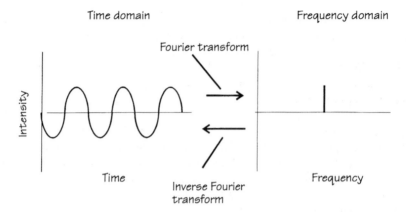

Figure 3.15 A single frequency wave displayed as a time domain and as a frequency domain spectrum.

mirror. The reflected beams from the two mirrors recombine at the beam splitter. Constructive and destructive interference takes place, depending upon the difference in path length between the path followed by the beam reflecting from the fixed and that from the movable mirror. This interference pattern is seen at the

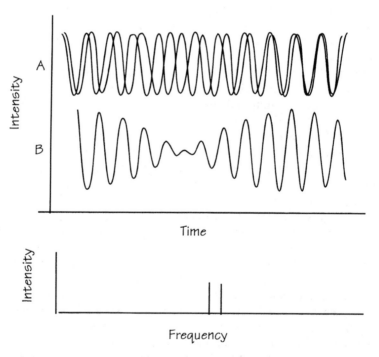

Figure 3.16 Waves of two different frequencies in the time domain (A), their sum (B), and the same waves in the frequency domain.

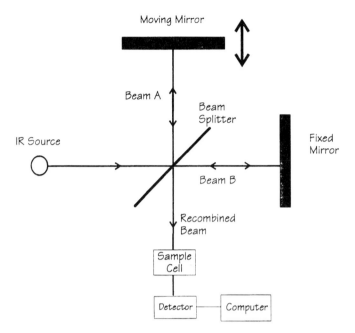

Figure 3.17 Michaelson interferometer used in Fourier transform infrared analysis. As the mirror moves the various wavelengths come into phase and out. The interferogram produced is the Fourier transform of the spectrum, which can be recovered mathematically.

detector. For example, assume that the incident beam is a monochromatic beam. As the movable mirror moves forward, constructive and destructive interference occurs alternately. The resulting detector signal is a sine wave, as shown in Figure 3.18. In other words, the interferometer converted the frequency domain spectra (the monochromatic light) into a time domain spectra. When many frequencies are present, the detector output is as shown in the second plot. The time domain spectra from the interferometer is called an interferogram. When the Fourier transform is performed on the interferogram, the frequency domain spectrum is constructed. The sample is placed in front of the detector, and some of the frequencies are absorbed by it. Fourier transform is then used to obtain the IR spectrum from the interferogram.

3.6.2.1 *Advantages of FTIR.* IR sources have low intensity, and, when a monochromator is used, the IR beam is further attenuated by the slit. The FTIR design eliminates the need for a monochromator, so there is no attenuation of the IR beam. Therefore, the power of the beam reaching the detector is significantly higher and a high signal to noise ratio is obtained. This is referred to as the **throughput** or **Jaquinot** advantage.

In a scanning instrument, it takes several minutes to scan the whole infrared region. In FTIR, there is no scanning to be done, and the spectrum of the whole

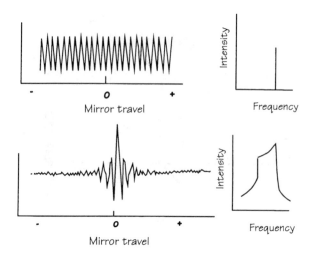

Mirror travel   Frequency

Mirror travel   Frequency

Figure 3.18 Detector signal as a function of mirror travel. The first plot shows monochromatic radiation, and the Fourier transform shows only a single frequency. The second plot is the detector output when many wavelengths are present. The Fourier transform shows the frequency domain spectrum.

region can be obtained in a second or less. FTIR has high wavelength accuracy and precision allowing many scans to be taken and their signals averaged. The signal has a fixed pattern, and the noise is random in nature. When scans are averaged, the signal increases while the noise is canceled out. The enhancement in signal to noise ratio is proportional to the square root of the number of scans averaged. For example, Figure 3.19 shows the result from a single rather noisy scan, and that obtained when 100 scans are averaged. While it may take 1500 s

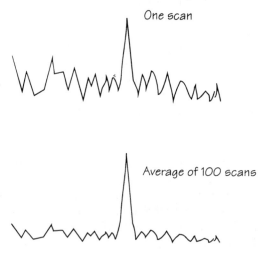

One scan

Average of 100 scans

Figure 3.19 When replicate scans of the same spectrum are done and averaged, the signal to noise ratio is improved.

to obtain a spectrum in a scanning IR, the same spectrum can be obtained by an FTIR in about 1 s. Therefore, in 1500 s, 1500 scans could be obtained and signal averaged, which would enhance the signal-to-noise ratio by a factor of the square root of 1500 or 39.

The FTIR depends heavily upon computer power to deconvolute the interferogram and reconstruct it into a spectrum. As computers have become more powerful and inexpensive, the cost difference between the FTIR and the grating IR instrument has steadily diminished. Since the FTIR can report an IR spectrum in a few seconds, compared to several minutes for a grating instrument, the small additional cost for the FTIR is usually justified. Resolution is also better than in comparable grating instruments. Finally, because the entire spectrum is produced in digital form in the computer, subsequent data analysis is easily done. Backgrounds can be subtracted, and a spectrum can be compared to thousands of standard spectra in a computer library, with a few keystrokes.

3.6.2.2 *Samples for infrared spectroscopy.* Because of the low power of IR sources, most solid or liquid IR samples must be rather thin. Gaseous samples, being of low concentration, need a longer path length. A common application of IR spectroscopy in environmental analysis is for long path sensing of pollutants in air. Gases can be placed in a cylindrical cell with IR transparent windows at the ends. The cells have inlets and outlets so that they can be purged with the sample gas. The windows are cut from a sodium chloride crystal or other IR transparent material. The pressure of the gas sample can be varied. Highly absorbing samples can be measured at pressures of a few millimeters of mercury, while low concentration samples can be used at higher pressures. For trace level samples, the required path length may be so long that the cell is impossibly bulky. In this case the light path may be folded, by placing front surface mirrors inside the cell. These are aligned so that the incoming beam of radiation is bounced back and forth through the sample several times before being brought to the exit slit. The adjustment and alignment of these cells are usually carried out using a visible laser beam. Figure 3.20 shows a long path cell.

For air pollution studies, it is sometimes easier to bring the instrument to the sample. Long range IR analysis is done using an IR source beam which is passed over a path many meters in length. Figure 3.21 shows such a system. A gold plated reflector is mounted some distance away from the source and

Figure 3.20 Long path cell for IR measurements of gases. The path length is increased by reflecting the beam back and forth using front surface mirrors inside the cell.

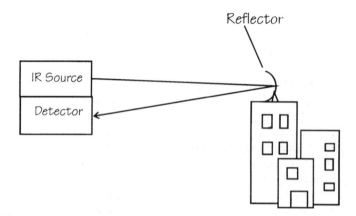

Figure 3.21 A long path IR system for air monitoring.

detector. The reflectors are gold plated mirrors, formed from a series of corner cubes. This arrangement of mirrors, lining the inside surface of three planes meeting at right angles, will always direct a beam that falls into it back along the same path as the incident ray. The beams are thus directed back to the source where they are intercepted and fed into an interferometer to be analyzed. The system is calibrated by using a gas cell of 10–15 cm path, filled with a known concentration of a standard gas mixture, and placed in the beam. The path length of the cell and the known distance to the reflector are used in the calculations.

The difficulties with the open long path systems are due to changes in high level components of the air, such as water vapor and carbon dioxide. This causes large changes in the IR spectrum at certain wavelengths. The complex mixture of organic compounds in the air also makes it difficult to determine specific compounds in the sample. However, if a single component is present at high concentrations compared to other possible absorbers at a particular wavelength, it is possible to follow the changes in that compound. This has been used to watch the components of automobile emissions over a roadway, and the results showed good correlation with the traffic density. It is also useful for industrial fenceline monitoring, where the compounds likely to be emitted are well known. Mathematical models involving the specific absorbances of compounds at different wavelengths are being used to develop algorithms to find the signature of specific pollutants in the mixture. This is possible in IR spectroscopy because of the information richness of the spectra. Another concern is that the concentration is integrated over the length of the path. A 500-m path system will show the same results if the entire air mass has a concentration of target compound of $1 \, \text{ml/m}^3$ or if a plume 5 m wide crosses the beam, carrying a concentration of $100 \, \text{ml/m}^3$.

Solid samples for IR analysis are handled by grinding the samples finely and mixing with refined mineral oil. This **mull** is smeared onto a transparent plate such as a sodium chloride plate. The absorbance spectrum is then taken. This technique is suitable for qualitative analysis, but is not suitable for quantitative studies because the sample thickness is not easily determined. Samples may also be ground finely with potassium bromide powder, and pressed into a pellet in a die. High pressure dies in which the samples are pressed while under vacuum can be used to form pellets which have little adsorbed water. Cells which have known spacings of a millimeter or so between the plates are available for liquid samples.

Quantitation is difficult in IR spectroscopy of liquids and solids primarily because setting the zero and 100% transmittance points is not readily done. The zero point, because of the low amount of energy reaching the detector, is difficult to set reproducibly. The 100% transmittance point cannot usually be set using a matched cell filled with the solvent because matched IR cells are not generally available. Therefore, a baseline method is usually used. The baseline is determined from the points of maximum transmission on either side of the peak of interest. The difference between the minimum of transmission and the zero point is considered I. The difference between the baseline and the zero point is I_O. Then the absorbance can be calculated as:

$$A = -\log (I/I_O) \tag{3.18}$$

Because of the relatively wide band passes used in IR methods, deviations from Beer's law are common. Quantitation by IR spectrometry is generally not as precise or accurate as that done by UV or visible spectrometry. However, it has a high capability of measuring a particular functional group in a complex mixture. This allows the estimation of, for instance, total ketone content or total aromatic content of a mixture, without separation.

In gas phase work, the cell path length is much greater and much easier to measure accurately. Therefore quantitative work in gases is more reliable and precise, if interferences can be avoided and if suitable standards can be obtained.

Example
An open path IR system is set up to study the concentration of SF_6 in the air. The SF_6 is to be used in a tracer study to see how gases emitted from a pollution source will be dispersed in the area under different weather conditions. The IR system is calibrated using a standard gas mixture containing 10 ml/m^3 of SF_6 in a 15.0 cm cell. The peak absorption occurs at about 945 cm^{-1}. The reflector is placed 18.0 m from the source/detector apparatus. The absorbance measured in the standard gas is 0.35. The absorbance measured in the atmosphere is 0.22. Assuming that Beer's law is obeyed, calculate the concentration of SF_6 in the atmosphere during the measurement time. (The calculated value will be averaged over the time of collection and over the distance between the mirror and the detector.)

The absorbances measured are assumed to follow Beer's law, $A = abC$. The extinction coefficient for the SF_6 is calculated from the standard as:

$$0.35 = a \ (0.15 \ M) \ (10 \ ml/m^3) \qquad a = 0.23$$

The path length for the open path system is 36 M, because the beam travels to the mirror and returns.

$$0.22 = (0.23) \ (36 \ M) \ C \qquad C = 0.027 \ ml/m^3 \text{ in the air.}$$

Question: Referring to a table of characteristic IR absorption bands, what interferences do you think should be considered, and how would you attempt to compensate or correct for these?

3.7 Atomic absorption spectroscopy

For elemental analysis, especially for the determination of metals, it is often preferable to decompose the sample molecules into atoms and to measure the absorption or emission of radiant energy due to these atoms. The advantage of these methods is that atomic spectra are line spectra, and do not include broad absorption and emission bands. This makes it easier to select individual elements from a complex mixture, with much less chance of interference.

Atomic absorption (AA) is mainly used for analysis of metals in air, water and solid samples. The absorption spectrum from AA is composed of narrow lines about 0.2–0.4 nm wide. Even the best monochromator does not produce such narrow bands. Consequently, it is not possible to select these bands from a continuous source by use of a monochromator. If the incident beam is wider than the absorption band, only a small fraction of the radiation is actually absorbed. Figure 3.22 shows the situation when a wide bandpass of light is being absorbed by a species which can only absorb a narrow segment of the light. The difference between no absorption and very strong absorption, or between low concentration and high concentration is very small, leading to very poor sensitivity. In addition, when the incident beam has a larger band width than the absorption band, deviation from Beer's law may occur.

This problem is overcome by using a line source which has a bandwidth similar to that of the absorbing species. The figure shows that a larger fraction of the radiation available can be absorbed by the species being measured, if the concentration is high enough.

For this reason, hollow cathode lamps, line sources, are used as the radiation source in atomic absorption spectroscopy. These contain a cathode filled with the element to be used in creating the line emission, and a wire anode. Figure 3.23 shows a diagram of a hollow cathode lamp. The body of the lamp is filled with a low pressure of an inert gas, usually neon or argon. When a voltage of 300–400 V is supplied to the electrodes, the fill gas atoms are ionized and fall into the cathode, sputtering off the element atoms. These atoms are in an excited state and, as they fall back to the ground state, they emit the characteristic

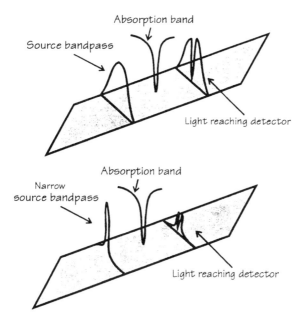

Figure 3.22 Absorbance of broad or narrow bandpass radiation by an atom. Since the atom can absorb only a very narrow line, in a broad bandpass spectrometer, there will be only a small percentage of the light absorbed, even with a very concentrated sample. Therefore, a narrow bandwidth source is used in atomic absorption measurements.

wavelengths of the particular element's spectrum. The width of these lines depends on the amperage passing through the lamp. If the current is too high a dense cloud of atoms is produced, raising the pressure in the lamp. This causes many intermolecular collisions, which spread out the energies of the atoms and thus increase the bandwidth of the line. In addition, the dense cloud will contain some ground state atoms, and these will absorb some of the emitted light before it exits through the window, lowering the output of radiation. Self absorption produces a broadened band with lower intensity in the middle of the band and higher emission at the sides.

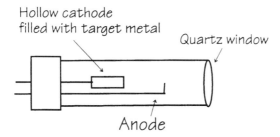

Figure 3.23 Hollow cathode lamp used in atomic absorbance spectrometry. The excited metal atoms sputtered from the cathode emit line spectra of the element to be determined.

The requirement for a specific lamp for each element usually limits atomic absorbance work to the analysis of one element at a time. Some instruments are capable of doing up to four elements simultaneously, by using multiple lamps and detectors, passing the beams through the same flame simultaneously.

3.7.1 Flame atomic absorbance spectroscopy

The atomic absorption spectrometer requires that the sample be atomized, broken down into individual atoms, before it is passed into the radiation beam for absorbance measurement. In flame AA, a liquid solution containing the sample is aspirated into a flame. This is achieved using a nebulizer, which mixes the sample solution with gaseous fuel and oxidant to form a uniformly mixed aerosol of the solution. Several different phenomena take place in the flame while the measurement is occurring. Each drop first dries to a small salt particle, then evaporates completely. The ion clusters heat further until they absorb enough energy to dissociate into free atoms in vapor state. The beam is passed through the flame and absorbance by the atomized species in the flame is measured. It should be noted that the absorbance is proportional to the concentration of ground state atoms in the flame.

The flame provides a complex and reactive atmosphere. Metal atoms can undergo chemical changes, forming, for example, refractory oxides or hydroxides. Atoms can also lose electrons to form ions. Any process which converts free ground state atoms to other forms lowers the sensitivity because the ground state atoms are the absorbing species.

Figure 3.24 shows a typical AA flame apparatus. The burner usually has a long narrow slot from which the flame emerges, and the light beam passes along the length of the slot. This allows for a longer absorbing path length, and better sensitivity. A commonly used flame is fueled with acetylene, with air for an oxidizer. When a higher temperature is needed, or where excess oxygen must be avoided, other flame gases are used. Table 3.4 shows some of the more common flame gas combinations and the reasons for their use.

Because the atomic absorption measurements in the flame are done in a dynamic system, it is especially important to be sure that the samples and standards are in a similar matrix. The viscosity of the solutions, the behavior of the mist in the flame, its drying and evaporation characteristics, and even the droplet size, can all have an effect on the rate of formation of atoms in the flame. In usual work, all standards and samples are made up in dilute acidic aqueous solutions. Where the characteristics of the sample are such that the standards may not be made in a similar matrix, the method of standard additions is often used.

3.7.2 Graphite furnace atomic absorption spectrometry

The electrothermal furnace, an alternative to flame atomization, may be used to atomize a sample for atomic absorption spectroscopy. This technique, often

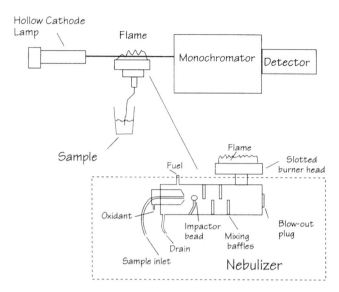

Figure 3.24 A flame atomic absorbance spectrometer includes a hollow cathode lamp, flame, monochromator, and detector. The nebulizer forms a fine aerosol of sample, well mixed with fuel and oxidant gases. The impactor bead helps to break up large droplets, and the baffles ensure that large droplets do not reach the flame.

called the graphite furnace method, minimizes sample preparation, because both liquid and solid samples may be used. The weighed or measured portion of sample is placed in a small graphite tube which is held between two electrodes. Some furnace tubes are manufactured with a small sample stage inside the tube, called a L'vov platform, which helps to ensure that the sample is atomized evenly. These systems have several different configurations but most allow the sample to be injected into the middle of a horizontal graphite tube. A current is passed through the walls of the tube, usually increasing the temperature in a programmed fashion. The initial stage heats the furnace to a fairly low temperature, usually just above the boiling point of the solvent, in order to dry the sample. Then the temperature is raised to a point at which the sample is ashed, destroying any organic material present. This temperature varies widely, depending upon the character of the sample matrix and the target metal.

Table 3.4 Some flames used in atomic absorption spectroscopy

Flame gases	Temperature, °C	Application
Acetylene–air	2100–2400	Useful for many metals
Hydrogen–air	2000–2100	Avoids molecular CH band interferences for metals whose analytical lines are in the same area
Acetylene–nitrous oxide	2600–2800	Higher temperature and less free oxygen in flame, so avoids formation of refractory oxides for metals such as aluminum

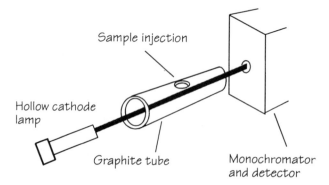

Figure 3.25 Graphite furnace spectrometer. The sample is injected into the furnace tube. It is heated by passing a current through it. The sample is atomized at temperatures of 3000–4000°C.

Finally the current is increased in a sharp pulse to volatilize the metals into the light beam. This current pulse raises the temperature rapidly to over 2000°C, in a matter of a few seconds. A puff of atomic vapor is produced, and its absorbance is measured. The signal is in the form of a peak, as the concentration increases and dies away. The graphite furnace is shown in Figure 3.25. Because the entire amount of sample is atomized and is measured immediately, the sensitivity of the method is usually higher than that obtainable with flame atomization. Also, the amount of sample needed is smaller than that needed for flame AA. A liquid sample of 2–100 μl is usually injected, using a syringe. Small amounts of solids may be weighed directly into the furnace. However, it is imperative that the solid is dried and ashed at a low enough temperature so the targeted metal is not vaporized. When samples can be atomized directly, a great deal of sample preparation can be eliminated. For instance, there are screening methods for lead in blood which require only a few microliters of sample injected directly into the graphite furnace for analysis.

There is a difference between flame atomized and furnace atomized samples. The flame aspirates a continuous flow of sample solution, keeping a constant concentration of absorbing atoms, for as long as is necessary to establish and note the absorbance. The furnace provides a puff of atoms in a small cloud, which is transitory. The concentration of the atoms increases then decreases, and the instrument must be designed to follow this signal rapidly and accurately. Modern atomic absorption spectrometers have microprocessor controlled signal acquisition circuitry which handles the recording and integration of these rapidly changing signals readily. Since the entire sample is atomized at once, and there is no rapid flow of gases to dilute the atom cloud, as there is in a flame, the residence time of the atoms in the light beam is much longer than in flame AA and the detection limits are usually lower.

The graphite furnace method is subject to matrix problems, as often happens with very sensitive methods. Some salts are quite volatile. For instance, lead chloride is more volatile than other salts of lead. Therefore lead in chloride-bearing samples may give a low value, because some of the lead is lost as chloride during the ashing and drying cycles. This problem may be addressed by addition of ammonium nitrate, which helps to release the chloride as ammonium chloride, before the lead is volatilized. Many of these **matrix modifiers** are useful in specific cases. It is always advisable to match the matrix of the samples and standards as closely as possible, and also to perform recovery studies by running spiked samples. These will usually show interferences, but may not detect the type of systematic bias which occurs when a sample component is tightly bound within a matrix and is not atomized efficiently. A spike of standard put into the same sample may be easily and efficiently recovered, because it is not bound in the same way as the sample component. The use of appropriate standard reference materials to test the method will be more likely to detect this kind of bias.

3.7.3 *Interference in atomic absorption*

While the method is generally quite specific, interferences can occur in atomic absorption spectrometry, both in flame and graphite furnace work. These may be classified as spectral, chemical, and ionization interferences.

3.7.3.1 *Spectral interference.*

Spectral interference arises when the absorbance (or emission) of an interfering species either overlaps or is so close to that of the analyte that they cannot be resolved by the monochromator. Since the emission line from the hollow cathode lamp is so narrow, spectral interferences are not usually a problem. Two lines have to be less than 0.1 nm apart for this type of interference to occur. Spectral interferences can be readily eliminated by choosing a line further removed from the interfering line. For example, the vanadium line at 308.211 nm interferes with the aluminum line at 308.215. This interference is easily avoided by measuring aluminum at 309.27 nm instead.

3.7.3.2 *Chemical interference.*

Species being measured may form refractory oxides which do not readily decompose into atoms. These require a higher temperature or a flame low in oxygen, so that oxide formation is reduced. For example, the formation of $Ca(OH)_2$ would interfere in the analysis of other metals. This problem can be eliminated by substituting nitrous oxide for air as an oxidant. The higher temperature decomposes the $Ca(OH)_2$ and eliminates the interference.

Chemical interferences occurs when chemical processes during atomization change the absorbance characteristics of the analyte. One of the most common types of chemical interference is the formation of compounds of low volatility. For example, in presence of sulfate or phosphate, the calcium absorbance can fall significantly due to the formation of sulfate or phosphate species which will not vaporize at the flame temperature. In the presence of aluminum, the absorbance of magnesium drops because complex oxides containing aluminum and magnesium form.

A **releasing agent**, for example, strontium or lanthanum, can be added to minimize the interference of phosphate in the determination of calcium. Here, the lanthanum or strontium ties up the phosphate and prevents it from combining with the calcium. A **protective agent** such as APDC (the ammonium salt of 1-pyrrolidinecarbodithioic acid) forms a stable, volatile complex which decomposes in the flame and prevents the formation of refractory compounds.

3.7.3.3 *Ionization interference.* The high temperature in the flame can cause ionization of atoms. Thus the concentration of free atoms, which are the species being measured, decreases resulting in a lower sensitivity. The metal atoms are in equilibrium in the flame.

$$M \Leftrightarrow M^+ + e^- \tag{3.19}$$

The equilibrium constant, K, for this reaction is given by:

$$K = \frac{[M^+][e^-]}{[M]} \tag{3.20}$$

From the above equation, it can be seen that the metal atom concentration can be increased by increasing the concentration of free electrons in the flame. This is achieved by adding an **ionization suppresser** such as lithium. Lithium ionizes easily in the flame, supplying a relatively large quantity of free electrons. The ionization equilibrium shifts, increasing the concentration of free atoms.

3.7.3.4 *Background correction in atomic absorption spectrometry.* If molecular species are formed in flames or in the cloud of atoms issuing from a graphite furnace, these will exhibit broad band absorbances rather than the line absorbances found with atoms. The bands may contribute substantial interference across a fairly wide area of the available spectrum. Absorbance due to molecules formed in the flame without any contribution from the sample can be zeroed out as the blank is being aspirated. However, if materials contained in the sample interact to form interfering molecules, a background correction technique must be applied. The absorbance on either side of the line of interest can be measured, and subtracted as a background. This, of course requires a light source which emits at the needed wavelengths, so a continuous source such as a deuterium lamp must be used in addition to the line source lamp. Figure 3.26 shows the apparatus for this background correction method. A chopper alternately focuses the beam from the hollow cathode lamp and that from the continuous source through the sample and on to the detector. The radiation from the deuterium lamp is absorbed by the broad band interferents, and is measured and subtracted.

Another background correction involves the use of the Zeeman effect. When an absorbing or emitting atom is subjected to a strong magnetic field, the line of radiation being emitted or absorbed is split into several lines. These are separated by only a few thousandths of a nanometer in wavelength. These extra lines

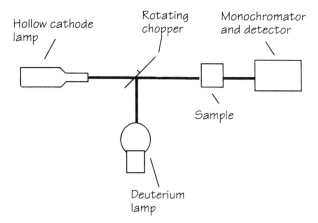

Figure 3.26 Hydrogen/deuterium lamp background correction system for atomic absorption. The deuterium lamp provides a continuous spectrum, allowing absorption measurements to be made at each side of the analytical absorption band, correcting for interfering broad band absorbances.

arise from a change in the magnetic quantum number of an atom, in addition to the usual changes in electronic energy level caused by the absorption or emission of a photon. These additional side absorbance bands will absorb radiation polarized perpendicular to the magnetic field. The line arising from absorbance with no change in magnetic quantum number absorbs light which is polarized parallel to the field. The molecules responsible for the interference absorb the polarized light equally, regardless of its direction of polarization. This effect may be used to correct for background absorption by subjecting either the emitting atoms that form the light source or the sample atoms to a magnetic field. The Zeeman effect, then, gives a way of determining which part of the absorbance is due to the analyte atoms and which part is unaffected by the field. Therefore, part of the absorbance can be attributed to interfering molecules and subtracted as background.

Another background correction, the Smith–Hieftje method, uses the self absorption phenomenon which occurs when the lamp current is raised. This reduces the available light at the center of the band, while increasing it on either side of the spectral line. As the current to the lamp is increased, the background absorbance on both sides of the line is measured. Then the current is returned to the normal setting, giving the analytical absorbance. The background absorbance can then be subtracted to obtain the corrected absorbance. Zeeman effect, Smith–Hieftje, and deuterium background correction are available on commercial instruments.

3.8 Inductively coupled plasma emission spectroscopy

In atomic emission spectroscopy, a sample is often excited by exposing it to high temperatures. With solid samples, an electrical arc or spark is used, but the spectrum obtained is difficult to use for quantitative work because it only lasts a second or so. However, if the sample is dissolved, the liquid solution can be aspirated into an argon plasma torch at 6000–8000 K. Because of the high temperature, the sample is very efficiently atomized, and the atoms are excited to an electronic level where they emit. Few molecular species are stable enough to survive this temperature and so interference from molecular emission bands is minimized.

To form the plasma, argon is passed at a high flow of 10–20 l/min through the torch. Surrounding the top of the torch tube is a water-cooled induction coil, powered by a radio-frequency generator and producing about 2 kW of energy, at around 27 MHz frequency. Ions formed by a spark passed through the argon are agitated by this strongly fluctuating field, and are forced to flow rapidly in a circular path. Friction heats the gas to the point at which it forms a plasma, which is a conducting gaseous mixture containing electrons and cations as well as undissociated atoms. Argon ions are stable enough that they will continue to absorb energy from the coil and stay hot enough to sustain the plasma indefinitely. The argon flow is directed through three concentric quartz tubes. The outermost one serves to cool the torch body, as it would easily melt at the plasma temperature. The argon flowing through the inner tube is used to aspirate the liquid sample into the plasma. The plasma torch is shown in Figure 3.27.

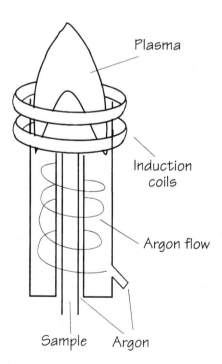

Figure 3.27 An argon plasma torch.

The central core of the plasma, which appears as a bright white flame, emits a strong argon spectrum. About 20 mm above the core, the argon spectrum is much less strong, and the plasma appears transparent. This area is quite free of background lines and is suitable for analysis. The droplets of sample are exposed to the plasma for a couple of milliseconds before they reach this area, by which time they have been dried, evaporated, and atomized.

Since the torch provides a much higher temperature than an atomic absorption flame, chemical interferences are less of a problem. Also, ionization is not as serious a problem as one might expect, since the plasma contains a high concentration of free electrons. This produces a high enough concentration of excited atoms to give a sufficiently intense emission spectrum. Oxides do not readily form, as the plasma is relatively free of any active species, including oxygen. Therefore, the spectra emitted from the sample elements are quite clean and free of most interference. Calibration curves tend to be linear over quite wide ranges.

The detectors and monochromators suited to atomic absorption spectroscopy are also used for ICP. Instruments for ICP are often combined with atomic absorption. This enables the use of the same monochromator and detector for both. Then a movable mirror either brings the beam from the hollow cathode lamp into the monochromator, or is switched to capture the emission from the plasma torch instead. The emitted lines are scanned as the sample is aspirated, or a diode array detector reads the lines simultaneously. Computer automated data collection on the ICP allows the comparison of emission at more than one wavelength for each targeted element, so that interferences can be detected.

Unlike atomic absorbance methods, ICP is an excellent tool for the simultaneous determination of multiple elements. The sensitivity is high, because the plasma gives a high atomization efficiency. In addition, molecular band spectra are much less of a problem because of the very high temperatures. Figure 3.28 shows an ICP spectrum of a sample containing copper, zinc and chromium.

Sample preparation is the same as that required for flame atomic absorption, except that modifiers for masking interference or buffering ionization are not often necessary. The applications are also similar. ICP analysis is used for determination of most metals in water, soil and other environmental sample extracts or digests. The plasma torch is also used in conjunction with mass spectrometry. The ions formed in the plasma can be sent directly into the source of the mass spectrometer for detection by mass/charge ratio, instead of measuring the concentration by the radiation emitted. This technique is discussed in the mass spectrometry chapter.

3.8.1 Comparison of atomic spectroscopic methods

Table 3.5 shows the detection limits for some elements in the various atomic spectroscopy methods. It should be remembered, however, that the true detection limits can be influenced by the other components of the sample, which may

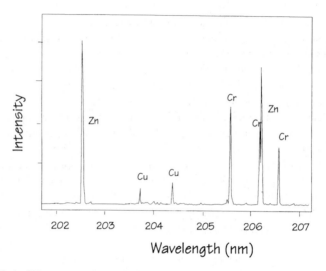

Figure 3.28 An ICP spectrum of a sample containing zinc, copper, and chromium.

interfere to some degree, and by the cleanness of the blanks, which can be the most difficult part of the analysis when ultratrace analyses are being done.

3.9 X-ray fluorescence

When atoms are bombarded by X-rays, interior electrons may be ejected. The remaining electrons cascade down to fill the vacancy. These electron transitions emit other X-rays with energies or wavelengths specific to the emitting elements. This phenomenon is known as X-ray fluorescence (XRF). It is especially suited to analysis of solids for their elemental composition. For screening for the presence of metals in reasonably high concentrations, for example for determining if lead is present in a painted wall, it is unsurpassed for ease of use, and is totally nondestructive. The great advantage of being able to analyze a solid

Table 3.5 Detection limits for some elements

Element	Flame atomic absorbance	Graphite furnace atomic absorbance	Inductively coupled plasma emission
Aluminum	30 ng/ml	0.005 ng/ml	2 ng/ml
Arsenic	100	0.02	40
Cadmium	1	0.0001	2
Chromium	3	0.01	0.3
Mercury	500	0.1	1
Manganese	2	0.0002	0.06
Nickel	5	0.02	0.04
Lead	10	0.002	2
Vanadium	20	0.1	0.2
Zinc	2	0.00005	2

directly without dissolving or extracting it, is counterbalanced by the difficulties of obtaining quantitative data from XRF and by its relatively low sensitivity. It is usually used for determining elements which are present in the 0.01 % range or above.

3.9.1 Wavelength dispersive XRF versus energy dispersive XRF

In the X-ray region, the grating is replaced by a crystal in which the planes of atoms serve to diffract the beam into monochromatic rays. Because of their much closer spacing, of the order of angstroms, these atomic planes can diffract radiation of very short wavelengths, such as X-rays. Some instruments are termed wavelength dispersive, and these contain a crystal which acts as a diffraction device to separate the various wavelengths of emitted X-rays.

However, since a photon in this region carries a relatively high amount of energy, it is possible to determine the energy of each photon as it strikes the detector. Therefore, one does not necessarily have to separate the rays of different wavelength in space, and scan them. All the emitted radiation can be directed to the detector, where the energy of each photon is determined as it impinges on the detector. These instruments tend to be simpler and less expensive than the wavelength dispersive X-ray spectrometers, but also provide lower resolution between closely spaced emission lines.

3.9.2 X-ray instrumentation

3.9.2.1 *Sources.* X-rays are generated by directing a beam of electrons at energies of 10–100 keV onto a metal target. The X-rays have a wavelength profile which depends on the target metal. A continuum of radiation is produced, with photons of discrete energies characteristic of the target superimposed on it. Higher atomic number targets give more intense radiation, and the spectrum extends to shorter wavelengths as the energies of the impinging electrons are increased. It is also important that the target has a high melting point, so that it can stand up to the electron bombardment.

The selection of a target material can change the emitted spectrum, since the continuum wavelenths of the excitation are not as intense as the characteristic emissions. If the intense characteristic line is near an absorption band in the sample, the fluorescence will be enhanced.

Another means of producing the rays to stimulate X-ray fluorescence is by use of radioactive sources. These can be limited to a specific band of radiation by passing the radiation through two filters, one to absorb the energies below the desired range, and the other above the range. The specific wavelength bandpass is tuned to the element to be detected, while the radiation returning to the detector is likewise selected. The radiation produced by the radioactive source is several orders of magnitude weaker than that of an X-ray tube. Therefore, the source, the sample and the detector must be located close together. However, the

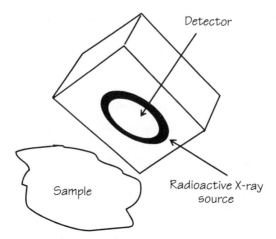

Figure 3.29 Hand-held X-ray fluorescence spectrometer. The ring source contains a radioactive substance, emitting X-rays. The instrument is pressed against the sample, and the specific fluorescence X-rays emitted by lead in the sample are measured by the detector in the center of the ring.

radioactive source, which needs no power, is ideally suited for hand held, portable screening instruments, tuned to a single element, often lead. The radioactive ring source with sample and detector is shown in Figure 3.29.

3.9.2.2 X-ray detectors. Geiger counters, proportional counters and scintillation counters are used to detect the presence and intensity of radiation in the X-ray region of the spectrum. Geiger counters are simple to operate and do not require very sophisticated electronics. They have a rather slow counting rate and a relatively long "dead time", losing much of the radiation, even when the intensity is moderate. This **dead time** is the time between counts, when any radiation falling on the detector is not counted. Also, the Geiger counter is not capable of determining the energy of the measured radiation.

Proportional counters are useful over a similar spectral area as the Geiger counters, but have several advantages. The dead time is short, so the detector is suitable for use at higher intensities. The output voltage is proportional to the energy of the input X-ray photon, so energy discrimination is possible without a dispersing device.

The scintillation detector has very good sensitivity for X-rays with wavelengths below 2 Å, and a very short dead time. It is capable of a rapid counting rate. The output voltage is proportional to the input ray's energy, so it can be used to reject radiation of energy bands outside that being measured. Its energy resolution is not as good as that of the proportional counter, but is useful for rejection of higher order bands of X-rays.

Many modern X-ray fluorescence spectrometers use semiconductor detectors. These are composed of silicon wafers, germanium crystals, or germanium crystals doped with lithium. The detectors are solids, in which electron–hole pairs are formed when the detectors are exposed to ionizing radiation. The electrons in the material are excited into the conduction band of the semiconductor by the absorbed radiation, where they move freely toward the positive electrode, producing a current pulse. The vacated hole travels in the opposite direction by repeatedly exchanging electrons with neighboring sites. Lithium-drifted germanium detectors require cooling with liquid nitrogen at all times after manufacture. The semiconductor is inherently unstable at higher temperatures and must be replaced if it is allowed to warm. Silicon based detectors are only useful down to wavelengths of 0.3 Å. Germanium-based detectors are necessary at shorter wavelengths. Pure germanium detectors require cooling only during use, to reduce thermal noise, but not constant cooling, and so these are somewhat more convenient and less expensive to use.

These detectors have much greater energy resolution than the earlier types. For each photon absorbed, a large number of electron–hole pairs are formed. The intensity of the pulse of energy formed in the detector is used to distinguish the wavelengths or energies of the radiation emitted by the samples. The resolution of semiconductor detectors enables discrimination between elements only one or two atomic numbers apart, making them useful for use in simpler, less expensive, qualitative and semi-quantitative analyzers. The basic layout of the X-ray fluorescence apparatus is shown in Figure 3.30.

3.9.2.3 X-ray fluorescence samples.

Sample preparation for X-ray fluorescence analysis is often very simple. Because the samples can be analyzed in the solid or liquid states, there is often no preparation at all. Samples such as airborne particulates can be captured on a filter and be inserted directly into the instrument for analysis. Liquids may be poured into holders, with a thin mylar film forming the bottom. The exposure is then done through this window. Since

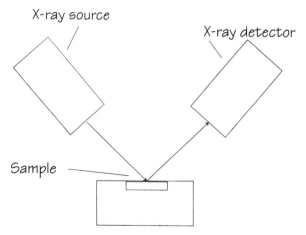

Figure 3.30 An X-ray fluorescence spectrometer. No monochromator is needed, as the detector differentiates among the energies of the emitted X-rays.

several peaks may be produced for each element, these can be used to confirm identities of analytes.

While these simple methods are adequate for qualitative and perhaps semi-quantitative estimations, quantitative work must include consideration of the serious matrix effects which occur in X-ray fluorescence. Because the radiation released from the sample, the fluorescence, is absorbable by other elements in the sample, variations in the matrix will cause differences in the amount of interelement interferences.

First of all, the X-rays, especially at longer wavelengths, do not penetrate deep into most samples. Therefore the surface must be prepared and be representative of the whole sample. For most environmental samples, the sample will be a particulate material. Particle size and particle size distribution have substantial effects on the results of the analysis. To avoid matrix effects, samples may be diluted in a solid material such as sodium tetraborate or borax. The sample may be mixed with the diluent and formed into a pellet in a high pressure press. Alternatively, the sodium tetraborate may be melted with the sample and the fused mass used as the sample. The latter technique gives somewhat better results than pellet formation, because the fused sample is more homogeneous.

When trace materials are being sought, some concentrating technique is usually required. Airborne particles on a filter are used directly. The background due to the filter may be substantial, and filters of paper or organic membranes are generally more suitable than quartz and glass fibers. Ion exchange membranes may be used to collect ionic species from water samples. When the ions to be determined are a small part of the total ionic loading, as in sea water, chelating resins may be more suitable for collection and concentration of the sample. Interelement interferences can be reduced by keeping the sample thin, so that the excited emission does not have to pass through any substantial amount of the matrix before it is detected. If this is not possible, then the matrix should be comparable between the sample and the standard. Standards can be prepared for filters by drying known amounts of standard solutions on a filter, or by dispersing a known amount of the element in powder form on the surface of the filter. Calibration plots are often not linear and so calibration may have to be done at several different concentrations to determine the curvature of the plot.

If matrix interferences are minimized, and the sample and standard are similar in composition, a simple ratio of intensities of the emission for the sample and standard is adequate for quantitation. Internal standards, standard addition, and comparison with standards made up in a very similar matrix are all techniques that have met with success in X-ray spectroscopy. However, each of these must be looked at with care before use. Analysis of standard reference materials similar to the sample under consideration is a good way of determining whether or not there are substantial matrix or calibration problems. Mathematical methods have also been developed which assist in the calibration and correction for interferences.

3.10 Hyphenated spectroscopic methods

The ability to determine both qualitative and quantitative information about a sample by examining its interactions with radiation make the combination of spectroscopy with separation techniques particularly valuable. Liquid chromatography uses both absorbance and fluorescence methods in detection, while infrared detectors have become useful in gas chromatography with the advent of Fourier transform methods. The complete IR spectrum of a peak can be obtained as it exits the GC column, and a match with a computerized library can be done. GC/FTIR is almost as useful as GC/MS for environmental analysis, although it is inherently less sensitive. Since the IR is a nondestructive detector, it is possible to use both FTIR and MS in series after the gas chromatographic column. This gives the power to characterize most compounds by their mass spectra, to distinguish between isomers with similar mass spectra, and to identify compounds by functional groups in the chromatogram, by IR.

Suggested reading

Haswell, S.J. (1991) *Atomic Absorption Spectrometry: Theory Design And Applications*, Elsevier, Amsterdam.

Johnston, S. (1991) *Fourier Transform Infrared: A Constantly Evolving Technology*, Ellis Harwood, New York.

Montaser, A. and Golightly, D.W. (1992) *Inductively Coupled Plasma In Analytical Atomic Spectrometry*, 2nd edition, VCH Publishers, New York.

Skoog, D.A. and Leary, J.J. (1992) *Principles Of Instrumental Analysis*, Saunders College Publishing, New York.

Willard, H.H. *et. al.* (1991) *Instrumental Methods Of Analysis*, 6th edition, Wadsworth, Belmont, CA.

Wolfbeis, O.S. (1993) *Fluorescence Spectroscopy: New Methods And Applications*, Springer-Verlag, New York.

Study questions

1. What are some reasons why a particular solvent in solution may appear not to obey Beer's Law?
2. Explain why fluorescence measurements tend to be more sensitive than absorbance measurements.
3. What are the major components of a spectrometric instrument?
4. The molar absorptivities ε of two compounds, A and B were measured with pure samples at two wavelengths. From these data determine the amount of A and B in the mixture.

	ε at 277 nm M^{-1} cm^{-1}	ε at 437 nm M^{-1} cm^{-1}
A	14780	5112
B	2377	10996

The absorbance of a mixture of A and B in a 1.0 cm cell was determined to be 0.886 at 277 nm and 0.552 at 437 nm.

5. Why is the hollow cathode lamp line source used instead of a broad band source for atomic absorption spectrometry?
6. What is the purpose of background correction in atomic absorption spectrometry? For what type of samples is it most important?
7. What are the advantages of ICP over flame AA?
8. Why is FTIR able to obtain useful spectra on samples where scanning IR will not?
9. What is the principal advantage of XRF analysis? What is its chief drawback?
10. A sea water sample is to be analyzed for cadmium by flame AA. To avoid matrix effects from the high salt content, the calibration will be done by standard addition. Samples are prepared by pipetting 20.00 ml of sea water into several volumetric flasks and adding 1 ml of nitric acid. After this, measured amounts of a standard containing 0.1 mg/μl cadmium are added to each and the solutions are diluted to the 25.00 ml mark. Each sample is run and the absorbances are as follows.

Sample number	μl of standard added	Absorbance measured
1	0.00	0.31
2	10.0	0.57
3	20.0	0.84
4	30.0	1.11

Calculate the concentration of Cd in the sea water.

11. Acetone has a molar absorptivity of $2.75 \times 10^3 \, l \, cm^{-1} \, mol^{-1}$ at 366 nm, in ethanol. What range of concentrations can be measured, so the absorbance remains within the range of 0.08–0.3, using a 1.0 cm cell?
12. A 0.2×10^{-4} M solution of a chromium complex, with a molecular weight of 284, has an absorbance of 0.4 in a 1.0 cm cell.

   a. Calculate the molar absorptivity of the complex at this wavelength.
   b. Calculate the absorptivity when the concentration is expressed in mg/l.
   c. Calculate the concentration of a sample which shows an absorbance of 0.8.

13. Why are the absorption bands in the spectrum of $Ca(OH)_2$ much wider than the absorption bands in that of beryllium?
14. Why is better sensitivity usually achieved using graphite furnace atomization rather than flame atomization.
15. Why is spectral interference more of a problem in flame emission AA than in ICP?
16. What are the advantages of FTIR over scanning IR?
17. A FTIR system was set up to monitor emissions at a factory fenceline. When 20 scans were collected and averaged, the signal to noise ratio for the C—H stretching band was 5:1. How many interferograms should be averaged to obtain a 50:1 signal-to-noise ratio.

4 Chromatographic methods

Chromatography, a group of methods for separating very small quantities of complex mixtures, with very high resolution, is one of the most important techniques in environmental analysis. The ability of the modern analytical chemist to detect specific compounds at ng/g or lower levels in such complex matrices as natural waters or animal tissues is due in large to the development of chromatographic methods.

The science of chromatography began early in the twentieth century, with the Russian botanist Mikhail Tswett, who used a column packed with calcium carbonate to separate plant pigments. The method was developed rapidly in the years after World War II, and began to be applied to environmental problems with the invention of the electron capture detector (ECD) in 1960 by James Lovelock. This detector, with its specificity and very high sensitivity toward halogenated organic compounds, was just what was needed to determine traces of pesticides in soils, food, and water and halocarbon gases in the atmosphere. This happened at exactly the time when the effect of anthropogenic chemicals on many environmental systems was becoming an issue of public concern. Within a year, it was being applied to pesticide analysis. The pernicious effects of long lived, bioaccumulating pesticides, such as DDT, would have been very difficult to detect without the use of the ECD. The effect of this information on public policy has been far-reaching.

The basis of all types of chromatography is the partition of the sample compounds between a stationary phase and a mobile phase which flows over or through the stationary phase. Different combinations of gaseous or liquid phases give rise to the types of chromatography used in analysis, namely gas chromatography (GC), liquid chromatography (LC), thin layer chromatography (TLC), and supercritical fluid chromatography (SFC).

Chromatography has increased the utility of several types of spectroscopy, by delivering separate components of a complex sample, one at a time, to the spectrometer. This combination of the separating power of chromatography with the identification and quantitation of spectroscopy has been most important in environmental analysis. It has enabled analysts to cope with tremendously complex and extremely dilute samples.

4.1 Principles of chromatography

All chromatographic systems rely on the fact that a substance placed in contact with two immiscible phases, one moving and one stationary, will equilibrate between them. A reproducible fraction will partition into each phase, depending on the relative affinity of the substance for each phase. A substance which has affinity for the moving or **mobile phase** will be moved rapidly through the system. A material which has a stronger affinity for the **stationary phase**, on the other hand, will spend more time immobilized in that phase, and will take a longer time to pass through the system. Therefore, it will be separated from the first substance. By definition, chromatography is a separation technique in which a sample is equilibrated between a mobile and a stationary phase.

Gas chromatography employs an inert gas as the mobile phase, and either a solid adsorbent or a nonvolatile liquid coated on a solid support as the stationary phase. The solid or coated support is packed into a tube, with the gas flowing through it. Separation depends on the relative partial pressures of the sample components above the stationary phase. Liquid chromatography uses similar packed tubular columns and usually a pump to force a liquid mobile phase through the column. Supercritical fluid chromatography occupies a middle ground between gas and liquid chromatography. The mobile phase is a super-critical fluid, i.e., a fluid above its critical temperature and pressure. This allows the use of GC type detectors, since the mobile phase has gas-like properties, but also allows continuous variation in such mobile phase properties as viscosity and density, by changing temperature and pressure. Finally, chromatography may be done on a planar surface. The sample is transported over a solid surface such as cellulose or silica gel, coated on a plate. The sample components are moved over the surface by the mobile phase which is usually allowed to travel through the adsorbent layer by capillary action.

The reason that all molecules of a certain type tend to exit the system at the same time is that they are always re-equilibrating between the phases. Over a large number of such equilibrations, the molecules spend, on average, the same amount of time in each phase. Let us look at one point in the chromatographic column. When the analyte achieves or approaches equilibrium, the mobile phase moves on, leaving the stationary phase with too much of the analyte and the mobile phase with too little. To attempt to reestablish equilibrium, more sample dissolves in the mobile phase and moves along. Figure 4.1 shows a mixture of three substances as they move through a chromatographic column.

The movement of analytes in the column can be described mathematically. The basis of chromatography is the equilibrium of each analyte between the mobile and stationary phase. This can be expressed by a simple equilibrium equation, where K_x is the partition coefficient

$$K_x = [C_s]/[C_m] \tag{4.1}$$

that is: the concentration in the stationary phase ($[C_s]$) is directly related to the concentration in the mobile phase ($[C_m]$), at least when the concentrations are low. Chromatographic separations are best done with a small amount of analyte, which keeps either phase from becoming saturated with analyte, so that the concentrations in the two phases are directly proportional. Overloading the

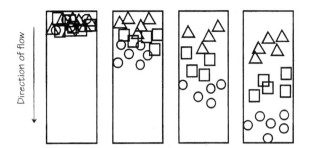

Figure 4.1 Separation of three substances in a chromatographic column. Note that the bands of the individual compounds become wider as they move down the column.

column with sample causes one of the phases to become saturated with sample, leading to a loss of column efficiency, and poorly shaped peak profiles.

The quantities in the equilibrium expression for K_x, $[C_s]$ and $[C_m]$ are not easy to measure. We can define a new constant, the capacity factor, k':

$$k'_x = \frac{\text{(moles of } X \text{ in stationary phase)}}{\text{(moles of } X \text{ in mobile phase)}} \qquad (4.2)$$

Since the number of moles can be expressed as the concentration multiplied by the volume, Equations 4.1 and 4.2 can be combined and reduced to:

$$k'_x = K_x(V_s/V_m) \qquad (4.3)$$

All sample molecules spend the same amount of time in the mobile phase. If they were completely unretained by the stationary phase they would exit the column in the time it takes for one volume of mobile phase to pass through the column. This is equal to the void volume of the column. Molecules pass through the column in the time equal to the passage of one void volume, V_m, plus the time spent in the stationary phase, expressed by k'_x. Therefore, the volume of eluent which will pass through the column before the sample elutes (the retention volume, V_r) can be expressed as:

$$V_r = V_m(1 + k'_x) \qquad (4.4)$$

The retention volume, V_r, is related to the column flow F_c, and the retention time, t_r. Likewise, the volume of the mobile phase, V_m, is related to the flow and the time the void volume takes to pass through the column, t_o.

$$V_r = t_r F_c \qquad (4.5)$$

$$V_m = t_o F_c \qquad (4.6)$$

Substituting these into Equation 4.4 and rearranging gives:

$$k'_x = (t_r - t_o)/t_o \qquad (4.7)$$

Values for all these variables can all be obtained from the experimental chromatogram, as shown in Figure 4.2. The term $(t_r - t_o)$ is called the adjusted retention

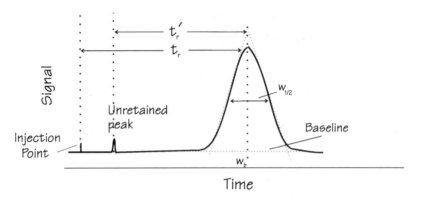

Figure 4.2 A chromatogram showing the peak for an unretained substance and an analyte. The retention time, adjusted retention time, peak width, and peak area can all be determined from measurements made on the peak.

time and is often expressed as t_r', so k_x' can also be expressed as t_r'/t_o. This is then the basis for separation of any two analytes. The separation is directly related to the difference in the k' values for the two substances. If the k' for the sample components is very small, there is so little retention of the compound that separation is not possible. If the difference between the k' factors for two compounds is small, separation of them will be difficult. The selectivity, α, of a column for a particular separation, say of substances A and B, is expressed as a ratio of their retention times or retention factors:

$$\alpha = \frac{t_{r_B} - t_0}{t_{r_A} - t_o} = \frac{k_B'}{k_A'} \tag{4.8}$$

One should notice that the sample bands tend to spread out as they move through the column. The narrower the initial band of sample, and the less the individual compounds are spread out as they traverse the column, the more efficient the column is. An efficient column can separate a greater number of individual compounds in a given time.

4.1.1 Column efficiency

By analogy to the process of distillation, the separating power of a column is expressed in terms of "theoretical plates". This term refers to the length of column in which the analyte equilibrates between the two phases. The more efficient the column is, the smaller the height of the plate will be, and the more equilibrations will occur in the length of the column. An increase in efficiency is most easily seen as a decrease in the width of each sample peak, showing that the bands of sample have not spread much as they passed through the column. A chromatography peak can be approximated to be a normal or Gaussian peak, and "height equivalent of a theoretical plate" H is defined as the variance per unit length. This is a measure of band spreading per unit length of the column. The separation capability of a column is expressed as the number of "theoretical plates" or N, i.e., number of plates in a column. The number of plates contained in the column, is made evident in the peak width of the sample components. Therefore, the number of plates, N, and H can be calculated from a test chromatogram.

$$N = 16(t_r/w)^2 \quad \text{or} \quad N = 5.54(t_r/w_{1/2})^2 \tag{4.9}$$

where $w_{1/2}$ is the peak width at half height, and

$$H = L/N \tag{4.10}$$

where L is the column length.

The efficiency of a column is a function of several parameters. These include the size of the column packing particles, the uniformity of the packing, the flow of eluent, and the rapidity with which equilibrium is established between the two phases.

Two molecules, moving through the column in the eluent flow, may find different paths, especially if there are gaps in the packing which allow eddies or swirls in the eluent flow. Molecules caught in these eddies will be slowed in their movement through the column, and will therefore elute on the back tail of the compound peak. If molecules can take a variety of paths of different lengths through the column, the peak will be broadened. Figure 4.3 shows the phenomenon of eddy diffusion.

Band broadening is also due to the fact that a solute in a gas or liquid stream has a natural and unavoidable tendency to diffuse both forward and backward in the stream. The only factor which can change this is the viscosity of the mobile phase, which is not usually very easy to change. A faster flow, however, gives the peak less time to diffuse.

The mass transfer kinetics between the two phases also have an effect on the band width. Equilibrium between the mobile and stationary phases is never quite achieved in a chromatographic system. The further from equilibrium the system is operating, the poorer the efficiency will be. Diffusion of the sample band in the mobile phase and in the stationary phase both have an effect. Solute molecules may be held up excessively when they become trapped in deep pools of stationary phase, or in stagnant portions of mobile phase. If a substantial

Flow

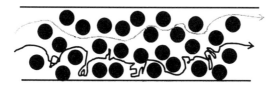

Figure 4.3 Eddy diffusion. If packing is not uniform, molecules may find different paths through the column. Void spaces contribute substantially to band spreading.

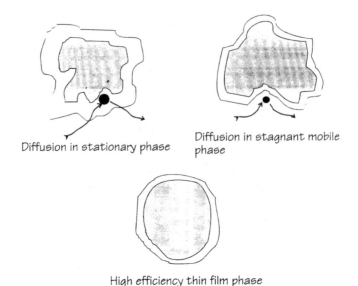

Diffusion in stationary phase

Diffusion in stagnant mobile phase

High efficiency thin film phase

Figure 4.4 Diffusion processes in stationary phase and in stagnant mobile phase. These processes contribute to band spreading. Regularly shaped particles coated with a thin film give highest efficiency.

fraction of the molecules encounters such delays, the result is a spread-out peak. Figure 4.4 shows cross sections of support particles coated with a liquid stationary phase, and how molecules may be excessively delayed. These pools and backwaters may be avoided by spreading a very thin film of support on a fairly regular shaped particle. Of course, this thin phase has little volume, which makes k' smaller, and leads to easy overloading of the column. Figure 4.5 shows how the maximum concentration of the solute in the mobile phase is slightly ahead of that in the stationary phase. The bigger the difference in the location of these peaks, the wider the final exiting sample band will be. If the flow is too

Mobile Phase

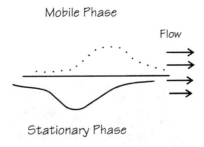

Flow

Stationary Phase

Figure 4.5 Concentration distribution between the mobile and stationary phases. Faster mobile phase flow contributes to the separation of the regions of maximum concentration in the two phases, leading to band broadening.

fast for the equilibration rate, it will cause band broadening due to mass transfer effects.

The plate height, H, is a function of the flow rate of the mobile phase expressed as linear velocity, v, and can be calculated from the equation:

$$H = \frac{B}{v} + C_s v + C_m v \tag{4.11}$$

The first term describes the longitudinal diffusion, and is related to diffusion of the analyte in the mobile phase. This term becomes smaller as the velocity increases, because less time is available for the solute to diffuse. The C_s term is related to mass transfer in the stationary phase, and C_m to mass transfer in the mobile phase. Both of these terms increase with flow, because the solute is less able to reach equilibrium between the two phases as the flow becomes faster. The contribution to the overall H of the column by each of these factors can be plotted against flow velocity. Figure 4.6 shows such a plot. The H line shows the combined effect of the factors. The minimum in the curve indicates the flow velocity which will give the maximum efficiency in that particular column. Using a carrier gas flow which is higher than the optimum will cause some loss of efficiency. However, this may be a satisfactory compromise, since it will also reduce the retention times, and so may shorten the time required for an analysis. Using a flow slower than the optimum is, however, never a good idea, since it lengthens the analysis time and also causes a loss in efficiency.

Although information on the optimal flow rate is often supplied by the column manufacturer, the best flow can be determined experimentally. The value of H can be found at several different flows, and a plot constructed for the column.

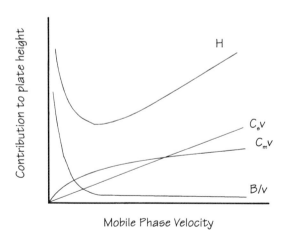

Figure 4.6 Plot of the contributions of longitudinal diffusion and mass transfer in the mobile and stationary phases to the height of a theoretical plate, H. The minimum in the upper curve indicates the flow which will yield the minimum plate height and the most efficient separation on the column.

If there is a difficult separation to be done, there are two approaches. The column material may be changed, which will change both k'_A and k'_B, in order to produce a larger difference between them. On the other hand, if a more efficient column is used, the same amount of difference in retention time will give a more complete separation, because each peak will be narrower. Figure 4.7 shows a poor separation which is improved in one case by increasing the value of α, which moves the peaks apart, and in the second case, by improving the efficiency of the column, which makes the peaks narrower, while not making the retention time difference any greater. Increased efficiency can take the form of either lengthening the column, which has the drawback of also increasing analysis time, or decreasing H by improving the quality of the column itself.

The degree of separation between any two peaks, A and B, can be expressed as the resolution of the peaks. This is defined as the difference between the two retention times divided by their average peak width.

$$R = \frac{t_{r_B} - t_{rA}}{w_A/2 + w_B/2} = \frac{2(t_{r_B} - t_{rA})}{w_A + w_B} \tag{4.12}$$

Alternatively, resolution can be expressed in terms of the selectivity factor and the capacity factors of the two substances being separated:

$$R = \frac{1}{4} \sqrt{N} \left(\frac{\alpha - 1}{\alpha} \right) \left(\frac{k'}{1 + k'} \right) \tag{4.13}$$

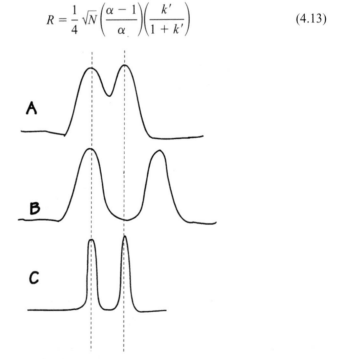

Figure 4.7 Improvement of resolution. In case B the selectivity is changed, moving the peaks apart. In case C, the efficiency is improved, without changing the retention times.

where α and k' are averages (since the values for these factors are similar, if the retention times are close, as a consequence of Equations 4.7 and 4.8).

From this equation, the interaction of the three factors which influence resolution can be seen. A low value of k' for the compounds being separated will lower resolution, since separation cannot take place efficiently if the analytes are not retained long enough to be separated. If α, the selectivity, is not sufficiently large, the compounds are only slightly different in their retention properties and are therefore very hard to separate. Finally, the resolution depends on N, the number of theoretical plates in the column, a measure of the efficiency of the column, indicating that a more efficient column can accomplish a separation even if α or k' are lower than one would wish.

The same relationships can be expressed in terms of retention time rather than resolution. The retention time of the second of two peaks, A and B, is given by

$$(t_r)_B = \frac{16R_s^2 H}{v} \left(\frac{\alpha}{\alpha - 1}\right)^2 \frac{(1 + k_B')^3}{(k_B')^2} \tag{4.14}$$

where v is the carrier gas velocity.

Example

Heptane and toluene were separated with retention times of 15.4 and 16.5 min, respectively, on a 1.0-m packed column. An unretained species passed through the column in 1.8 min. The peak widths measured at the base were 1.15 min for heptane and 1.20 min for toluene.

a. Calculate the resolution between the peaks:

$$R = \frac{2(16.5 - 15.4)}{(1.15 + 1.2)} = 0.94$$

b. Calculate the average number of plates for the column.

$$N = 16\left(\frac{15.4}{1.15}\right)^2 = 2869 \text{ for heptane}$$

$$N = 16\left(\frac{16.5}{1.2}\right)^2 = 3025 \text{ for toluene}$$

$$N_{\text{Average}} = \frac{2869 + 3025}{2} = 2947$$

c. Calculate the average plate height.

$$H = \frac{L}{N} = \frac{1.0\,M}{3025} \times \frac{1000\text{ mm}}{M} = 0.33 \text{ mm for toluene}$$

$$H = \frac{1.0\,M}{2869} \times \frac{1000\text{ mm}}{M} = 0.35 \text{ mm for heptane}$$

$$H_{\text{Average}} = 0.34 \text{ mm}$$

d. Calculate what column length will be necessary to achieve a resolution of 1.5 on this column.

Since k' and α do not change with column length, Equation 4.13 indicates that the resolution is proportional to $\sqrt{N}$. So

$$\frac{R_1}{R_2} = \frac{\sqrt{N_1}}{\sqrt{N_2}}$$

and

$$\frac{0.94}{1.5} = \frac{\sqrt{2947}}{\sqrt{N_2}} \qquad N_2 = 7504.$$

Since $L = NH$, the length required to generate this number of plates is:

$$L = 7504 \times 0.34 = 2551 \text{ mm} \qquad \text{or} \qquad 2.5 \text{ m}$$

e. Calculate the capacity factor for toluene:

$$k' = \frac{t_r - t_m}{t_m} = \frac{16.5 - 1.8}{1.8} = 8.6$$

4.1.2 The general elution problem

There is a problem which arises in all types of chromatography, when samples of widely differing retention properties are present in the same sample. If the elution conditions are correct for the early eluting compounds, the late ones will remain in the column too long. They will be so broadened that it will be difficult to determine their area accurately. Indeed, they may not elute at all. On the other hand, if we set up the system so that the later eluting compounds spend less time in the stationary phase, the early peaks will come out so quickly that they will not have sufficient time in the stationary phase to allow adequate separation. This problem is called the "general elution problem" and is solved in different ways in different types of chromatography. Usually, some type of "programming" is done. This involves a gradual or stepwise change in one of the operating parameters. The best conditions for the separation of the early compounds are set up. Then these are changed during the run to conditions better for the elution of more retained compounds. In gas chromatography, the column temperature is raised over time until a temperature favorable for the separation and elution of the later peaks is reached. This is called **temperature programming**. In liquid chromatography, where retention is more dependent upon strength of the mobile phase, the composition of the mobile phase is changed as a function of time. This is called **gradient programming**.

4.2 Quantitation in chromatography

The amount of each component in the sample is measured as it issues from the column, by passing the effluent through a detector and integrating the detector signal over time. For quantitation in both gas and liquid chromatography, the detector signal is usually fed into a digital or analog recorder. The strip chart recorder produces the typical chromatogram, a plot of signal versus time, as a series of peaks. The peak area is related to the quantity of each material, and the location on the time axis or the retention time is used to identify the compound. Modern chromatographs have high precision in retention time as well as peak area. Consequently, chromatograms provide excellent qualitative and quantitative information. Digital processing of the signal allows the peak areas to be integrated automatically, with the beginning and ending points of peaks indicated. Baselines are usually set automatically and may sometimes be adjusted by the operator. Peak areas obtained on strip chart recorders may be determined graphically by drawing tangent lines to the peak sides and measuring the widths and heights, as was indicated in Figure 4.2. The area (A) is calculated from:

$$A = h_{max} \times w_{1/2} \quad \text{or} \quad (h_{max} \times w_b)/2 \quad (4.15)$$

In addition, a peak area may be measured by cutting the peak out of the chart paper with sharp scissors, and weighing it on an analytical balance.

In order to calculate the quantity of sample from the peak area or peak height, the detector must be first calibrated by running standards. Quantitation may be done by either peak height or peak area, with most accurate results being achieved with use of areas, although, when chromatograms are being manually interpreted, heights may be more readily measured. External standards, internal standards or a combination of both can be used.

4.2.1 External standard method

A series of standards containing known quantities of the analytes are prepared and run. The peak heights or areas are plotted versus the quantities or concentrations, and the samples are run in the same way. A major source of error in this method is the reproducibility of the injection, especially if manual injections are being made using a syringe. Since most chromatographic samples are only 1 µl or less, accurate measurement is difficult. Automatic injectors and sampling valves can reduce this error to a few percent.

The concentration of the sample is calculated from the calibration curve or from:

$$C_{unk}/C_{std} = A_{unk}/A_{std} \quad (4.16)$$

where C is the concentration and A is the peak area for the standard (*std*) and the unknown (*unk*). This simple calculation is usually good enough if the sample and standard are of the same order of magnitude and the detector response is linear over the range of concentrations covered by the sample and standard. If sample concentrations vary over a wider range, a calibration curve should be constructed over the entire range.

Since it is not always possible to obtain standards over a suitable range or for every compound one needs, detector response factors may be used. These allow calibration with one compound as a surrogate for others.

$$C_{unk}/C_{std} = f_{unk} A_{unk}/A_{std} \qquad (4.17)$$

where f_{unk} is the detector response factor which relates the detector response to a quantity of the sample compound to its response to the same amount of the surrogate standard compound. These factors can be determined experimentally. Then the chromatograph can be calibrated daily against the surrogate compound, and the detector response factors used to calibrate for other compounds. It is best to use compounds for standards which resemble the target compounds as much as possible, being of similar molecular weight and polarity, to reduce possible errors.

4.2.2 The internal standard method

An internal standard is added to the sample before analysis. This is composed of one or more compounds with sufficient similarity to the target analytes, so that they behave similarly, but must surely not appear in the samples. Internal standard compounds must also be readily separated from any of the compounds in the sample. If an internal standard is added before such steps as concentration, extraction, or dilution, it is then not as necessary to make accurate volume measurements, and the injection volume is also not as critical. By examination of the peak area of the internal standard in sequential injections, the reproducibility of the injection volumes can be seen. Small changes in the volume show up as changes in the internal standard peak. These can be compensated for in the calculations, by working with the ratio of the sample peak/internal standard peak, rather than just the peak areas. Again, it is important to keep the concentrations of the internal standard and the target compounds in a similar range, to keep linearity problems from arising. The area of the internal standard (A_{is}) is used to normalize the areas of all other sample peaks, thus eliminating the effect of differences in injection volumes or dilutions.

$$C_{unk} = (A_{unk}/A_{is})C_{is} f_{unk} \qquad (4.18)$$

4.3 Gas chromatography

In gas chromatography, the eluent is an inert gas, often helium, hydrogen, or nitrogen. The eluent actually has little effect on the separation process, which is governed more by the volatility of each sample component and its interaction with the stationary phase. Stationary phases are either solids or liquids and are contained in columns which range in internal diameter from 100 μm to 4 mm. The chromatographic system consists of three essential elements: an injection system, a temperature controlled column, and a detector. To this basic system, many enhancements can be added. The column oven may be equipped to carry out temperature programming and subambient cooling. Additional detectors may be added, and an automated injection system, computerized instrumental control and data analysis systems may also be installed. For a chromatograph to be used in environmental analysis, specialized injection systems, such as concentrators or thermal desorbers for air analysis, or purge and trap apparatus for water analyses are often useful. In addition, complex environmental samples often require a detector or array of detectors to assist in identification of the sample components as well as to determine their concentrations.

4.3.1 *Injection devices*

An injector which can place a small, narrow, and reproducible band of sample on the head of the column is absolutely essential to good chromatography. The more efficient the column is, the more demand is placed on the injector to keep the initial sample band narrow. The sample band will only become wider as it traverses the column. If it is wide to begin with, the efficiency of the column will be overshadowed by the band broadening which takes place outside the column, and which contributes nothing to the separation. This is known as **extra-column band broadening**.

A syringe is the traditional injection device in GC. A heated injection port, equipped with a soft polymeric septum, is located at the head of the column. The sample is injected through the septum into the heated carrier gas stream, and vaporizes. The carrier gas flows through the injection port and sweeps the sample onto the column. The injection port must be well designed so that the sample is quickly and efficiently moved onto the column, with all interior areas rapidly swept out by the carrier gas. Unswept areas allow the sample to become trapped, and diffuse slowly back into the gas stream, causing peak tailing. A glass liner is often incorporated into the injection port, so that it may be removed and cleaned or replaced if samples leave a non-volatile residue.

Since a high efficiency column may require a sample of only a few nanograms, a syringe may not produce adequately reproducible samples. A splitting injector port may be used. In this case the sample is injected into a chamber, where it is mixed with a volume of carrier gas. Then a small part of this mixture is allowed to flow into the column, and the rest is vented. An injection port and a splitter are shown in Figure 4.8.

When the sample to be injected is a gas, a **gas sampling valve** is often used to inject a measured portion into the column. This valve contains a loop of known volume which is filled in the "load position" by allowing the sample gas to flow through it. When the valve is turned to the "inject position", the carrier

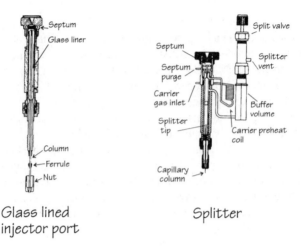

Figure 4.8 A GC injection port and a splitting injection port for use with capillary columns.

gas is diverted through the loop, and the sample is carried onto the column. In Figure 4.9, you can see that the carrier gas always flows to the column, whether the valve is in the load or inject position. The sample size is usually changed by installing a different sized loop, although, if a pressure gauge is mounted on the system, the sample pressure may be changed to change the amount injected. Since the sample is a gas, the temperature of the loop will also make a difference, if it is subject to wide variations.

More elaborate injection systems are found in instruments dedicated to a particular determination. A thermal desorber, for instance, is employed for transferring samples from an adsorbent trap into the column. A purge-and-trap apparatus is used for stripping volatile compounds from water or sludge samples, and injecting them into the column. These will be described in the chapters on air and water analysis.

4.3.2 Columns

There is a vast selection of GC columns available. Many of them are so slightly different from others that they are readily interchangeable. The columns differ in physical dimensions. Large diameter packed columns, having a larger volume of stationary phase, are able to accommodate large samples. The smallest columns, open tubular capillary columns, have the highest efficiency, but are limited to tiny samples.

Chromatographic stationary phases may be either solid or liquid. When a solid stationary phase is used, adsorption is the retention mechanism, and the technique is called **gas solid chromatography**. It is used mainly for low molecular weight volatile organic compounds and gases such as CO, CO_2, SO_2, or H_2S. The stationary phase may also be a viscous liquid coated or chemically

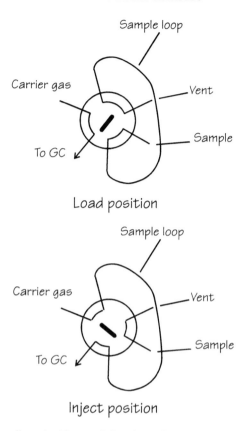

Figure 4.9 A gas sampling valve. The sample loop is usually made of stainless steel tubing, and the volume is adjusted by using different lengths of tubing for the loop.

bonded to the surface of a packing material, or on the inside wall of a open tubular capillary column. This is called **gas liquid chromatography (GLC)**. The mechanism of retention in a liquid stationary phase involves partitioning of the analyte between the gas and the stationary phase. Liquid phases must have good solvent characteristics, and should be inert and stable at relatively high temperatures.

Table 4.1 shows a selection of liquid stationary phases, with the kinds of samples for which they are suited. As gas chromatography developed, hundreds of stationary phases were described in the literature. To characterize these in a quantitative fashion, Rohrschneider developed a series of constants to describe the selectivity of a column toward different types of sample. A suitable phase for a particular sample can be selected on a rational basis by using these constants. This work was extended and refined by McReynolds, and the McReynolds constants for stationary phases are listed in the catalogs of almost every supplier of chromatographic columns.

Table 4.1 Some liquid stationary phases for GC columns

Phase composition	Polarity	Retention indices for probe compounds						Uses
		x	y	z	u	s	Av.	
Dimethyl polysiloxane	Nonpolar	657	648	670	708	737	684	Solvents
5% Diphenyl, 95% dimethyl polysiloxane	Nonpolar	672	664	691	745	761	707	Volatile organics, aromatics
20% Diphenyl, 80% dimethyl polysiloxane	Slightly polar	720	736	744	826	830	765	Volatile organics, alcohols
35% Diphenyl, 65% dimethyl polysiloxane	Moderately polar	754	717	777	871	879	803	Pesticides, PCB's, amines, nitrogen-containing pesticides
14% Cyanopropyl, 86% dimethyl polysiloxane	Moderately polar	726	773	784	880	852	803	Pesticides, PCB's, alcohols, oxygenated compounds
50% Phenyl, 50% methyl polysiloxane	Moderately polar	777	764	806	911	917	835	Phthalate esters, phenols
Trifluoropropylmethyl polysiloxane	Selective for lone pair electrons	740	755	882	980	980	898	Environmental samples, solvents, halocarbons, ketones, alcohols
Polyethylene glycol	Polar	956	1142	987	1217	1185	1097	Acids, amines, solvents, xylene isomers
90% Biscyanopropyl 10% cyanopropyl polysiloxane	Very polar	1061	1232	1174	1409	1331	1241	cis- and trans-dioxin isomers

The McReynolds constants are based on the assumption that different molecular interactions, such as dispersion, orientation-induced dipole, and donor–acceptor complexation, between functional groups on the column material and those of the sample are additive. A series of probe compounds was chosen to represent each type of interaction, and the retention time for each probe is determined on the column in question. Since retention times change with phase loading, temperature, and flow, a retention index for each probe is determined by injecting the probe compound along with a set of normal hydrocarbons which will bracket the probe compound's retention time. The retention index is calculated from the following equation:

$$I = 100z + 100 \frac{\log(t'_{R(x)} - t'_{R(z)})}{\log(t'_{R(z+1)} - t'_{R(z)})} \tag{4.19}$$

where each t'_R is an adjusted retention time. The unknown is indicated as x, and z is the number of carbons in the normal hydrocarbon eluting just before the sample. The hydrocarbon, $z + 1$, is that which elutes just after the sample. This index is not entirely independent of column conditions, but is still useful. The McReynolds constant is expressed as the retention index of each probe compound on the tested column minus that on squalane. The probe compounds, and the interactions they indicate, are listed in Table 4.2.

These constants are useful in comparing columns, and in selecting a new column to improve a particular separation. For instance, if alcohols are to be

Table 4.2 Probe compounds for characterization of GC column phases

	Probe compound	Classes of compounds represented
X	Benzene	Aromatics, hydrocarbons
Y	Butanol	Alcohols
Z	2-Pentanone	Ketones, aldehydes
U	Nitropropane	Nitro groups
S	Pyridine	Amines, basic groups

separated, one would look for a column which has a large Y constant, indicating good retention and selectivity for alcohols. When a column described in the literature for a particular analysis is not available, a suitable substitute can be chosen, by selecting one with similar McReynolds constants.

4.3.2.1 *Packed columns.* Packed columns vary in internal diameter, with most being constructed from 1/4 or 1/8 in OD tubing. They also vary in the particle size of the packing, and the amount of stationary phase coated on the support. One of the parameters usually listed is the percent loading. The loading is determined by the mass of stationary phase coated on a given mass of support. Loading ranges from a few percent to 20 or 30%.

Stationary phases are usually coated onto an inert support material, which should be of uniform size and good mechanical strength. The efficiency of the finished column depends on both the particle size and uniformity of the size. Smaller particles give more efficient columns, but if the particles become too small, the flow through the column at reasonable gas pressures is hindered. Usually column packings in the 60–80 mesh or 80–100 mesh ranges are used. These are often made from treated diatomaceous earth, a silica based material.

Supports vary in their surface activity and many treatments have been developed to modify the surface. The support may be extracted with acid to remove surface metal ions which may form active sites on the packing, adsorbing sample molecules too tightly and causing tailing. The support may also be treated with a silanizing agent to reduce surface activity further. For extremely polar compounds like water or sulfur dioxide, an extremely inert support is made from granulated fluorocarbon polymer.

The packing is covered with the stationary phase, generally by mixing the support with a solution of the stationary phase, a high boiling liquid, in a volatile solvent. As the solvent is evaporated the packing is coated with an even layer of stationary phase. When the solvent is dried off, the packing should be free-flowing and easy to pack into the column without gaps or loose spaces.

Stainless steel or glass tubing is used to form the columns. Stainless steel is considerably easier to handle, as it is not breakable and can be gently bent to fit the instrument at hand. Glass columns are usually deactivated by being sila-nized, and have a lower surface activity, so may be needed for especially sensitive samples, such as very labile organic compounds. It is important to handle packed columns fairly gently, and bend them carefully. Rough treatment will fracture the packing particles, increasing the non-uniformity of the packing by making smaller particles.

For gas–solid chromatography, several types of solid column packings are available. These may be porous polymeric materials, silica, or carbon based solids. Solid phases may be more durable, and able to stand up to some fairly aggressive samples, since they have no surface coating. They are usually used for small organic molecules and inorganic gases. Molecular sieves are used for gases such as CO, CO_2, and O_2. Some packings are designed for specific types of samples, such as water and alcohols, amines and nitro compounds, or sulfur gases.

Polymeric packing materials differ in their surface polarity and their pore sizes. They are especially useful for gaseous samples and for aqueous samples. Recently, carbon based materials have become quite popular. These are manufactured specifically for chromatographic separations and vary in such parameters as the pore size, the surface area per unit mass, and the particle size. Table 4.3 lists some of the more common types of solid packings.

4.3.2.2 *Open tubular columns.* Open tubular columns are often used for complex mixtures, where large numbers of components must be separated. These are made of fused silica tubing, coated with a thin film of stationary phase on the interior, and covered with a polyimide coating on the outside surface. The polyimide is necessary to provide mechanical strength as fused silica can break easily if it is scratched. Recently, open tubular columns are also being made with silica lined stainless steel tubing. Choosing the best stationary phase is done in the same way as for packed columns. Selection of column diameter and phase thickness requires consideration of the sample to be separated. The narrowest diameter columns give the highest separating power, but the sample size is very limited. When trace components are to be determined, there may not be enough sample to detect. The thinnest coatings of stationary phases also give

Table 4.3 Some solid phase GC packing materials

Packing	Applications
Graphitized carbon black	Solvents and chlorinated organics with low to moderate boiling points
Carbon molecular sieves	Permanent gases, light hydrocarbons, low MW organics
Divinylbenzene-ethylene glycol dimethacrylate copolymer	Permanent gases, methane, C2 hydrocarbons, water
Divinyl benzene-acrylonitrile copolymer	Polar organic compounds
Divinylbenzene polymer	Light gases, CO, CO_2, acetylene, water, H_2S
Porous silica	C_1 to C_4 organic gases, sulfur gases, acetates
Diphenyl-*p*-phenylene oxide	High boiling alcohols, diols, phenols, ethanolamines, chlorinated aromatics, ketones
Zeolite molecular sieves	Permanent gases such as H_2, O_2, Ne, Ar, N_2, CH_4, CH_4, Ne. Traps water
Activated alumina	C_1 to C_4 hydrocarbons

highest efficiency, but again limit the size of the sample which can be analyzed.

Many available capillary columns have chemically bonded phases. In these, a chemical reaction takes place between the liquid phase and the surface after the column is coated. The liquid forms chemical bonds to the hydroxyl groups found on the surface of the fused silica. These phases have several advantages. They are mechanically stable, and will not tend to creep towards the lower part of the column, even at high temperatures. This helps to extend the useful life of the column. If a bonded column becomes contaminated, which may happen especially easily when it is used for direct on-column injection, it may be washed out with solvent. This will often return the column to its original efficiency.

Internal diameters of capillary columns range from the highest efficiency ones of 0.15 mm to wide bore columns of 0.53 mm. Typical values for H range from 0.13 mm for a 0.15 mm ID column to 0.45 mm for a 0.53 mm ID column. The column length selected again depends on the complexity of the sample and the number of plates needed to effect the separation. The shortest column which will accomplish the separation is best, since a longer column will require a longer analysis time without adding to the information gained. One virtue of capillary columns is the ease of dividing a long column into two shorter ones. Sometimes a long column, which has been shown to have an excess of theoretical plates for the use, may be cut in half.

The usual liquid stationary phase layers are in the range of 0.12 to 0.2 μm. The thickest stationary phase coatings may range up to 5 μm, and find their most usual usage in the analysis of very low boiling substances, especially samples which are gases at room temperature. Figure 4.10 shows the effect of film thickness on separation, and Figure 4.11 shows the effect of column diameter. The best separation is that which gives an adequate separation in the shortest time.

Porous layer open tubular (PLOT) columns are somewhat of a hybrid between packed and capillary columns. These are open tubular columns with a solid stationary phase layered on the inside surface of the column. These provide a means of doing gas/solid chromatography, with the advantages of an open tubular column.

The considerations which go into the selection of a column for any particular analysis can be summarized as follows:

- Is there a column offered which is designed for this analysis? Suppliers often design and sell columns which have been optimized for a certain commonly done sample. For example columns are made for light hydrocarbons, gasoline, pesticides, fatty acid methyl esters, and amines.
- Is there a column specified in a reference in the literature for a similar analysis? If so, the same column or one with similar McReynolds numbers should be satisfactory.

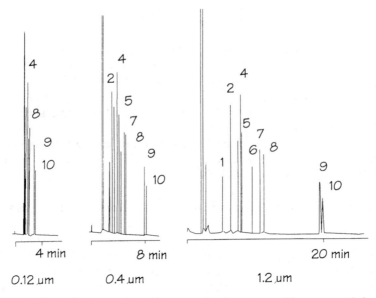

Figure 4.10 Effect of stationary phase thickness on the chromatogram. The same sample is run at the same temperature on three columns. A thicker phase increases the capacity factor and the retention time. Note that the separation is not optimized for these runs, in order to show the effect of coating alone.

- A rule of thumb for column selection relies on the fact that the separation takes place in the stationary phase. Therefore, a phase which has a higher affinity for the sample components will be better for the separation. Relying on the maxim "Like dissolves like", a nonpolar column will do best for nonpolar samples and polar samples should be separated on a polar phase.
- Column length, diameter, and film thickness depend on the complexity of the sample, as well as the amount of sample which will be injected and the boiling range of the sample.

4.3.2.3 *Column temperature.* A column has a minimum and maximum use temperature, which depends mostly upon the stationary phase. The minimum temperature is usually the freezing point of the stationary phase, since the solid has quite different properties from the liquid, and does not always partition the sample adequately. The maximum temperature is usually that at which the column bleed becomes intolerable. This is usually due to thermal breakdown of the stationary phase, or to the boiling off of the lower molecular weight fraction of the phase.

The temperature for the analysis must be optimized in the development of the method. The usual method is an educated trial and error approach. Equations exist which attempt to predict retention times based on column temperatures and thermodynamic properties of the sample components, but, in practice, these are

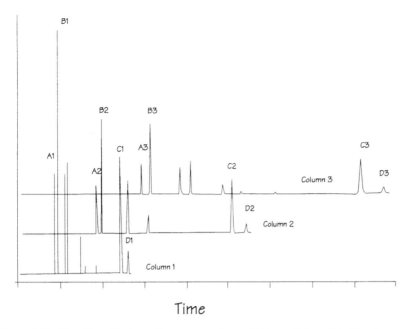

Time

Figure 4.11   The effect of column diameter on column efficiency. Column 1 is 10 m long, and 0.15 mm in diameter, column 2 is 16 m × 0.25 mm, and column 3 is 21 m × 0.32 mm. These columns all have the same number of plates. The smaller diameter column has a smaller value of H and provides more plates per minute.

seldom used. Generally, a standard or sample containing the components of interest is run on the selected column, at a temperature selected on the basis of the range of boiling points of the sample. It is not necessary to bring the column up to the boiling point of the higher boiling components. Compounds will migrate through the column at temperatures much lower than their boiling point. Thinner phase columns will require lower temperatures to elute peaks than those with thicker films. Referring to Figure 4.10, we can see that the 0.4 μm film column eluted the test peaks much more rapidly than the 1.2 μm film column. If the temperature on the 1.2 μm column was raised, the chromatogram would begin to appear more like that produced on the 0.4 μm column.

When the last peaks in a chromatogram elute late and are very broad, the problem may be corrected by raising the column temperature. However, this often causes the earliest peaks to be unresolved because they are not retained long enough. In this case, a temperature program must be used. The sample is injected with the temperature optimized for the separation of the earliest peaks. The temperature may be held until the first peaks are eluted, and then the temperature is ramped up to a temperature which will bring out the last peaks, separated, but not excessively broadened. The rate of temperature rise depends on the number of peaks eluting during the ramp. If few peaks are present, then a rapid ramp might be chosen, while a complex sample with many components requires a slower ramp rate. Figure 4.12 shows a standard containing four hydrocarbons. In the isothermal run, the early peaks are sharp and well resolved but the late ones are broad. When a temperature program is used, a better chromatogram is obtained, and the time to resolve the four compounds is halved.

Generally, a few test runs will be needed to determine the best temperature program. Some chromatographic instruments are capable of multiple ramps and pauses during a program. These can be used to fine tune specific parts of a separation. For example, let us suppose that both the early and late parts of a sample are well separated, but there is a pair of peaks in the middle which are overlapping. One might slow the rate of temperature rise before these peaks elute, to retard them a little longer in hopes of improving the separation. After they elute, the oven temperature is raised to the same final temperature as in the original program.

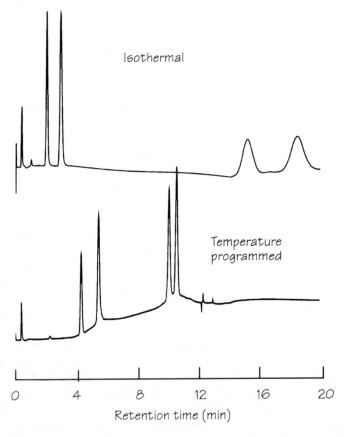

Figure 4.12 A mixture of four hydrocarbon gases run at constant temperature and with a temperature program. Note the improvement in the later peaks in the programmed run.

4.4 GC detectors

The eluent from the column is directed to one or more detectors. These produce signals which are proportional to either the amount of sample present in the detector at any moment, or the concentration of sample in the detector. The detector signal is usually displayed as a plot of signal magnitude versus time, giving the classic chromatogram. Detectors vary in their response to different classes of compound, from the thermal conductivity detector, which is universally responsive to all compounds, to such specialized detectors as the flame photometric detector, which detects only sulfur or phosphorous containing compounds, depending on the way it is set up. The selectivity of a detector is usually expressed by the ratio of the response to the desired analyte divided by that towards an interfering compound. For instance, the selectivity of a sulfur specific detector may be given by its response to a nanogram of sulfur divided by that for a nanogram of hydrocarbon. Since the interfering material may be present in much greater quantity than the analyte, selectivity factors of 10^4 or more are desirable.

The characteristics sought in a gas chromatographic detector are high sensitivity, a linear dynamic range of 4 orders of magnitude or more, a favorable signal to noise ratio, and good long term stability. A small detector dead volume is also important. If the sample has an opportunity to mix with a volume of carrier gas before the detection process is completed, the peak will be broadened and efficiency lost.

4.4.1 Thermal conductivity detector (TCD)

The TCD is a truly universal detector. It consists of a heated sensor in a thermostated chamber, through which the effluent flows. Helium is usually used as a carrier gas, as it has the highest thermal conductivity of any gas, except for hydrogen. As the peaks elute, the thermal conductivity of the gas in the chamber changes. This changes the heat flow from the heated sensor, through the gas, to the walls. Since the sensor is being heated at a constant rate, it becomes hotter as the thermal conductivity of the effluent drops. The change in temperature of the sensing wire filament or thermistor changes its resistance. The sensor is wired into a Wheatstone bridge circuit, and the change in resistance produces an unbalance, which produces a signal. The circuit for the TCD detector is diagrammed in Figure 4.13.

The filament is sensitive to oxidation while heated, and therefore must not have current flowing unless the carrier gas is passing through the chamber. The detector is limited by its relatively low sensitivity, compared to other detectors, and usually has a fairly large dead volume. It is, therefore, not very suitable for capillary work. Because of these limitations the TCD is little used in environmental work, except for the determination of major constituents of air.

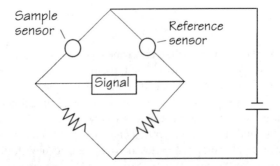

Figure 4.13 Thermal conductivity detector. The sample flows over a heated wire in one arm of the Wheatstone bridge, while pure carrier gas flows over the reference wire.

4.4.2 *Flame ionization detector (FID)*

The FID is a major workhorse of environmental analysis. It is nearly universally sensitive to organic compounds, and shows good sensitivity and excellent linearity. The column effluent is fed into a flame fueled by hydrogen, with a forced air flow. Figure 4.14 shows a typical FID. A potential of several hundred volts is imposed between the tip of the flame burner and the collector which surrounds the flame. As the sample components burn, they produce a burst of ions. These produce a tiny current between the flame tip and the collector. The current is amplified by a high impedance electrometer and measured. The background current flowing in the detector is in the region of

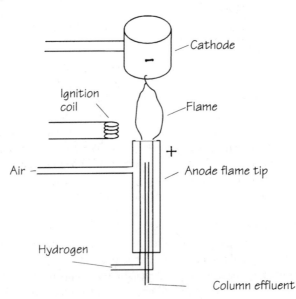

Figure 4.14 The flame ionization detector. Ions produced by the sample in the flame are collected on the cathode and the current amplified.

2×10^{-14}–10^{-13} amp. In the presence of organic compounds the current will rise to 10^{-12}–10^{-9} amp. The response of the detector will change if the flows of air and hydrogen to the flame change. These flows should be checked for consistency when the detector is calibrated and used.

The FID detector has a number of advantages. The response is roughly proportional to the number of carbon atoms in the flame at any time, although certain substituent atoms, such as chlorine, reduce the response. The detector is insensitive to inorganic gases, water, carbon dioxide, sulfur dioxide, nitrogen oxides, and other non-combustible gases. The detector has a very wide linear range, over about 7 orders of magnitude, has a low dead volume of about 1 μl, and is relatively noise free and easy to operate. Its major disadvantage is that it destroys the sample, so it cannot be passed on to another detector. Also, the fact that it requires both compressed air and hydrogen, as well as carrier gas, can be an inconvenience, especially when instrumental portability is an issue.

4.4.3 Electron capture detector (ECD)

This is probably the most used of the compound class specific detectors for environmental analyses. It has been used to trace the fate of such pesticides as DDT, as well as the halocarbon gases in the atmosphere. Its response depends on the electron-capturing properties of the sample. The detector is highly sensitive to the presence of electron capturing substituents, such as halogens, peroxides, and nitro- groups, on the sample molecules. When one looks at lists of regulated compounds, it is striking that so many of these are electron capture active.

The detector (Figure 4.15) consists of a chamber containing a β-emitting foil, usually nickel-65. This radioactive source emits electrons which ionize the carrier gas, and form a small current between the electrodes in the chamber.

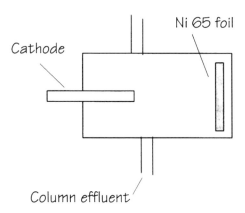

Figure 4.15 Electron capture detector. The β particles (electrons) emitted by the Ni[65] foil are collected on the cathode. As these are captured by sample components, the current drops, and this drop is monitored as the signal.

Table 4.4 Relative sensitivity of ECD detector to various compounds

Compound	Sensitivity relative to 1-chlorobutane (1.0)	Compound	Sensitivity relative to 1-chlorobutane
2-Chlorobutane	2	1-Bromobutane	280
1-Chloropentane	1.0	Benzene	0.06
1-Chlorohexane	1.1	Chlorobenzene	75
1-Chlorooctane	1.6	Bromobenzene	450
1,4 Dichlorobutane	15	1-Iodobutane	9×10^4
1,1-Dichlorobutane	110	Carbon tetrachloride	4×10^5

When electron-capturing species are present, they reduce this current. The reduction is detected by the electronics and measured.

The detector response varies widely, depending on the electronegativity of the species being detected. Table 4.4 shows the relative sensitivity of the detector towards a variety of substances. We can see that the response to the individual electronegative groups depends upon their number in a molecule, as well as their location. The response to chlorinated hydrocarbons increases by about a factor of ten with each additional chlorine atom present in the molecule, for example.

If the carrier gas forms metastable ions, it may cause undesirable collision reactions. Therefore, helium cannot be used with the ECD detector. Nitrogen is suitable, and a mixture of argon with 5–10% methane is also sometimes used. If a capillary column is being used, extra gas, called make-up gas must be added to the detector to bring the flow up to that for which the detector is designed. This helps to minimize band broadening. In this case, if nitrogen is used for the make-up gas, helium may be used for the column carrier gas.

Since ECDs contain a radioactive isotope, they are subject to governmental regulation. Depending on the design of the unit, a license may be necessary before one can be purchased. There are tests, called wipe tests, for leakage of radioactivity which must be done at specified intervals, to comply with regulations. The company which sells the unit will supply information on required licensing and routine testing.

The sensitivity of the ECD toward halogenated hydrocarbons and many pesticides is extremely high. The response to a particular compound is usually quite temperature sensitive, and the detector's linearity varies with conditions and analyte. Therefore, a calibration curve should be generated for each compound which is to be quantitated, and the samples should fall within the range of the calibration. Extrapolation is especially risky when the detector may be non-linear, and a large error may easily arise if the calibration line is extended beyond the last measured point.

4.4.4 *Photoionization detector (PID)*

A detector which has much the same response as the FID, but which requires no support gases is the photoionization detector. This detector exposes the effluent

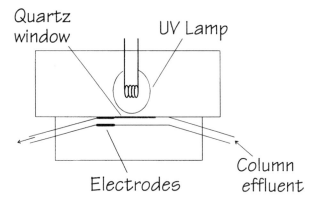

Figure 4.16 The photoionization detector. Sample flows through the irradiating chamber and ions are collected on the electrodes, forming a current which is amplified.

stream to ultraviolet light, thus ionizing the sample. The ions are collected on an electrode, with the resulting current amplified and measured with an electrometer. Figure 4.16 shows the detector. The range of compounds to which the detector is sensitive depends on the wavelength of the lamp used in the detector. Lamps can be purchased with wavelength peaks at 9.5, 10.0, 10.2, 10.7 and 11.7 eV. The 10.2 lamp is the most commonly used. The detector will respond to substances having ionization potentials below the lamp energy, and up to about 0.4 eV above it.

The PID is about 35 times more sensitive to aromatic compounds and somewhat more sensitive to alkanes than is the FID. It has a linear range of about 10^7. The response toward various compounds seems to be most closely related to their ionization potentials, with those with the lowest potentials giving the highest response. In general, as the carbon number increases, the sensitivity increases.

The chief advantages of the PID are that it is nondestructive, so it can be used in series with other detectors, and does not require support gases, as does the FID. This makes it ideal for portable instruments, which may use air for the carrier gas, and not require any cylinder gases to be carried. A portable PID detector, without a gas chromatograph, is available and can be used for screening for organic emissions, without speciation. The main drawback is a deposit which may form on the window separating the UV lamp from the gas stream. Some sample components react under the UV light and form solid products, which contaminate the lamp window.

4.4.5 *Electrolytic conductivity detector (ElCD)*

The electrolytic conductivity detector (Figure 4.17) may be set up to determine compounds containing chlorine, sulfur, or nitrogen. The detector is reconfigured,

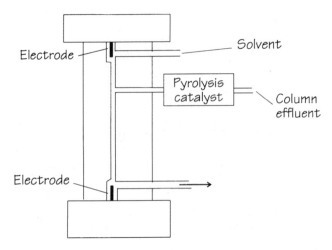

Figure 4.17 The electrolytic conductivity detector. Sample components are pyrolyzed and the resulting soluble gas is dissolved in the flowing solvent. The change in solvent conductivity is monitored.

depending which of the elements are to be determined. When configured for chlorine, the effluent from the column is passed over a catalyst which converts any Cl to HCl. In sulfur mode, SO_2 is formed, and in nitrogen mode, NH_3. To reconfigure the detector, the catalyst is changed and reactor temperature is adjusted.

The reacted gas is scrubbed into a flowing aqueous or alcohol stream, and passed into a conductivity cell. The response of the detector is proportional to the number of Cl, N, or S atoms passing through the cell. The solvent is usually recycled through ion exchange resins. The detector must be watched for dips in the baseline which begin to occur as the solvent becomes exhausted. Detection limits of 10^{-12} g nitrogen/s, 5×10^{-13} g chlorine/s, and 10^{-12} g sulfur/s are possible, and the selectivity ranges from 10^4 to 10^9 over hydrocarbons. The sensitivity is similar to that of the ECD, and the fact that the response is fairly consistent with the amount of the target heteroatom gives it an advantage over the ECD for some compounds.

4.4.6 *Flame photometric detector (FPD)*

This detector is designed for the specific detection of sulfur and phosphorous. It is similar in construction to the FID, but a cooler flame is produced by altering the hydrogen/air ratio. Instead of measuring the ions formed in the flame, the radiation emitted by the sulfur S_2 and phosphorous HPO species formed when the sample components enter the flame is measured. A filter photometer is used to detect the radiation emitted at 394 nm for sulfur or 526 nm for phosphorous. Figure 4.18 shows a schematic of this detector. Since the sulfur or phosphorous

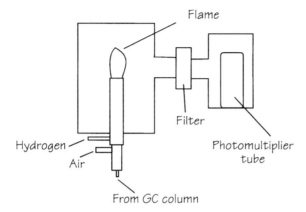

Figure 4.18 The flame photometric detector. Sample components containing sulfur or phosphorous emit UV radiation as they pass through the flame. The detector is set up for either P or S detection by changing the filter.

are measured in the same form for each component the response is governed by the total amount of the element in each sample component.

The detector is subject to negative interference from hydrocarbons, which quench the emission if they are present in the flame at the same time as the sulfur or phosphorous compound being measured. A good GC separation will reduce this difficulty by eliminating the coelution of the interferent and the analyte.

The square root of the detector response is proportional to the sulfur concentration. For phosphorous, the response is directly proportional to concentration and is linear over 2–3 orders of magnitude.

4.4.7 Thermionic ionization or nitrogen–phosphorous detector (NPD)

The nitrogen–phosphorous detector is a modification of the flame ionization detector. In this detector, a bead of a rubidium salt is placed at the tip of the flame. This gives a selectivity of $10^3–10^4$ for N and P compounds over hydrocarbons. Helium carrier gas gives a better response with phosphorous compounds, while nitrogen is better for nitrogen containing compounds. The actual mechanism of the selective response is not entirely clear. It is believed that free radicals, formed in quantity by the nitrogen or phosphorous compounds in the flame, cause the vaporization and ionization of rubidium from the bead, adding to the signal. While the NPD shows selectivity toward the N- and P-containing compounds the response to these is not a great deal higher than that of the unmodified FID. This detector gains its selectivity as much by repressing the response to hydrocarbons as by enhancing the response to N and P compounds. Its most common environmental use is in detection of nitro-PAH and other nitrogen-containing compounds in petroleum products.

4.4.8 *Mass selective detector (MSD)*

The mass selective detector (MSD or GC/MS) is probably the most powerful tool in the hands of the environmental analyst. The detector, essentially a mass spectrometer, is a universal detector, as well as a very specific one. The MS can detect any molecule. Because each effluent component is fragmented and a mass spectrum is generated, plotting the intensity of a single mass fragment or group of fragments will generate a compound-specific chromatogram. These detectors will be discussed in Chapter 5.

4.4.9 *Comparison of detectors*

The advantage of a universal detector is obvious, in that it responds to any component of the sample. Class specific detectors are also useful, since they can be used to simplify complex chromatograms. For most environmental work, the mass selective detector is the detector of choice, as it not only identifies but also quantitates the sample. It is mandated in many of the standard methods, but does not match the sensitivity of the ionization detectors. All the ions produced in an ionization detector are collected and measured, while each ion fragment in a mass spectrum is collected for only a small fraction of the time. This makes mass spectrometry inherently less efficient. However, one of the most difficult parts of the analysis with a general purpose detector is the identification of peaks. Because of the difficulty of reproducing the temperature exactly, and because the amount of loading on the column, especially of water vapor, may cause small shifts in retention time, retention times are not always sufficient to determine the identity of individual peaks unequivocally. A mass spectrum, combined with the retention time, is a much better way of making a reliable identification.

4.5 High performance liquid chromatography (HPLC)

When a material is non-volatile or when it is so thermally fragile that it cannot be analyzed by gas chromatography, then liquid chromatography may be appropriate for the analysis. High performance columns are constructed of packing materials with very small particle diameters, on the order of 3–10 μm. Therefore, the eluent cannot flow through under gravity flow as was done in early liquid chromatographic analyses. A high pressure pump is necessary to force the eluent through the column at the rate which delivers the maximum number of theoretical plates. High pressure, high performance liquid chromatography is one of the essential tools of environmental analysis. It can readily handle high molecular weight compounds, such as polynuclear aromatic hydrocarbons, highly polar compounds such as phenols or organic acids, and even inorganic ions.

Choice of detector is somewhat limited in HPLC compared to GC. Since the eluent is usually an organic liquid, ionization detectors, which are so useful in GC, are not applicable. Detection is therefore restricted mostly to spectroscopic methods, which are limited by the absorbance characteristics of the analytes.

HPLC can be divided into several related techniques, depending on the separation mechanism and the column type. The most useful types in environmental analysis are reverse phase, normal phase and ion chromatography. Reverse phase liquid chromatography is probably the most frequently used, and the most versatile. It is called "reverse" because of the comparison with "normal phase", which is only called normal because it was invented first.

4.5.1 *Reverse phase liquid chromatography*

Reverse phase columns have a packing composed of solid silica support particles with an organic coating bonded to their surface. The bonded phase is produced by reacting a halogen substituted organosilane with the surface $-OH$ groups present on the silica support. This leaves hydrocarbon chains, which may contain two, eight, or 18 carbons, bonded at their ends through $Si-O-Si$ groups to the surface of the support. Figure 4.19 shows the bonding of octyl groups to the silica surface. These columns are designated by the carbon number of the chains attached, with the most frequently used column being the bonded octadecyl type, called C18.

Since these coatings are very non-polar in nature, the chief mechanism of retention is dispersion forces. This makes them useful for separation of organic compounds based on slight differences in their backbone or side chain configuration. The mobile phases commonly used are fairly polar in nature, with alcohols and water being common constituents. Since these are weaker eluents than the non-polar solvents which have a strong affinity for the highly non-polar

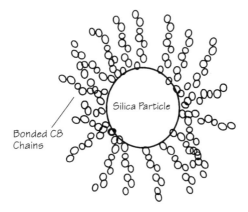

Figure 4.19 A C8 bonded particle for HPLC packing. The organic molecules are bonded to the surface to prevent them from being dissolved in the organic eluent.

column surface, sample components are retained long enough for good separation to take place. Components are eluted with the most polar ones being least retained and the least polar ones being held the longest.

Figure 4.20 shows a chromatogram of a sample containing five compounds. In the first run, an isocratic eluent was used, a mixture of 30% methanol and 70% water. The first peaks are poorly separated while the later ones are too broad and take a long time to elute. In the second case a gradient elution from 10% methanol to 100% methanol is done. The early peaks are better separated, since the initial eluent was weaker, while the late peaks are moved through the column more rapidly, as the eluent increases in strength. When gradients are done, it is important to begin with a weaker mobile phase, in this case one with a substantial amount of water. This allows the earliest peaks to remain in the

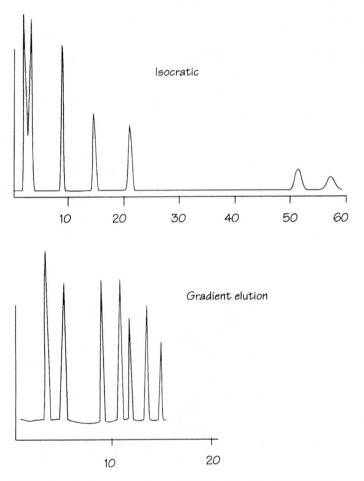

Figure 4.20 The isocratic chromatogram is run with 30% methanol in water. Note the improvement when the same sample is run with an eluent gradient from 10% methanol to 100%.

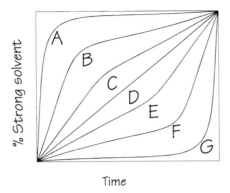

Figure 4.21 Examples of solvent programs. In addition to linear gradients, eluent composition may change rapidly at first, then slower, or vice versa.

column sufficiently long to achieve separation. Then the strength of the eluent is increased, by adding more of the less polar acetonitrile.

Gradients may be linear or curved. Figure 4.21 shows some gradient profiles. If one had a situation in which several peaks were eluting close together at the beginning of the chromatogram, one might want to select a concave gradient similar to the one labeled F in the figure. This would allow the eluent to increase in strength very slowly, as these early peaks are being separated. Then, toward the end of the run, the strength is increased rapidly to bring out the later peaks. It is usually better to begin by experimenting with a linear gradient, then see if the beginning or end of the chromatogram could benefit from being stretched out a bit. A curved elution profile might then be tried.

4.5.2 *Normal phase liquid chromatography*

Normal phase chromatography relies on such column packings as silica and alumina. Modern silica packings with polar bonded coatings are also available and are more reproducible and easy to use than is the bare silica. The difficulty with silica is its high affinity for water. Any trace of water in the solvent will be adsorbed onto the column, thereby changing its characteristics. This makes reproducible chromatography harder to achieve. Characteristically, normal phase columns have a polar surface, and eluents are rather nonpolar, to achieve reasonable separation. In contrast to reverse phase separations, the strongest eluents used in this system are the most polar.

Solvent gradients, in this case, would begin with the least polar solvent, and gradually increase in polarity to bring out the most retained, most polar compounds. Samples best separated on normal phase columns are those comprised of different classes of compounds. Homologs are better separated on reverse phase columns.

4.6 HPLC instrumentation

A complete apparatus consists of a pumping system, either for a single eluent or for a gradient, an injector, a column, one or more detectors and a data handling system. Typical setups are shown in Figure 4.22.

4.6.1 *Solvent delivery systems*

Before being fed to the pumping system, the solvents should be filtered and degassed. Any particulate material in the solvent must be removed, because particulates may damage the pumps, and will, in time, collect at the top of the column and cause plugging. Degassing is important, because the dissolved gases may form bubbles when the pressure drops as the solvent enters the detector. Many detectors are severely disrupted by bubbles. Dissolved gases can be removed by purging the solvents with helium, which is quite insoluble in most solvents, or by passing the solvent through a microporous filter under vacuum. Vacuum filtration is the most common technique, since it accomplishes both the filtration and degassing processes at the same time. When solvents have been standing for some time, re-filtration is a good idea.

The characteristics of an HPLC pump which are of highest importance are its ability to deliver a constant, pulse-free flow, over a wide range of different flows. The materials of the pump system must be resistant to attack by the wide variety of mobile phases to be used. An additional desirable feature is the ability to generate a gradient of two or even three solvents, in a reproducible fashion. Pressures of up to 10 000 psi are generated by the pump. Another consideration

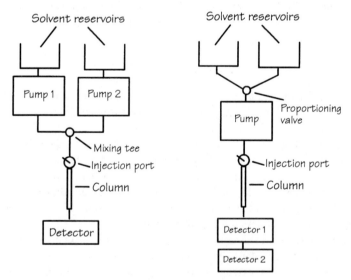

Figure 4.22 Two HPLC setups. The solvent gradient may be mixed before the high pressure pump, or may be mixed after being pressurized by two separate pumps.

is the ease of changing solvents, which is related to the hold-up volume of the system.

Reciprocating dual piston pumping systems are the most common type. In these, one piston chamber is filling while the other is pumping. The pistons move in small chambers, each of which contains less than 0.5 ml of eluent. A system of check valves keeps the solvent flowing in the correct direction. Pulses are kept to a minimum by elaborate design of the piston stroke cycle. As one piston begins to slow at the end of its stroke, the second one, newly filled, begins to deliver solvent, keeping the pressure and flow as constant as possible.

When using reciprocating piston pumps one must be careful to rinse the pump before turning it off. A solvent which contains no solids is pumped for several minutes, to remove any buffer salts which may remain from running an analysis. Salts can dry and crystallize on the surface of the piston when it is idle. Then, when the pump is restarted, the solids will cause abrasion of both the piston rods and the seals through which they pass.

A simpler, less expensive pumping system uses an eluent-filled syringe, driven by a constant speed motor, turning a screw drive. High pressure, pulseless flows can be generated by a syringe pump, but the volume is limited by the capacity of the syringe. This causes some downtime, as the pressure must be brought down, and the flow stopped, each time the syringe is refilled. These pumps are particularly useful for analyses using microbore columns which require very low flows. A constant pressure supplied by a compressed gas cylinder can also be used to drive a pump piston, giving a very smooth, pulseless flow. However, this system suffers from the same inconveniences as does the syringe pump.

Connections between solvent containers, pumps, columns, and detectors are usually constructed of narrow bore stainless steel tubing. Extra-column band broadening is highly dependent on the radius of the tubing through which the sample and eluent pass. The smallest bore tubing which is practical without causing undue plugging should be used. Commonly, tubing around 0.01 inch i.d. is used. Increasing either the diameter or length of tubing through which the sample is passed will have a deleterious effect on the separation efficiency. The effect is more serious as smaller diameter columns are used.

When narrow-bore tubing is cut, it is important to avoid plugging it, and to produce a smooth, square end for attachment of fittings. If the end of the tubing is cut raggedly or at a slant, a void space will usually occur when the tubing is placed into a compression fitting. Special cutting wheel tools are available to make good cuts.

4.6.2 Solvent gradient systems

Solvent gradient systems require a method of mixing a constantly changing amount of solvents from two or three separate reservoirs. The mixing may be done either before the high pressure pump, or after it. The low pressure mixing method uses low pressure metering pumps to deliver the components of the solvent mixture to the inlet of the high pressure pump. One must be careful to have no dissolved gases in the solvents, because the gases may have lower solubility in the mixture, and bubbles can form.

High pressure gradient systems use separate high pressure pumps for each component, feeding into a small mixing chamber just before the injector. These are usually controlled by a microprocessor, which gradually increases the speed of one pump, while slowing the other. This is the most commonly used method, although it is somewhat more expensive, since it requires additional high pressure pumps. In addition to the ability to run samples which require gradient elution, another advantage of having a gradient system available is the ability to change the elution mixture easily. When setting up a new method, even one which will be done isocratically, it is faster and easier to make successive runs while having the microprocessor prepare different mixtures for trial runs, than it is to make the solvent mixtures by hand. In a laboratory with several instruments available, the gradient instrument should be used to set up methods, even for isocratic systems.

4.6.3 Sample injectors

The requirement for an injector in HPLC is the same as it is for other types of chromatography. It must put a very narrow plug of sample into the eluent stream. One difficulty is that the eluent is under high pressure, which makes the use of a syringe impractical. The usual method is the use of a sampling valve, containing a small sampling loop, with a volume of a few microliters. An excess of dissolved sample is flushed through the sampling loop, to fill it completely, with the excess passing out to waste. When the valve is rotated, the eluent flow is diverted through the loop, picking up the sample and moving it on to the column.

Some of these injection valves may also be used in a partial-filling method. In this case, a volume of sample, smaller than that of the loop, is injected into the loop, using a syringe to measure the sample. This displaces some, but not all of the eluent in the loop. The accuracy is less than can be achieved by complete flushing of the loop, but it has the advantage that the sample volume can be readily adjusted. Figure 4.23 shows a typical injection valve.

4.6.4 HPLC columns

4.6.4.1 *Precolumns and guard columns.*   The analytical column is expensive and can be damaged by particulate material depositing at the head of the column, as well as by attack of the eluent on the packing itself. Eluents with pH outside the 2–7 range may dissolve the silica support of the column. To lengthen the useful column life, guard and precolumns are used.

Precolumns are short segments of tubing packed roughly with similar material to that used in the column. The precolumn will pick up any particulates which are present at the exit of the pump. These particles may arise from poorly filtered eluents, or from wear fragments from the pump. More importantly, if the eluent is aggressive, and is dissolving the silica backbone of the packing, the precolumn serves to saturate the eluent with silica. Since it is located before the injector port, it cannot contribute to band broadening. Therefore, it is not necessary to have this column packed as carefully as an analytical column, or to be concerned about having very low dead volume fittings used to install it. It can be made of rather inexpensive components and packed in the lab.

Another source of particulate matter which may damage columns is the sample. To protect the column from materials in the sample which may deposit at the entrance of the column or which may be irreversibly adsorbed on the column packing, a guard column may be used. Guard columns can contribute to loss of separation efficiency because the sample passes through them. Therefore, they must be packed as carefully as the analytical column, and connected with low dead volume fittings. These columns are placed immediately before the analytical column, and can be considered to be the first part of the analytical column. Guard columns are commercially available, usually in 2–5 cm lengths. Some column systems are available which allow a replaceable cartridge to be

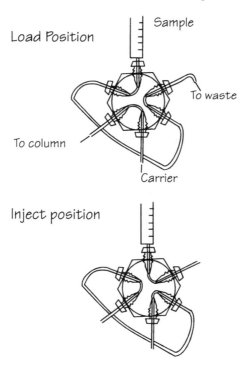

Figure 4.23 Injection valve for HPLC use.

placed in the inlet fitting of the analytical column, to serve as a guard column. Figure 4.24 shows such a system.

4.6.4.2 *Analytical columns.*   Columns are usually constructed of stainless steel tubing with inner diameters of 4 or 5 mm. Microbore columns, with diameters of 1 and 2 mm are also available, as are columns with inside diameters above 10 mm, which are mainly used for preparative scale work. The end fittings on the column contain a frit to hold the packing in place, and a flow distributing plate which spreads the flow from the pump over the end of the column, thus helping to achieve constant flow through the entire cross section of the column.

Installation of columns is achieved by use of threaded fittings, usually using a stainless steel compression ferrule. Many column manufacturers use similar fittings, so that the same threaded nuts and fittings can be used. However, once a steel ferrule is swagged onto a tube, it is often not possible to move this tube to another fitting. The length of tubing which protrudes from the ferrule may be different in each case. If the tubing is too long, the ferrule cannot seat properly, and a good seal will not be achieved. On the other hand, if the tube is too short, a gap will be present inside the fitting and efficiency will be lowered. Figure 4.25 shows this problem. Unless a system which uses a ferrule which can be

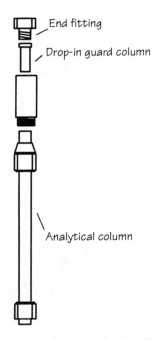

Figure 4.24 An integral guard column designed to be installed directly on the end of the analytical column (courtesy of Keystone Scientific).

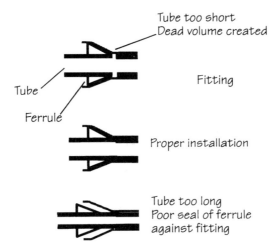

Tube too short
Dead volume created

Fitting

Tube

Ferrule

Proper installation

Tube too long
Poor seal of ferrule
against fitting

Figure 4.25 Ferrules must be properly installed on column ends so that leak free seals are made, and dead volume is not created.

readjusted on the inlet or outlet tube is used, the ferrule should be cut off and a new one fitted whenever columns are changed.

4.6.4.3 *Eluents.* The choice of eluent depends on the column and the sample. In reverse phase chromatography, a more polar eluent will move the sample slowly, and allow time for separation. A less polar solvent will elute late peaks more quickly and prevent excessive band broadening. There are several measures of eluent strength, including the polarity index, P'. A higher value of P' indicates a more polar eluent. Often solvents are mixed to produce an eluent of a suitable strength for a particular separation. For instance, various mixtures of methanol and water are used to produce a variety of different polarities, with an increase in the water content making a less strong, more polar eluent for reverse phase work. The polarity index of a solvent mixture P_m composed of solvents "a" and "b" is computed as:

$$P_m = P_a \times x_a + P_b \times x_b \qquad (4.20)$$

where P_a and P_b are the polarity indexes of a and b, and x_a and x_b are their volume fractions. The effect of eluent polarity on the capacity factor k' of a compound is given by the equation:

$$\frac{k_2'}{k_1'} = 10^{(P_2' - P_1')/2} \text{ for reverse phase chromatography, and}$$

$$\qquad (4.21)$$

$$\frac{k_2'}{k_1'} = 10^{(P_1' - P_2')/2} \text{ for normal phase.}$$

where P_1' and P_2' are the polarity indices of the two eluent mixtures.

Table 4.5 Some HPLC eluents

Solvent	Boiling point, °C	Viscosity (centipoise)	Polarity index, P'
Cyclohexane	81	0.90	0.04
n-Hexane	69	0.30	0.1
1-Chlorobutane	78	0.42	1.0
Toluene	110	0.55	2.4
Tetrahydrofuran	66	0.46	4.0
Ethanol	78	1.08	4.3
Methanol	65	0.54	5.1
Acetonitrile	82	0.34	5.8
Nitromethane	101	0.61	6.0
Water	100	0.89	10.2

The eluent must be able to keep the sample components in solution. The viscosity of the eluent is of concern, because a less viscous solvent can be used at a higher flow, without requiring very high pump pressures. Purity of the eluent, as well as its availability, cost and ease of disposal or recycling are other important considerations. Table 4.5 lists some common eluent solvents and their physical characteristics important for HPLC.

Example

In a reverse phase separation of a pesticide, the retention time was 15.5 min, with an eluent composed of methanol/water at a volume ratio of 30:70. An unretained peak eluted at 0.25 min. Calculate k'.

$$k' = \frac{15.5 - 0.25}{0.25} = 61$$

What water/methanol eluent composition will reduce k' to 5? Substituting values for methanol and water into Equation 4.20:

$$P' = 0.3 \times 5.1 + 0.7 \times 10.2 = 8.7$$

and

$$\frac{5}{61} = 10^{(P'_2 - 8.7)/2} \qquad \text{so } P'_2 = 6.52$$

To find the composition, let V = volume fraction of methanol.

$$6.52 = V \times 5.1 + (1 - V) \times 10.2$$

$V = 0.78$. Therefore, the eluent is 78% methanol and 22% water.

4.7 HPLC detectors

There is no sensitive universal detector available for use in HPLC. The only really universal, bulk property HPLC detector is the refractive index detector, which cannot be used with gradient elution, requires excellent temperature control, and is as much as 10^3 times less sensitive than other detectors. Therefore, it finds little use in environmental work. The detectors most often used are those such as absorption spectroscopic detectors, which respond to some property of the sample which is not exhibited by the mobile phase.

4.7.1 Ultraviolet absorption detectors

Ultraviolet detectors are fairly general in application, since most organic compounds absorb some wavelengths in the UV spectrum. However, the spectral region of wavelengths below 210 is usually not useable for analysis, because most solvents which would be used as eluents would also absorb in these areas. The response of this detector depends on Beer's Law, and therefore gives a linear response over four to five orders of magnitude. The detection limits vary widely, depending on the sample component and its extinction coefficient at the wavelength being used. In the most favorable cases, 1 ng or less of a compound may be detected.

Fixed wavelength detectors, using filters to isolate a single band of radiation, are inexpensive and stable. Light is passed through the filter, then through a flow cell containing the effluent from the column. Finally, it is allowed to impinge on a photocell, where the light is measured. Generally, these are single beam instruments, but dual beam systems are possible. They lack versatility, since the only compounds which can be analyzed are those which absorb at the fixed wavelength. However, for standardized, repetitive analyses, these detectors may be ideal since their reproducibility is often slightly better than that of variable wavelength detectors.

Variable wavelength detectors, are, however, much more versatile. These use a continuum source and a monochromator to select the wavelength desired. A manually adjusted grating disperses the light and passes the target wavelength through the flow cell.

Detectors which can rapidly perform a complete scan over a range of wavelengths can give qualitative as well as quantitative information. This can be done with a rapid scanning instrument, but, more commonly a diode array detector is used. The photo diode array (PDA) detector uses an arrangement of diodes positioned so that each diode intercepts a different band of wavelengths. The signal from each diode is recorded, and a spectrum of the effluent at any moment is obtained. This is very useful in confirming identity of components, and even more, in determining the efficiency of separation. Figure 4.26 shows the basic layout of a diode array detector.

The purity of a peak may be determined. Coelution of components can be confirmed or ruled out by comparing spectra taken on the leading edge, the top, and the trailing edge of a peak. It is difficult to identify a totally unknown compound from the UV spectrum. These are relatively simple spectra, and the solvent, mixed with the sample, has an effect on the spectrum. However,

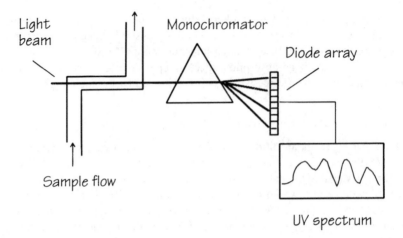

Figure 4.26 A diode array detector uses all the output of the monochromator simultaneously, giving a complete spectrum, without causing a time delay

comparison of samples and standards run in the same solvent, gives retention time and spectral information, and strong identification confirmation. Figure 4.27 shows part of a chromatogram of a sample of polynuclear aromatic

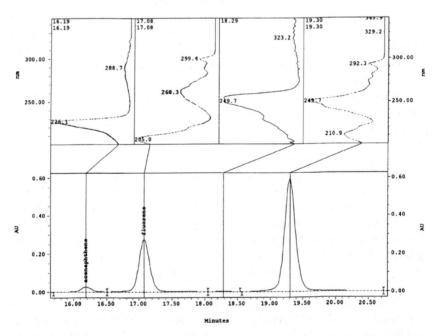

Figure 4.27 Output of a photodiode array detector from a separation of some PAH compounds. The peaks are detected by the data system, and a UV spectrum of each peak is displayed. Peaks whose spectra are found in the data library are identified.

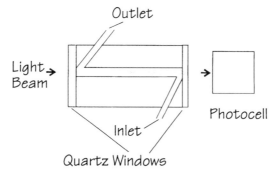

Figure 4.28 A cell for visible or UV absorbance detection. The path is made sufficiently long without increasing the cell volume by use of the Z configuration.

hydrocarbons run using a PDA. The peaks which have spectra and retention times matching those of the standard are identified, and the spectrum of each peak is printed on the report.

The flow cells used for absorption detectors are designed to give the maximum length of sample for the light to pass through, while keeping the volume as small as possible, to ensure that the resolution is not compromised. Figure 4.28 shows a UV detector cell. The cell has a Z shape to provide the maximum path length with as little cell volume as possible. The principal source of noise in absorbance detectors using a flowing sample is due to slight changes in refractive index. These are due to slight inhomogeneities in the composition of the eluent, changes in temperature, or turbulence in the flow. The change in refractive index diverts some of the light from the path to the detector, momentarily decreasing the signal.

4.7.2 *Fluorescence detectors*

Fluorescence detectors are among the most sensitive available. These are most suitable for, but are not limited to, compounds which fluoresce. Non-fluorescent compounds may be derivatized by adding a reagent after the column, which supplies a fluorescent tag to the sample molecules. Alternatively, the eluent may be made fluorescent and the sample peaks detected by the decrease of fluorescence as the peak elutes. This is known as "vacancy chromatography".

Fluorescence detectors need an intense high energy source, either line or continuous, to excite the fluorescence. Mercury lamps are used for line source excitation, and deuterium or xenon arc sources for continuum source. A monochromator is used to select the wavelength for excitation and for emission. The wavelength selection can also be done with filters, at the expense of versatility and sensitivity. The wavelength at the absorbance peak may not be available in a filter instrument, so that the highest sensitivity cannot be achieved. A photomultiplier is used to capture and amplify the weak emission from the fluorescent molecules. Figure 4.29 shows a fluorescence detector.

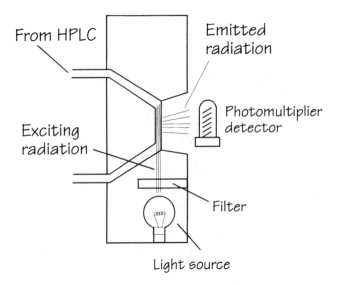

Figure 4.29 Fluorescence detector for HPLC.

For dilute solutions the equation which relates the emission to the concentration is:

$$I_f = I_o \phi_f (2.3 \, abC) \tag{4.22}$$

where I_f is the measured emission intensity, I_o is the excitation beam intensity, ϕ_f is the number of photons emitted per photon absorbed (the quantum yield), a is the molar absorption coefficient, b is the cell path length, and C is the sample concentration. The response is linear over about two orders of magnitude. Sensitivity varies widely, depending on the amount of light scattering in the optical system, the intensity of the excitation radiation, and the fluorescence quantum efficiency of the sample. Mobile phase composition is also important, since fluorescence is readily quenched. Oxygen is a particularly efficient fluorescence quencher, so solvents must be well degassed. Fluorescence is also temperature dependent, and, at higher concentration, self absorbance can be serious.

4.7.3 *Mass spectrometric detection*

The most informative detector is probably the mass spectrometer. Interfacing between the HPLC and the ion source is even more difficult than it is with GC/MS. The eluent is a liquid, and therefore, must be eliminated in some way before the sample is injected into the vacuum system.

4.8 Ion chromatography

Ion chromatography is used for separation of ionic species. The stationary phase is an ion exchange resin, and retention of the ionic species occurs as they are exchanged onto and off the resin surface. A cation exchange resin has $R-H^+$ groups on its surface and, a cation such as Zn^{2+} is retained because it exchanges with the hydrogen ions on the resin:

$$R-H^+ + Zn^{2+} \Leftrightarrow R-Zn^{2+} + H^+$$

Similarly, an anion exchange resin $R-OH^-$, will exchange OH^- ions for anions such as NO_3^- in the sample:

$$R-OH^- + NO_3^- \Leftrightarrow R-NO_3^- + OH^-$$

The partition coefficient K for the cation exchange is:

$$K = [R-Zn^{2+}]/[Zn^{2+}]$$

where $R-Zn^{2+}$ is the concentration on the ion exchange resin, and Zn^{2+} is the concentration in the mobile phase. The partition coefficient for the anion exchange is calculated in a similar fashion. The most common anion exchange column incorporates a quaternary amine group, while cation columns usually bear sulfonate groups. The packings are prepared by sulfonating or aminating the surface of the polymer core, with the active sites located close to the surface, to improve the mass transfer between the eluent and the stationary phase.

The instrumentation used for ion chromatography is similar to that used for HPLC and is shown in Figure 4.30. The conductivity detector which measures the electrical conductivity of the eluting mobile phase is commonly used.

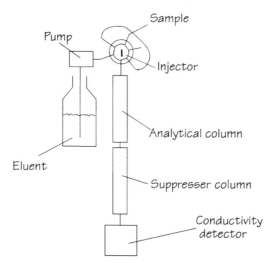

Figure 4.30 An ion chromatography system, using a suppresser column and conductivity detection.

If the mobile phase has a high ionic strength, the background electrical conductivity will be high and the detector will have low sensitivity. There are two methods used to reduce this difficulty: the suppresser column technique and the single column technique. In the **suppresser column** method, a fairly strong eluent is used to move the sample through the analytical column. Then the eluent is passed through a second column, the suppresser column. This neutralizes the eluent and allows easy detection of the sample ions. For instance, in the analysis of cations, a dilute HCl solution may be used as the eluent. The analytical column is a low capacity cation exchange resin, and the suppresser column is a high capacity anion exchange resin. The large excess of H^+ ions displaces the sample cations, with each cation establishing its own equilibrium between the eluent and the surface. The suppresser column is an anion exchange resin in the hydroxyl form, and the H^+ ions from the mobile phase react with the OH^-, forming water. This leaves the sample cations in the eluent stream, with a very low background conductivity, facilitating conductivity detection. The suppresser column eventually becomes exhausted and must be regenerated to replenish the OH^- on the surface.

For analysis of anions, the analytical column is an anion exchange resin, while the suppresser is a high capacity cation exchange resin. The eluent is, for instance, a dilute solution of NaOH. In the suppresser column, the OH^- ions in the eluent are neutralized by the H^+ from the column. This leaves only the sample anions in the solution, and high sensitivity is obtained.

The single column method, a more recent development, uses a low capacity ion exchange resin designed especially for chromatographic purposes. Since the resin has such low retention, the eluents of very low ionic strength can be used. Buffers of such weak acids as boric acid have very low conductivities and the detection of the sample ions can be done without the use of a suppresser column. This simplifies the system, and allows the usual HPLC equipment, with only a conductivity detector added, to be applied to ion chromatography.

Ion chromatography can be used for the detection and quantitation of many species: Inorganic anions such as chloride, fluoride, sulfate, nitrate, and nitrite; cations such as sodium, calcium, copper, lead, ammonium ions, as well as ionizable organic species such as carboxylic acids and amino acids can be determined using this technique. While many metals may be more easily determined by atomic spectroscopy, ion chromatography has the ability to distinguish between species having different oxidation states, such as Fe(II) and Fe(III). This is not possible if atomic absorption spectroscopy is used for the analysis. Figure 4.31 shows an ion chromatogram of an extract of anions from an air filter, obtained with a single column system.

4.9 Supercritical fluid chromatography (SFC)

A substance cannot exist in the liquid state at a temperature above its critical temperature. However, if a material is above its critical temperature, and is

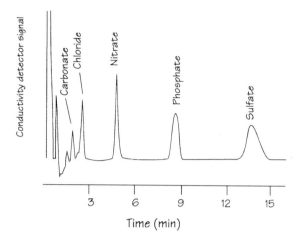

Figure 4.31 An ion chromatogram of an aqueous extract of an air particulate. The sample was analyzed using a borate–gluconate buffer on a non-suppressed anionic column, with conductivity detection.

subjected to sufficiently high pressure, it becomes much more dense than ordinary gases, and takes on some liquid-like properties. This is then referred to as a supercritical fluid. Figure 4.32 shows the phase diagram for CO_2, a commonly used supercritical fluid. The properties of supercritical fluids can be continuously varied between those of the gas and those of the liquid by changing the temperature and pressure. These fluids can be used as mobile phases in chromatography. Properties which can be varied include the viscosity, solvent

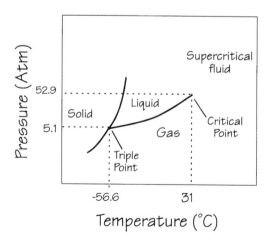

Figure 4.32 The phase diagram for carbon dioxide. At any temperature above the critical temperature, a supercritical fluid exists. At lower temperatures and high pressures, the compound exists as a liquid.

properties, and diffusivity, all of which are important chromatographic properties.

The properties of these fluids are usually closer to those of liquids than gases. The solubilizing power of a supercritical fluid is much greater than that of a gas. Therefore, nonvolatile and slightly volatile compounds may be separated by supercritical chromatography, while this would be impossible to do with GC. There is also an advantage over HPLC analysis for these compounds, since the solute diffusion coefficients in supercritical fluids are much greater. This means that the eluent velocity required for the maximum column efficiency is 5–10 times greater than that for HPLC. Equally efficient separations can therefore be done in much less time than is needed for HPLC. Finally, the viscosity of these fluids is much lower than that of liquids, making them much easier to pump through columns at a faster flow. Both packed and open tubular columns are used.

Any substance stable above its supercritical temperature might be used for eluent in SFC, but only a few are used routinely. Supercritical fluids which have been used are carbon dioxide, nitrous oxide, sulfur hexafluoride, Freon-13, ethane and ammonia. Of these, CO_2 is the most common, since its critical temperature, 31°C, is readily attained, it is non-toxic, and is readily available. Between the pressures of 72 and 400 atmospheres, and temperatures of 40–140°C, the density of CO_2 can be varied from 0.1 to almost 1 mg/ml. The only practical supercritical fluid which is reasonably polar is ammonia. It is, however, quite reactive and difficult to use. Modifiers are therefore used to improve the separation of more polar substances in nonpolar supercritical fluids such as CO_2. Modifiers are added to the eluent to improve peak shape and shorten retention times. These modifiers, including methanol, water and formic acid, are used in low concentrations, below 2% by volume. Modifiers at this low level may be thought of as deactivating silanol groups on the column, rather than increasing the solubility of the sample compounds in the solvent.

Programming in SFC is quite flexible, since temperature, pressure, and density all affect the retention of samples. The most common method of programming is density or pressure programming although temperature programming has also been used.

4.9.1 *SFC instrumentation*

The components of an SFC system are similar to those of HPLC, since a pump is used to produce the high pressures required, but GC and HPLC detectors can be used. A typical system is shown in Figure 4.33.

A syringe pump is the most commonly used pump, although reciprocating pumps have been used for packed column work. The mixing of modifiers complicates the system. Cylinders of eluent with modifier already mixed may be purchased, but then the amount of modifier cannot be adjusted. A second pump to add modifier is useful.

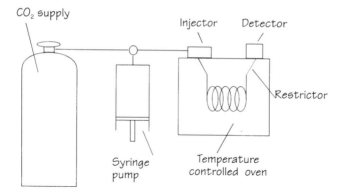

Figure 4.33 A supercritical chromatograph configured for use with a FID detector. If a liquid chromatographic detector, such as an UV adsorption detector, is used, the restrictor would be placed after the detector, rather than before.

Samples are injected into the system using high pressure rotary sampling valves. Sample volumes as small as a few nanoliters are needed for capillary work, so sample splitting techniques may be required. However, the sample is injected at room temperature, where the eluent may not be supercritical. The sample also may not be homogeneously mixed into the eluent quickly enough before the split is made. Therefore, the operation of a splitter is not always simple, and quantitation may be poor. Sample volumes for packed columns are larger, in the microliter range, so injection is a much easier task.

Columns used in SFC are usually small bore columns packed with 5–10 μm bonded phase particles similar to those used for HPLC. Short capillary columns of 1–10 m in length are also used, with stationary phases which are often more crosslinked than those used in GC. All bonded stationary phases have some contribution from unreacted silanol groups, and these can be a problem in SFC, because of the relatively nonpolar nature of CO_2. This is why polar modifiers are effective.

At the end of the column, a restrictor is required to keep the fluid in the column at the required pressure. The restrictor is positioned before the detector when the detector is a GC type detector, and after the detector, when an HPLC type detector such as an UV absorption cell, is used. A restrictor may be simply a short length of narrow bore fused silica tubing of 5–15 microns i.d. However, the sample may precipitate as fog droplets when the solvent suddenly decompresses at the end of the restrictor. These droplets, when fed into a flame ionization detector, cause signal spikes. This may be avoided by decompressing the eluent more gradually in a tapered or conical restrictor, which is kept warm at the tip so that the sample has a chance to evaporate.

The flame ionization detector (FID) is probably the most common detector for SFC, as it is compatible with the usual fluids. While organic modifiers interfere, water and formic acid can be used with FID detectors. The sensitivity of the FID is somewhat lower than its GC counterpart. The UV detector is also often used, especially when wide bore columns or organic modifiers make the FID unsuitable. The volume of the absorbance cell must be very small in SFC, on the order of 50 nl or less, to avoid band broadening when capillary columns are used. Standard HPLC UV cells may be used in packed column work, but modifications may be necessary to allow the cells to be used at the much higher pressures common in SFC.

4.10 Applications of chromatography in environmental analysis

Gas chromatography is the most widely used separation technique. Many of the target pollutants are volatile enough to be analyzed by GC. For semivolatiles such as PAH, PCBs and some pesticides HPLC is widely used. GC has several advantages over HPLC. GC columns provide a larger number of plates, and a variety of highly sensitive and selective detectors are available. Since environmental samples are complex, the high separation capability is very important. If there is a choice between GC and HPLC, GC is usually preferred. An example of a difficult separation of some metabolites of benzo-a-pyrene, too nonvolatile to be separated easily by GC, is shown in Figure 4.34.

Supercritical chromatography has still not become a standard technique in environmental analysis. Most samples can be done by either GC or HPLC, both of which are much more mature techniques. There are some advantages, especially the ability to use high sensitivity GC detectors for relatively nonvolatile samples, but the equipment has not yet reached the sophistication and ease of use of GC or HPLC. Supercritical extraction, on the other hand, will probably develop into a major technique for environmental sample preparation, because it readily replaces toxic solvents.

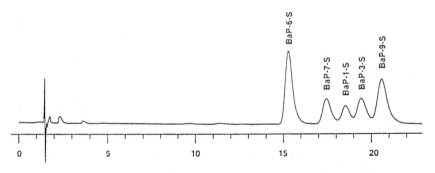

Figure 4.34 Separation of hydroxy-benz-a-pyrene sulfate isomers (metabolites of BaP) on a Vydac 201TP column. Mobile phase is an aqueous solution of 7 parts 1% triethylamine adjusted to pH 4 with acetic acid, and 3 parts acetonitrile (courtesy of C. Johnson, A Greenberg, University of North Carolina, Charlotte, L. Sander, NIST).

Suggested reading

Bruner, F. (1993) *Gas Chromatographic Environmental Analysis: Principles, Techniques and Instrumentation*, VCH Publishers, New York.

Poole, C.F. and Poole, S.K. (1991) *Chromatography Today*, Elsevier, Amsterdam.

Skoog, D.A. and Leary, J.J. (1992) *Principles of Instrumental Analysis*, Saunders College Publishing, New York.

Snyder, L.R. and Kirkland, J.J. (1980) *Introduction to Modern Liquid Chromatography*, 2nd edition, John Wiley and Sons, New York.

Weiss, J. (1995) *Ion Chromatography*, 2nd edition, VCH Publishers, New York.

Willard, H.H. et al. (1991) *Instrumental Methods of Analysis*, 6th edition, Wadsworth, Belmont, CA.

Study questions

1. What are the common factors which exist in all types of chromatography?
2. What is the significance of t_o in gas chromatography?
3. A gas chromatographic peak from a 25-m long column has an adjusted retention time of 11.2 min, and a width at half height of 20 s. How many theoretical plates are present in the column? What is the value of H?
4. A sample of pesticide is analyzed by gas chromatography. A 0.1 µl injection of a standard containing 0.234 µg/l gives a peak of area 34 873. The same size injection of an unknown sample solution gives an area of 39 945. What is the concentration of the pesticide in the sample solution?
5. In gas chromatography, predict the effect on the efficiency of a separation when the following changes are made:
   a. Particle size of the packing is increased.
   b. Gas flow is increased.
   c. Thickness of liquid phase in a capillary column is increased.
6. In an HPLC analysis, a reverse phase column is being used with a solvent gradient, starting with solvent A and increasing concentration of solvent B over time. The program starts with 50% A and 50% B and runs linearly, reaching 100%B at 20 min. Explain which of the two solvents is more polar.
7. What are some of the properties of a good chromatographic detector?
8. What is the advantage in using a halogen specific detector such as the ECD in environmental analyses?
9. Define the following terms:
   - Gas–liquid chromatography
   - Bonded phase
   - Normal phase HPLC
   - Gradient elution.
10. What is the function of the suppresser column in ion chromatography? Under what conditions is it not needed?
11. Predict the elution order of the following in reverse phase and in normal phase chromatography:
    a. *n*-heptane, heptanol, and toluene
    b. nitrobenzene, benzene, and phenol.
12. What parameters can be used to improve resolution in GC? in HPLC?
13. What are some advantages of SFC over GC? Over HPLC?

14. In a normal phase LC separation, nitrobenzene had a retention time of 28.0 min, while an unretained compound eluted in 0.9 min. The mobile phase was a 50:50 mixture of toluene and hexane. What mixture of toluene and hexane will reduce the k' to 9.0?

15. The following data were obtained for a separation using a 0.5 mm i.d., 10 m open tubular column:

	Retention time (min)	Peak base width (min)
Methane	2.0	
Cyclohexane	8.0	0.65
Methylcyclohexane	8.17	0.72
Toluene	10.2	0.95

Draw the chromatogram and label the peaks.

a. Calculate the number of plates for each compound

b. Calculate the capacity factor for toluene

c. Calculate the length of the column needed to separate cyclohexane and methylcyclohexane at a resolution of 1.6.

16. In a reverse phase separation of chlorinated phenols, trichlorophenol has a retention time of 15.0 min, when 30:70 acetonitrile-water mixture is used as the mobile phase. What mixture of these two solvents will reduce the retention time to 12.0?

5 Mass spectrometry

The mass spectrometer, in combination with a chromatographic inlet system, is probably the most powerful and most useful tool available to the environmental analyst. As we have seen, chromatography is an efficient separation technique which uses a variety of sensitive and selective detectors to analyze a wide array of compounds. However, these detectors do not provide much information about the molecules being detected. In chromatography, the retention time is the major piece of information used to identify a compound. If two compounds have very close retention times, identification becomes very difficult. The mass spectrometer, however, provides information about the structure of the molecule. In a mass spectrometer, a molecule is ionized and fragmented, then the molecular mass of each fragment ion is determined. Each molecule forms a unique set of fragments, so the mass spectrum shows a pattern which can be considered as a fingerprint of the molecule.

Modern mass spectrometers have digital libraries containing thousands of compounds with which the mass spectrum of an unknown compound can be compared. The combination of GC or HPLC with the mass spectrometer makes an even more powerful analytical tool. The chromatograph separates the components in the complex mixture and the mass spectrometer identifies each compound as it elutes. Compounds can be identified by their retention time, the molecular weight information provided by the mass spectrometer, and by their fragmentation pattern. The mass spectrometer adds an additional dimension to the analysis. Mass spectrometers are also used in conjunction with ICP for elemental analysis.

The mass spectrometer traces its roots back to the experiments of the English scientist J.J. Thompson. He developed an apparatus which would deflect an ion beam with a magnetic field, and used it to determine the masses of the two isotopes of neon. For many years, a mass spectrometer was a very complicated, expensive instrument which usually required a dedicated technician to operate it. Recently, both more sophisticated, powerful research instruments and simpler, much more reasonably priced ones have been developed. Small mass spectrometers have found immediate application to routine environmental analyses. Their usefulness arises from their ability to provide both quantitative measurements and qualitative confirmation of the sample component identity, in a single analysis. Improvements in vacuum systems, computer control of instrument adjustment and data collection, and increased detection sensitivity have all contributed to the ease of use and general utility of this instrument.

5.1 Interpretation of spectra

Mass spectra are commonly displayed as vertical lines, located at the unit mass values along the x-axis, with the height of each line proportional to the intensity of the ion current found at that mass. While we often speak of these peaks as

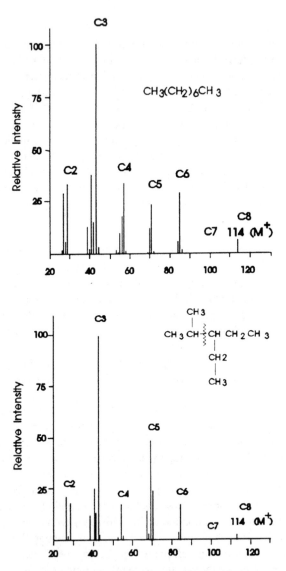

Figure 5.1 Comparison of branched and nonbranched isomers of octane. Note the large C5 peak in the branched compound because of the preferred fragmentation at the branched carbon. Also notice the absence of a C7 fragment in both.

occurring at certain masses, they are actually mass/charge (m/e) ratios. However, the most common charge is +1, so the mass and m/e are usually equal. The highest intensity line is usually referred to as the base peak and is given an arbitrary value of 100. All other ion peaks are scaled relative to the base peak. The ions are formed in large numbers by fragmenting the analyte molecules. The distribution of fragments is statistical, with each particular fragmentation having a certain probability of happening, depending on the strengths of the bonds being broken and the stability of the ions formed. This yields a reproducible spectrum for each molecule.

The heaviest mass peak, except for any isotope peaks, will probably be the molecular ion, the unfragmented molecule, simply missing an electron. In some cases, however, the ionized molecule is so unstable that it disintegrates immediately, and no molecular ion is found. To decide if the largest mass peak is truly the molecular ion, look at the next-to-the-highest mass peak. If this is between 4 and 14 mass units lower than the supposed molecular ion, the largest peak is probably **not** the molecular ion. If it were, the molecular ion would have to lose more than four hydrogen atoms, which is unlikely. The smallest mass to be lost, other than a hydrogen atom, is likely to be a CH_2 group of mass 14.

The spectrum of hydrocarbons is quite distinctive, with groups of peaks occurring at intervals of 14 mass units, corresponding to losses of CH_2 units. Normal hydrocarbons give fragments with decreasing intensity as the fragment mass increases. There is usually a large peak at 43, from the C_3H_7 ion, and for higher hydrocarbons, no peaks at 14 mass units lower than the molecular ion, as the molecules seldom lose just a single methyl group. The molecular ion is usually present, although it is small. Branched hydrocarbons will be preferentially cleaved at the point of the branch, so the most intense ions can give a hint about where the branch is located and what the structure is. Figure 5.1 shows the mass spectrum of normal octane and its isomer 1-methyl-2-ethyl-pentane, showing the preferential cleavage in the branched compound which produces fragments containing three and five carbons. Also note that the fragments corresponding to seven carbons are missing from both spectra, but the molecular ions at 114 are present, although small.

In mass spectra, the ion masses correspond not to the masses found on the periodic table, but rather to the sums found by adding up the actual masses of the isotopes contained in the ions. Isotopes are very useful in identifying fragments in mass spectra. The isotope ^{13}C, for instance, has a natural abundance which is 1.1% of the common ^{12}C isotope. This fact can be used to determine the number of carbons in a molecule, using the equation:

$$\text{number of C atoms} = \frac{\left(\dfrac{M+1}{M}\right)}{0.011} \tag{5.1}$$

where $M + 1$ is the intensity of the isotope peak at 1 mass above the molecular ion, and M is the intensity of the molecular ion peak. For instance, in the mass spectrum of benzene the molecular ion at mass 78 has a relative intensity of 100. The peak at 79 has an intensity of 6.6. Using Equation 5.1, we find that the number of carbons is six, as it should be.

For ions containing elements such as chlorine which have more than one major isotope, very easily identified patterns are produced. Since isotope distributions are constant, several peaks are produced, corresponding to different combinations of isotopes. Chlorine, for instance, has two isotopes, ^{35}Cl which comprises about 75% of natural chlorine, and ^{37}Cl, almost 25%. The spectrum of trichloroethylene, C_2HCl_3 (Figure 5.2) is a good example. The molecular ion, made up of the most abundant isotopes of each of the elements, has a mass of $2(12) + 1 + 3(35) = 130$. The ion contains three chlorine atoms. If the three are all ^{35}Cl, the mass is 130. If one of the chlorines is ^{37}Cl, then the mass is 132, while if all three are the heavier isotope, the mass is 136. Because the ^{35}Cl is three times more abundant than the ^{37}Cl, we expect that the mass of most of the $C_2HCl_3^+$ ions will be 130. However, the 132 peak will be nearly as large, because there are three different ways to obtain this mass. Each one of the three chlorines has a 1 in 4 chance of being a heavy one. A combination of two ^{37}Cl and one ^{35}Cl is considerably less likely, because fewer ^{37}Cl are available, and the ion containing three ^{37}Cl is even less probable. The spectrum confirms this analysis, showing a strong peak at 130, a slightly less intense peak at 132, and much

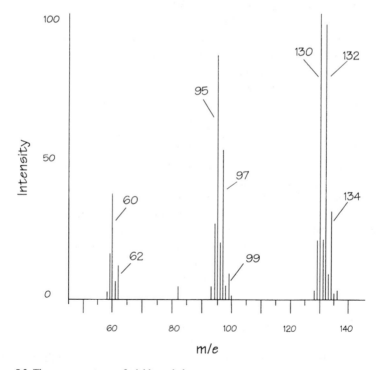

Figure 5.2 The mass spectrum of trichloroethylene.

smaller ones at 134 and 136. The pattern of four peaks for a species containing three Cl atoms will always show the same relative intensities. Notice also the smaller peaks between the four we have discussed. These are due to the molecular ion which has lost a H atom, and to the presence of ^{13}C.

Lighter fragments are produced by the loss of a chlorine atom. The ion produced by this loss contains two chlorines, giving three isotope peaks, with masses of 95, 97 and 99, from the ions: $C_2H\ ^{35}Cl\ ^{35}Cl$, $C_2H\ ^{35}Cl\ ^{37}Cl$, and $C_2H^{37}Cl\ ^{37}Cl$. Again, they form a relative intensity pattern characteristic of all species containing two chlorines. Finally, the ions containing one chlorine atom will show two isotope peaks at m/e of 60 and 62, with relative intensities of 3 to 1, because of the 3:1 ratio of the abundances of the atoms.

As can be seen, the interpretation of mass spectra is logical, but requires a good deal of knowledge of the fragmentation patterns of various combinations of atoms, and their isotopes. Generally, in environmental mass spectroscopy, the sample is separated by a chromatographic process before the mass spectrum is obtained, and the retention time is an added piece of information. The spectra produced are compared in the data system with library spectra, and the best matches are displayed for the analyst to examine. More sophisticated spectral analysis is necessary when compounds not present in the library are encountered, but for most environmental work this is not necessary. There are several good reference books on mass spectral analysis which can be referred to when needed.

5.2 Basic instrumentation

All mass spectrometers consists of certain essential components. Because all types of mass spectrometers operate by producing and sorting ions according to their mass and charge, the space through which these ions flow must be free of other gases. In other words, the mass spectrometer must operate with a vacuum within the analyzer section. Therefore a **vacuum system** is essential. The system must also provide a means of converting the molecules of the sample into ions which comprise recognizable fragments of the original molecule. This takes place in the **ion source**. The ions must then be sorted by mass and charge in the **analyzer** and be brought to a focus on the **detector**. A mass spectrometer system is shown in Figure 5.3.

5.2.1 Vacuum system

The most important parts of the vacuum system are the pumps. Usually, a mechanical pump is used as a roughing pump or forepump, to reduce the pressure to a few millimeters. Then the high vacuum pumps reduce the pressure still further to the range of 10^{-5}–10^{-8} torr required in the analyzer. Until recently, diffusion pumps were most commonly used for achieving high

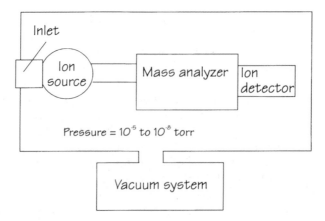

Figure 5.3 Essential components of a mass spectrometer.

vacuum. The turbomolecular pump, which requires less maintenance and which provides much cleaner backgrounds than the diffusion pumps, has become much more popular, in spite of the higher cost. The diffusion pump contains a high boiling liquid, a highly purified oil or mercury, which is heated to provide a stream of heavy vapor molecules. The vaporized pump liquid is condensed at the top of the pump, and carries molecules of gas from the space being pumped, down into the pump. Despite chilled traps between the pump and the analyzer, some of the oil vapor can reach the analyzer and contribute to the background spectrum. Mercury gives a somewhat cleaner spectrum than does an oil pump, with only a single mass peak, but mercury is heavy, toxic, and expensive.

The turbomolecular pump contains a bladed turbine which rotates at high speed and sweeps molecules down into the throat of the pump, from where the backing pump removes them. These pumps, because of their very high speeds, can be readily damaged if they are activated when the system is at a pressure of more than a few millimeters or if the system pressure rises suddenly. Because it takes a minute or two for the turbine to slow down, a sudden venting may do damage, even if a safety circuit cuts off the power when a pressure rise is detected. The turbomolecular pump is shown in Figure 5.4. At set intervals, the pump bearings must be lubricated or replaced. A record of hours of operation must be kept and servicing done at the specified times.

5.2.2 *Inlet*

An inlet system is required to take the sample from its high pressure gas, liquid or solid form, vaporize it if necessary, and reduce its pressure to the point at which it may be injected into the source without overloading the source. Common inlets are batch inlets for gases and liquids, direct injection probes for liquids and solids and interfaces to gas or liquid chromatographs.

Rotating blades

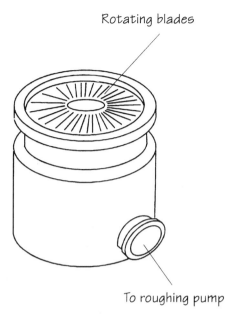

To roughing pump

Figure 5.4 A turbomolecular pump. The pumping action is generated by rapidly rotating blades in the pump.

The batch inlet consists of several chambers of known volume, usually heated, and with interconnecting valves. The inlet is evacuated with a roughing pump, and the sample, gas or volatile liquid is introduced into one chamber. From there, it may be reduced in pressure by being admitted to another chamber, with the pressures being monitored. Finally, it is allowed to leak slowly into the source through a very small orifice.

For less volatile liquids or solids, a direct injection probe may be used. This is a rod which carries a thimble a few millimeters in diameter at its tip. Solids or liquids are placed in the thimble and the rod is slipped through an airlock and ball valve into the source. The probe is usually heated, and the temperature of the tip may be raised slowly if a mixture is present, giving, in effect, a rough distillation.

Most often, in environmental work, the samples are so complex that a mass spectrum of only a single compound from a sample is not usually called for. To obtain mass spectra of the many components in a water extract, a soil sample, a concentrate of air volatiles, or any one of the many environmental samples presented for analysis, a separation must be done before the spectrum is taken. The most efficient way to do this is to separate the sample using a GC or LC column, passing the effluent directly into the mass spectrometer. As the compounds elute, their spectra are recorded. However, the mass spectrometer requires its samples at very low pressures, while the chromatographic effluent is an atmospheric pressure gas or a liquid. To make this sample compatible with the requirements of the mass spectrometer without increasing the peak width, or disposing of so much of the sample that the sensitivity is seriously reduced, is a problem which has been addressed in several ways. These are discussed in the GC/MS and LC/MS sections.

5.3 Ion sources

The function of the source is to produce ions from the vaporized sample. The sample molecule loses an electron and is given an excess of energy. This destabilizes the molecule sufficiently to cause it to break apart into fragments. The ionization process may be characterized as **hard** or **soft**. In a hard ionization, a greater amount of energy is deposited in the molecule, causing the rupture of several bonds and forming many fragments. A soft ionization leaves many of the ionized molecules intact, so that there is a significantly large peak at the molecular mass of the sample molecule. The harder ionization yields a rich spectrum with many peaks, and therefore is useful in the identification of compounds. The soft ionization gives a spectrum with a large molecular ion peak. By keeping much of the mass in a single peak, the sensitivity is enhanced and quantitation is somewhat easier. There are several common ionization methods used in environmental work.

5.3.1 Electron impact ionization

Electron impact ionization (EI) is the most common ionization method. As can be seen in Figure 5.5, the vaporized molecules from the inlet flow into the source. There they encounter a beam of electrons produced by a heated filament

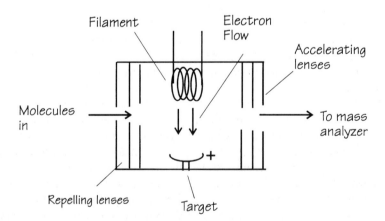

Figure 5.5 Electron impact ionization. The electron beam crossing the incoming stream of molecules forms ions in collisions.

and accelerated across the source chamber to a positively charged target at the opposite side. The energy of the electrons can be controlled by changing the voltage drop across which the electron beam flows. The accelerating voltage is usually set to produce electrons with an energy of around 70 eV, substantially higher than the ionization potential of the sample molecules, which is generally near 10 eV. These electrons collide with the sample molecules with sufficient energy to eject an electron from one of the higher energy molecular orbitals. This leaves the molecule unstable and may cause it to fragment.

When all the molecules which ionize are considered, these have a reproducible distribution of energies. When they fragment, they produce a reproducible group of ions under constant ionization conditions. There are such a large number of molecules in a sample that the production of fragments is statistically governed. Lowering the accelerating voltage of the electron beam will produce a different distribution of energies, less fragmentation, and more unfragmented **molecular ions**. It should be noted that the ionization process is rather inefficient, with only about one molecule in 1000 actually being ionized. The rest are pumped out by the source vacuum pump.

The EI source can be used for any sample which can be volatilized at the reduced pressures found in the source. It provides a rich spectrum with many different fragments, ideal for identification of molecules. Most extensive libraries of mass spectra are compiled of EI spectra.

5.3.2 *Chemical ionization*

Sometimes, the molecular ion is nearly absent, especially when the molecule is easily fragmented. This can make determination of the molecular weight more difficult. To obtain spectra with much larger molecular ion peaks, **chemical ionization (CI)**, is often used. This method injects a stream of reagent gas into the source along with the sample. The sample may account for 10^{-3}–10^{-5} torr in the source, while the reagent gas brings the source pressure to 0.3–1.0 torr. The source is not much different from the EI source, except that the source chamber is more tightly enclosed, to allow the higher pressure. A large capacity, differential pumping system is usually required, which has separate pumps on the source and analyzer. A very narrow slit is placed between the source and analyzer, leaving just enough space for the ion beam to pass. In addition, the electron emitting filament is kept at 10^{-4} torr or less, to keep the filament from burning out. Combination sources which can be readily switched from CI to EI are commercially available, with the main difference between the EI source and the convertible source is that the convertible source instrument has a higher pumping capacity.

Methane, ammonia, isobutane, and water are some of the reagent gases used in this process. Because there are so many more reagent gas molecules than sample molecules, the electron beam ionizes reagent molecules almost exclusively. These excited ions undergo ion-molecule reactions with the sample molecules and with other reagent molecules. With pressures up to 1 torr in the source, the 70 eV electron beam usually used for ionization has a relatively short range, and so instruments are usually equipped to produce higher energy electrons, up to 400 eV. Most reagent gases do not have a seriously negative effect on the service lifetime of the filament, but if strongly oxidizing reagent gases such as nitric oxide are used, a different ionization method may be required.

The most common ions formed in a high pressure source when methane gas is used as the reagent are CH_5^+ and $C_2H_5^+$. These strong proton acids react with the sample molecules by transferring a proton to the neutral sample molecules, yielding many positive ions with a mass equal to $M + 1$, where M is the original sample molecular mass. Some of the reactions which can take place in a source with methane reagent gas, CH_4 and sample molecules (denoted as S) are:

$$CH_4 + e^- \rightarrow \bullet CH_4^+ + 2\,e^-$$

$$\bullet CH_4^+ + S \rightarrow CH_4 + \bullet S^+$$

$$\bullet CH_4^+ \rightarrow CH_3^+ + \bullet H$$

$$\bullet CH_4^+ + CH_4 \rightarrow CH_5^+ + \bullet CH_3$$

$$CH_5^+ + CH_4 \rightarrow C_2H_5^+ + H_2$$

$$CH_5^+ + S \rightarrow SH^+ + CH_4$$

$$C_2H_5^+ + S \rightarrow SH^+ + C_2H_4$$

$$C_2H_5^+ + S \rightarrow SC_2H_5^+$$

Because ion–molecule reactions occur more easily than electron–molecule reactions, the ionization efficiency of CI is higher than that in EI sources. The amount of energy transferred to the sample molecule in the collision is smaller than with EI, and, in addition, excited molecules can be relaxed to lower states by collisions with the abundant neutral molecules in the high pressure source. Both these factors contribute to the increased sensitivity and better quantitation obtained when CI is used. It should be noted that while there are generally fewer peaks in a CI spectrum, the spectrum of the reagent gas is always present along with that of the sample.

In most cases, the positive ions formed in the source are accelerated into the analyzer and separated. However, both negative and positive ions are formed in the source. Therefore it is possible to perform negative ion chemical ionization mass spectrometry. All molecules contain the high energy level electrons which are ejected to form positive ions, but some molecules have lower energy vacant orbitals which are able to pick up an extra electron, forming a negative ion. Therefore, a strong positive ion spectrum can be obtained from any molecule, while a select group of molecules will also give a negative ion spectrum. Generally, molecules which give strong negative ion spectra are those with positive electron affinities, often classed as oxidizing agents or alkylating agents.

Interestingly, the substances which are of high interest in the environment are often those which give strong negative ion spectra. These are generally the same compounds which are detected by electron capture, since the mechanism on a molecular level is similar for both processes. Because many fewer compounds give negative ion spectra, it can be used to selectively measure those environmentally interesting compounds. The switch from positive to negative ion detection is done by reversing the voltage to accelerate negative rather than the positive ions out of the source into the analyzer.

5.3.3 *Atmospheric pressure ionization sources*

Atmospheric pressure chemical ionization sources (APCI) have found a place in air monitoring because of their sensitivity and because the introduction of sample is simple. In these sources, chemical ionization takes place at much higher pressures. These sources operate at even higher pressures than the usual CI sources, and the primary ions are formed by either a β particle emitter, ^{63}Ni, or by a corona discharge. A carrier gas flows through the cell and serves as the reagent. There are two basic types of APCI sources. One is of small volume, similar in size and configuration to an electron capture detector, and the ionized sample is allowed to flow into the analyzer chamber through an orifice of about 25 µm. These sources are generally heated to avoid adsorption on the cell walls. With the small cell volume and the requirement for a flow of carrier gas, these sources are well suited as GC or LC mass selective detectors.

The second type of APCI source is of considerably larger volume, with a corona discharge placed in the chamber. The orifice is typically about 100 µm, larger than that used in the small volume source. This larger orifice requires very efficient pumping of the analyzer, to cope with the large amount of gas flowing in. A commercial API mass spectrometer of this type uses cryogenic pumping because of its very good pumping speed. This consists of chilling the analyzer walls with liquid nitrogen and liquid helium, so that gases present in the source rapidly freeze to the walls. The sampling orifice is protected from particles in the carrier gas stream by a "curtain" gas stream which sweeps around the edges of the orifice. The large flow through the source and the curtain gas around the orifice prevent sample from contacting the walls and therefore the source can be operated at ambient temperature. This source, with its high sample intake, is well suited to ambient air studies, and can be tuned to detect selectively an ion which is characteristic of a particular target compound. A van-mounted APCI unit can be used to monitor an emitted pollutant in real time, using the air as the carrier gas as well as the sample. As the van is driven back and forth across an impacted area, the location and concentration can be plotted, giving a map of the pollutant plume.

There are several other common ionization sources, mostly used to ionize less volatile and more thermally fragile molecules, such as biologically generated molecules. These sources are less used in environmental analyses.

5.4 Mass analyzers

After the sample molecules are ionized they are accelerated into a mass analyzer section of the instrument. Three types of mass analyzers are commonly used in environmental work. These are the **quadrupole**, the **magnetic sector**, and the **ion trap**. Each of these has the function of separating the ions according to their mass/charge ratio, or their momentum/charge ratio. Ions of each individual mass are brought to the detector sequentially, producing a mass spectrum. The spectrum consists of peaks whose location along the axis represents the mass to charge ratio (m/e) of the ions producing the signal and whose height indicates the relative number of ions of that mass which were produced. Mass spectra are conventionally displayed with the largest peak in the spectrum, the **base peak**, set at 100 units and all other peaks normalized to this one. The amount of separation between adjacent masses in the spectrum is described by the **resolution** of the mass analyzer. Resolution, R, is defined as:

$$R = \frac{m}{\Delta m} \qquad (5.2)$$

where Δm is the mass difference between two separated peaks and m is the nominal mass of the peaks. Low resolution instruments are adequate for the identification of compounds by matching with libraries, which usually only requires unit mass resolution. High resolution instruments can be used to calculate the atomic composition of molecules and fragments, and are often used in identification of new organic compounds.

Example
The nominal masses of CO^+ and N^+ are both 28. The actual masses of the isotopes are: $N = 14.003070$; $C = 12.0000$ and $O = 15.9949$. What resolution would be required for a mass spectrometer to distinguish between these two ions?
   The mass of CO^+ is 28.00614. The mass of N_2^+ is 27.9949.

$$R = \frac{28}{28.00614 - 27.9949} = \frac{28}{0.01124} = 2491$$

so a mass spectrometer with a resolution of about 2500 would be required to distinguish between the two species.

5.4.1 Quadrupole mass analyzer

The **quadrupole** detector is probably the most widely used mass analyzer for environmental work because of its inherent sensitivity, its ease of use, and its relatively attractive cost. While it may not serve well for a research instrument because it is not suited to high mass samples, for most environmental purposes the measurement of molecules of over 500 AMU is not necessary. Figure 5.6

shows the quadrupole mass analyzer. It consists of four rods in a parallel bundle. The rods or poles are subjected to a DC voltage as well as a radio frequency field. The opposite poles change their polarity at twice the radio frequency. The ions are injected at the end of the bundle and travel down the center axis to the detector located at the end. Because of the changing fields, the ions follow an oscillating path through the analyzer. At each particular applied voltage and frequency, there is only one mass of ion which will hold a stable path. All others with larger or smaller masses will oscillate widely enough to encounter the poles and be lost. The masses are scanned by varying the frequency and the voltage on the rods, bringing each ion mass to the detector in sequence.

The quadrupole is a true mass analyzer, and does not depend on the initial momentum of the ions as they are injected into it. Therefore, the voltage which accelerates the ions into the analyzer does not have to be as high as in a magnetic sector instrument, and ions are transmitted efficiently through the instrument. Because only voltage changes are required, the instrument is capable of rapid scanning. The rods are only a few inches in length, and therefore the

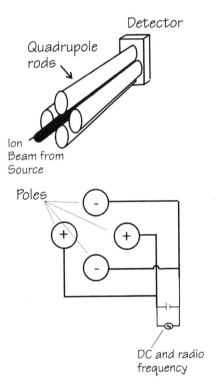

Figure 5.6 The quadrupole mass analyzer. The ions oscillate and move along the center of the space between the four poles towards the detector. At any time the applied voltages allow only one ion mass to follow a stable path, and reach the detector. Other masses will oscillate too widely, hit the poles and be lost.

whole unit can be made quite compact. Commercial versions of the quadrupole are produced as dedicated detectors for gas chromatographs, as well as full function mass spectrometers. The dedicated detector, often referred to as a **mass selective detector**, consists of the source, mass analyzer, and detector, with the GC column as the sole inlet.

5.4.2 Magnetic sector mass analyzer

The **magnetic sector** mass analyzer is the most direct descendent of the original mass spectrometers. It consists of a curved vacuum tube surrounded by either an electromagnet or a permanent magnet. The ions are fired out of the source by the **ion gun**, and they enter the tube all carrying the same momentum. The magnetic field bends the path of these ions, deflecting the lightest ones the most and the heavier ones less. The curvature of the ion beam is scanned by changing the magnetic field or by changing the accelerating voltage of the ion gun. The kinetic energy of the ions leaving the source is given by:

$$\frac{mv^2}{2} = Vq \tag{5.3}$$

where m is the ion mass, v is its velocity, V is the accelerating voltage and q is the charge on the ion. In the magnetic field of the analyzer, the ions which are able to travel through the curved path are those for which the centrifugal force tending to push them into a straight line path is balanced by the centripetal force of the magnetic field, pulling them into the curve.

The balance of these two forces is expressed as:

$$\frac{mv^2}{r} = Bqv \tag{5.4}$$

where r is the radius of curvature of the analyzer tube, and B is the magnetic field. These two equations can be combined and rewritten in a form with m in atomic mass units, the charge expressed as z, the ionic charge number, B in Gauss, r in centimeters, and V in volts.

$$\frac{m}{z} = \frac{B^2 r^2}{20\,740\,V} \tag{5.5}$$

Since all the other parameters are fixed, the different ions can be brought to the correct curve to reach the detector by varying either B, the magnetic field or V, the accelerating voltage. The magnetic sector mass analyzer is shown in Figure 5.7.

5.4.3 The ion trap mass analyzer

Figure 5.8 shows the schematic of an ion trap system. The usual vacuum, inlet, and detector systems are used. It is one of the more recent developments in commercially available mass spectrometers. The ion trap is composed of three

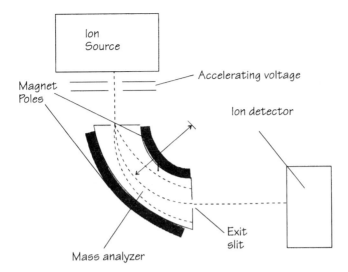

Figure 5.7 Magnetic sector mass analyzer. Beams of different mass ions are bent along different radii. Only the ones which are bent at radius *r* reach the ion collector.

electrodes. The central electrode is in the form of a toroid, a doughnut shaped ring. The inner surface of the electrode has a hyperbolic cross section. The other two electrodes are perforated end caps which are placed above and below the central hole in the ring. These are also hyperbolic in shape. When a radio frequency (*rf*) voltage is applied to the ring, and DC voltage on the end caps, an electrical field is established within the central chamber, which will trap a range of masses of ions. This range can be wide or can be as narrow as a single mass,

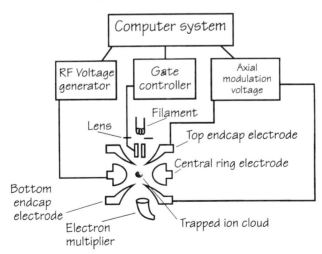

Figure 5.8 Ion trap mass spectrometer. Ions are trapped by oscillating fields inside the electrode assembly. As the endcap voltage is swept, the ions are ejected one mass at a time.

depending on the combination of DC and *rf* fields. The ions can be trapped in stable orbits for many seconds.

Volatilized sample molecules, along with helium, are allowed to flow into the trap. The trap pressure is usually kept relatively high, at about 10^{-3} torr. An electron stream from the filament flows through the endcap and forms ions in the trap. The DC and *rf* voltages are held at a constant value which will hold the desired range of ions in the trap. The helium atoms damp the ionic oscillations, helping to stabilize the ions in the trap. The electron gate is then closed to interrupt the flow of electrons through the trap. The voltage is then increased rapidly up to the point at which the desired ions will begin to be ejected from the trap. There the scan begins and the paths of the ions within the chamber become more and more unstable. The ions begin to be ejected from the trap through the perforated end caps and drop into the detector. The lightest ions are most easily ejected, then the heavier ones. As the *rf* amplitude is ramped, the ions are ejected in order of their m/z ratio, producing a mass spectrum.

The trap can also be used to do chemical ionization analyses, simply by changing the timing, levels, and sequences of voltages applied to the electrodes. The electron beam is allowed to form ions, while the *rf* voltage is kept at a level which will hold the low mass ions of the reagent gas but not any higher mass sample ions. Then the electron gate is closed, and the sample is allowed to mix and react with the reagent ions. The voltage at this point is kept for an appropriate time at a level to trap the desired sample ions which are forming by reaction with the reagent ions. Finally, the voltage is adjusted to the start of the scan point and the scan is started. Figure 5.9 shows the pattern of *rf* voltage for both EI and CI spectra.

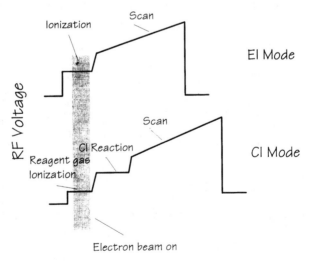

Figure 5.9 The ion trap voltages control the ionization and trapping cycles. If reagent gas is inserted, CI will take place during the ionization time. Extra time is allowed for reaction before scanning in this case.

The ion trap has several advantages and some disadvantages. The analyzer is compact in size and does not need a separate source chamber, and therefore is relatively inexpensive to build and requires a simpler vacuum system. The sensitivity of the ion trap analyzer is high because **all** the stored ions are eventually delivered to the detector. In contrast, in magnetic sector and quadrupole analyzers, an ion sent into the analyzer can only reach the detector when the fields are correct for that ion. At all other times, the ion is lost. As a detection system for gas chromatography, an ion trap spectrometer can detect a few $\mu l/m^3$ of an analyte in an injection of 1 ml of gas phase sample. Figure 5.10 shows the efficient throughput advantage of the ion trap compared to magnetic or quadrupole analyzers.

Problems with the ion trap are mostly due to the non-standard spectra produced by this analyzer under some conditions. When a large concentration of ions is allowed to build up in the trap, ion–ion reactions take place and the sample spectrum can be significantly changed. Samples which contain a great deal of water, for example, are likely to be subjected to this type of reaction, as the water ions react with the sample ions. This makes it difficult to identify compounds by reference to the standard libraries. The problem has generally been addressed by monitoring the ion buildup in the trap and emptying the trap at intervals to keep the concentration of ions low enough that they have little chance to collide and interact.

A rapidly developing use of the ion trap is for multidimensional mass spectrometry, often designated as MS/MS or MS^n. This method may, in time, replace or supplement GC/MS. A complex mixture can be fed directly into the source. After ionization, only the molecular ion of the compound of interest, a

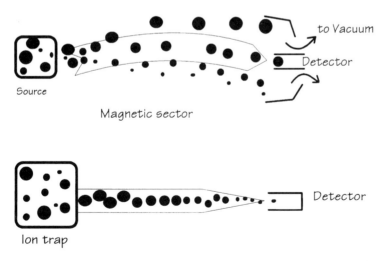

Figure 5.10 When an ion does not reach the detector in the quadrupole or magnetic sector MS, it is pumped out. In the ion trap, each ion is held until the proper time for it to be detected arrives. Therefore, the ion trap is more efficient.

single ion mass, is trapped. A reagent gas is allowed to react with the parent ions to form daughter ions. These can then be scanned into the detector so that the mass spectrum of the fragmented molecular ion is produced. Since no initial separation of the sample is required, this analysis can be done in a few seconds. The system can be used for triple or quadruple MS by selecting one of the daughter ions for further ionization before scanning.

5.5 Ion detectors

The usual method of detection of ions is the **electron multiplier**. This is a device which is similar in design and operation to the photomultiplier used in UV and visible region spectroscopy. An electron multiplier of 10–20 dynodes can multiply a single ion impact into 10^8 electrons. This detector has a very fast response time, low noise, and a wide dynamic range. It can respond to input ion currents ranging from 10^{-18} A, which corresponds to a single ion per second, up to 10^{-9} A. The **Faraday cup** detector is much less sensitive, but is easily calibrated. It consists of a cup which is placed in the ion beam. The charge from the collected ions is measured directly.

Detection of ions can also be done with a **channel electron multiplier**, a device which operates similarly to an electron multiplier, but which does not have discrete dynodes. This consists of a horn-shaped glass tube, coated on the inside with a lead oxide semiconducting material. A voltage difference is impressed on the cone, with the wide end being held at about -3 kV, and the back at near ground. As ions impact the oxide near the entrance, electrons are ejected. These bounce down the tube, producing more electrons at each encounter with the walls. The resulting pulse of electrons at the end of the tube is amplified by a factor of 10^8. These have an advantage over electron multipliers in that exposure to air will not harm them if the voltage is not on at the time. These multipliers have a limited lifetime and must be replaced when the sensitivity begins to decline, and higher voltages must be applied to keep the response at the same level. This detector is usually used in the ICP-MS. The two types of electron multipliers are shown in Figure 5.11.

5.6 Gas chromatography mass spectrometry (GC/MS)

Many current GC/MS instruments use capillary columns, so that the entire column effluent may be delivered into the ion source of the MS. The interfacing of the MS and the GC must be done with some care, since carrier gas is exiting the column at 1 or 2 ml per minute. The mass spectrometer source usually must be maintained at a pressure of about 10^{-4}–10^{-6} torr. The system must have sufficient pumping capacity so that the source is maintained at the proper pressure. This is especially important with environmental samples which may

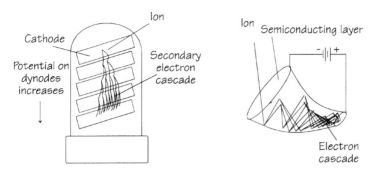

Figure 5.11 Discrete and continuous dynode ion multipliers. The initial ion impact generates secondary electrons, which are multiplied as they impact on successively more positively charged dynodes or at more positively charged areas in the continuous dynode.

contain high quantities of water vapor or background gases, such as nitrogen or oxygen. Since these gases are less mobile than the helium carrier, the large peak of, for example, water vapor, will cause a sudden increase of pressure in the source. This may cause the instrument's protective circuits to cause a shutdown, to protect the ionization filaments. Good pumping capacity is therefore vital, and should be a major concern when new instruments are being purchased.

The most common interfaces for capillary GC columns are the direct insertion and the open split systems. The direct insertion method takes advantage of the fact that the flow of carrier in the capillary column is rather low. If the pump on the mass spectrometer source has sufficient capacity, the entire sample and carrier may be allowed to flow directly into the source. This direct interface has the highest sensitivity, because all of the sample is delivered to the source. Also, there is little or no peak spreading, because there is no dead volume in the interface.

If the flow is too large, as might happen if a wide bore capillary column was used, the open split interface is useful. In this system, there is a small gap between the end of the column and the inlet capillary leading into the source. The gap is flushed with a sweep gas flow to avoid peak broadening, and the total flow does not enter the MS source.

If wide bore capillary or packed columns are used, the flow will usually be too large for the source pumps to handle. In this case, some sort of effluent splitter will be needed. The effluent may be simply split, with a portion going into the mass spectrometer and the rest either to waste or into another detector, such as a FID, for simple quantitation. The problem with splitting is that the sensitivity may be too greatly reduced, since much of the sample never reaches the MS.

More enrichment of the sample by separation from the carrier gas is required if packed columns are used. For these, a jet separator is often used. This involves the preferential removal of some of the carrier gas, to reduce the quantity of gas flow into the source, while preserving a greater portion of the sample. Helium is a smaller and more mobile molecule than most of the sample molecules are, so separators can be constructed to remove it. Probably the most popular of these is the jet separator. This separator consists of a jet from which the effluent flows, and an inlet tube positioned directly across from the jet. The effluent passes through a fine nozzle, across a small gap, and into the inlet tube to the source. The nozzle and gap are contained in a small, low-pressure chamber. The jet separator takes advantage of the ease with which the helium carrier gas atoms can be deflected and pumped away. The sample molecules, being heavier, preferentially move in a direct line from the jet into the source inlet tube. The light carrier gas molecules diffuse rapidly into the surrounding vacuum and are pumped out. The separator is usually constructed of glass and operated at a temperature higher than that of the GC column to prevent condensation, absorption, or decomposition of sample on the walls. The open split interface, jet separator and direct interface are shown in Figure 5.12.

Another consideration in the coupling of a GC and MS is the scanning speed of the MS. The MS must be able to complete a sufficient number of scans within the duration of a peak produced by the GC column. The scans must be fast enough to minimize changes in the concentration of the compound during the scan. Otherwise, the spectrum produced may be skewed. The higher mass peaks may be much too large or small in relation to the lower mass ones, depending on whether the scan is taken on the rising or falling slope of the peak. Scans which

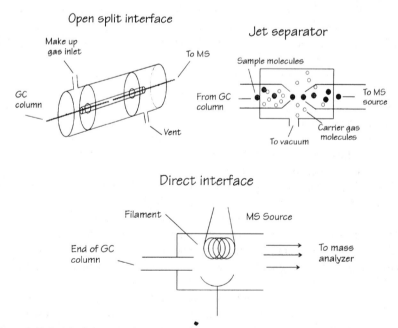

Figure 5.12 Interfaces between the GC and the MS. The open split and jet separator selectively remove some of the carrier gas, and concentrate the sample somewhat.

are too infrequent or slow will also not produce reasonable reconstructed chromatograms for quantitation. Of course, the fastest mass scans also collect the fewest ions at each mass, and so are less sensitive. When the number of ions collected becomes very small, the absolute limit of precision becomes governed by the randomness of ions arriving at the detector. One method of overcoming the problem of sensitivity is to skip the detector from one to another of a small, selected group of mass fragments, without doing a complete scan. This takes less time than a full scan, and allows more ions to be collected at each selected mass. This technique is called "selected ion monitoring". Of course, it is only useful in cases where one knows in advance which molecules are of interest.

For environmental GC/MS, two types of mass analyzers, the ion trap and the quadrupole, are preferred for their scanning speed and their sensitivity. They are also among the simpler and least expensive mass analyzers. When these are configured for GC/MS, they can be stripped of batch inlet systems and solid insertion probes. Since the sample must be introduced through the GC column, the scanning range need not go to very high molecular weights, as high molecular weight compounds will not be sufficiently volatile. These modifications allow the spectrometer to be made smaller and less complex than a full mass spectrometer.

5.7 Liquid chromatography/mass spectrometry (LC/MS)

For liquid chromatography, the interfacing of the mass spectrometer with the chromatograph is more difficult, since the eluent and the sample may both be more similar in their volatility than they are in gas chromatography. Easy separation between the eluent and the analytes is, therefore, not possible, except with very nonvolatile sample components. Several systems are in development and some are commercially available. The most common are the thermospray and the particle beam interfaces. In the thermospray device the sample-carrying eluent is passed into a heated tube, which flashes it into vapor. The vapor exiting the tube at high speed enters a vacuum chamber where much of the lighter solvent vapor is pumped off. The heavier sample molecules are undeflected and continue in a straight line into the source entrance. The particle beam works much like a jet separator, in which easily deflected lighter solvent molecules are diverted or skimmed off the stream of vaporized eluent, while the heavier sample molecules are preferentially passed into the mass spectrometer. Figure 5.13 shows these HPLC/MS interfaces.

5.8 Inductively coupled plasma mass spectrometry (ICP/MS)

The inductively coupled plasma source mass spectrometer (ICP/MS) is a rather recent development, which has spread rapidly since commercial instruments came on the market in 1983. It is now a standard regulatory method for many elemental analyses. For samples which are not easily volatilized, e.g., salts and metals, a high temperature source is needed. The inductively coupled plasma source begins with essentially the same apparatus which is used in atomic emission spectrometry. However, in this case, the ions formed in the plasma are aspirated into the mass analyzer, usually a quadrupole mass spectrometer. The heart of this system is the interface between the atmospheric pressure, very hot gases of the plasma torch, and the vacuum of the analyzer. A schematic of the ICP-MS system is shown in Figure 5.14.

The ICP-MS has several advantages over the ICP used in the emission mode. The most important difference is that when traces of material are sought in samples, the presence of high concentrations of the bulk constituents of the sample may have rich line spectra of their own which will interfere with the weak lines from the trace element. Calcium, a major component of many geological samples, is a particularly bad offender. In mass spectrometry, however, each element yields a single mass peak (or a few isotope peaks). This reduces but does not eliminate the chance of interference.

Interference can arise when isotopes of two elements have the same nominal mass, although it is usually possible to find an isotope of the analyte which does not suffer from such an interference. Molecular ions, while not produced in high

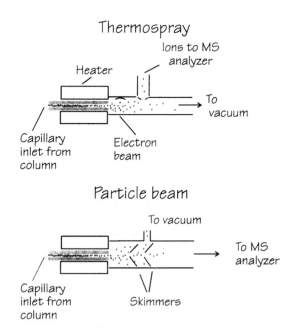

Figure 5.13 Two methods of interfacing HPLC and MS. The particle beam works rather like a jet separator, so some of the eluent is selectively pumped out before the sample reaches the MS source.

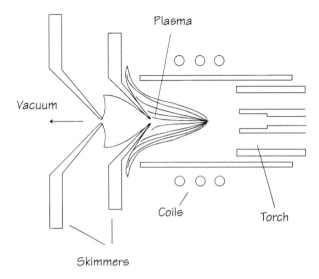

Figure 5.14 In ICP-MS, the argon plasma torch produces the ions for analysis.

amounts, can be a more serious cause of interference. These may form in the plasma, or as the ionized gases pass through the interface. They are usually species formed from argon and elements such as oxygen, hydrogen or nitrogen, which are present in the sample. Species such as $^{36}Ar^{38}ArH$ or $^{40}Ar^{35}Cl$ at mass 75, for instance, interfere with the determination of arsenic, which has only one natural isotope, also mass 75. Fortunately, the natural abundances of ^{36}Ar and ^{38}Ar are very low. Nevertheless, the interference from the $^{40}Ar^{35}Cl$ species is a reason to avoid introducing chloride into samples to be analyzed for arsenic.

The chief requirement for a source for elemental analysis is that it dissociates the sample completely into its constituent atoms. Unlike most mass spectrometry, the ICP-MS seeks no information on the molecules in the sample, only on the elemental composition. Therefore the source must produce as few molecular species as possible and maximize the production of singly charged atomic ions. The ICP allows the sample to be injected into the plasma at atmospheric pressure, and ionizes different elements with relatively equal efficiency. Most of the ions produced are singly charged, and few molecular fragments remain, which reduces interference. Since ionization is quite efficient, the concentration of ions is high and the sensitivity is good.

The removal of the ions from the 5000 K torch is critical. The ions are not distributed evenly in the torch, with the sample ions being concentrated in the cooler center of the torch and argon ions being higher at the sides, where the gas is hotter. The sample ions are concentrated within the central 2–3 mm area of the torch, so alignment is crucial. The position of the sampling point along the axis of the torch is also important, as the sample species must reach the base of the torch, dry, evaporate and then be ionized, before entering the orifice to the

MS. The location of the highest concentration of analyte atoms in the plasma will vary with the sample and with the analyte. Samples with high dissolved solids will form larger microparticles when they dry, and so will require a longer exposure to the plasma before they completely volatilize. The time available for atomization and ionization to occur depends on the power to the torch and the flow of gas along the central channel of the torch. Elements which form oxides will require the orifice to be placed further along the torch axis, to prevent interference by large amounts of oxide.

The orifice must be large enough so that the sample does not cool before it can be transferred through. The orifice, about 1 mm in diameter, allows the ions to flow through into the pumped vacuum system, forming a supersonic jet. The center of this jet flows through a second similarly sized orifice, called a **skimmer**. Because of the rapid expansion of the gases into the vacuum, the ions pass through so rapidly that they are basically unchanged in state or in relative numbers. A boundary layer forms over the edge of the first orifice, and the oxides and other impurities formed in this zone are eliminated when the gases pass into the skimmer. By the time they reach the second orifice, the gases are cooler and much less concentrated, so that the skimmer is not exposed to such an aggressive atmosphere. The entire interfacing apparatus is water-cooled to keep the surfaces from melting in the high temperatures.

Because of the supersonic speeds reached by the ions, there is no discrimination between ions of different masses. All are accelerated to the same velocity before entering the skimmer. The ions are focused into a collimated beam by ion lenses, electrostatic fields which force the ions into a narrow beam, and are allowed to pass into the quadrupole mass analyzer. There they are separated. The mass spectra are much simpler than the usual molecular mass spectra, since only elemental masses are seen, so the analysis of a mixture of atoms is readily done. The ICP-MS is very well suited to such analyses as the determination of trace metals in water and wastewater samples, as well as for air particulates, soil, and biological samples, once these have been brought into solution. The techniques for dissolving samples are similar to those used for atomic absorbance studies, and such techniques as internal standards, standard addition and isotope dilution can be applied. A sample analyzed by ICP-MS is shown in Figure 5.15.

5.9 Data collection

The mass spectrometer, especially if it is attached to a chromatograph, is capable of generating an enormous amount of data in a short time. This makes computerized data handling essential. Each spectrum will have perhaps 100 peaks in it. As each peak is detected, two numbers, the mass and the intensity must be recorded. In a typical GC/MS run, a scan may be done every 2 or 3 s over a span of 15 min or more. This run would produce a total of 60 000 numbers, to record each mass and its intensity. The computer handles this data, and also carries out

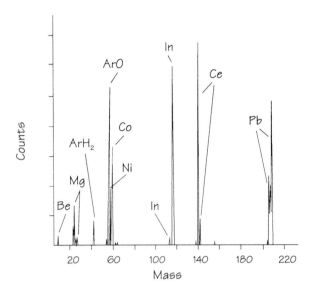

Figure 5.15 The ICP-MS sample output is simple to interpret. Each element shows peaks at its isotope masses, with their heights relative to each other related to their isotopic abundances. Note, for example, the three Pb peaks, corresponding to the isotopes of lead of mass 206 (25% abundance), 207 (22%) and 208 (52%). Note also the presence of some interferent peaks from argon species formed in the plasma.

such tasks as automatically tuning the mass spectrometer, and processing the data into the desired output format.

The data may be displayed by the computer in several ways. Individual spectra are selected and a background spectrum may be selected and subtracted. The background spectrum is usually selected at a point as near to the peak of interest as possible, to correct for any column bleed or impurities in the vacuum system, such as nitrogen or water, which may be present in that area of the chromatogram. The selection of a background spectrum is important, because subtraction of a spectrum which is not the same as the background actually found in the peak can result in a distorted spectrum which may be hard to match with the library. In general, quantitation is done on the reconstructed chromatogram, comparing peak areas with those of a standard. Background subtracted mass spectra of individual peak components are used for qualitative identification.

The spectrum is usually compared with a spectral library for identification, and several reasonable matches will be displayed for the operator's final judgment. The chromatogram can be reconstructed mathematically by addition of all the ions produced in each scan and plotting this total ion intensity against time.

A selected ion mass chromatogram can be plotted to help correct problems in the chromatogram. To bring a very small component out of the baseline noise, a single ion, characteristic of the compound or class of compounds being sought, may be plotted. A range of ions, perhaps a large interfering solvent peak, can be eliminated by plotting only a mass range greater than that of the solvent. Figure 5.16 shows part of the separation of a standard gas mixture of VOCs listed in the EPA TO-14 method. Two components which are only partially separated, in this case benzene and carbon tetrachloride, can be better quantitated by plotting the most prominent ion for each, mass 78 for benzene, and mass 117 for carbon tetrachloride, separately. The full chromatogram and each of the two individual ion chromatograms are shown. Notice that the y-axis is scaled differently for each chromatogram.

5.10 Library searching techniques

The usual computer matching algorithms reduce a sample spectrum to a group of major peaks, giving more weight to the higher mass peaks, which are more

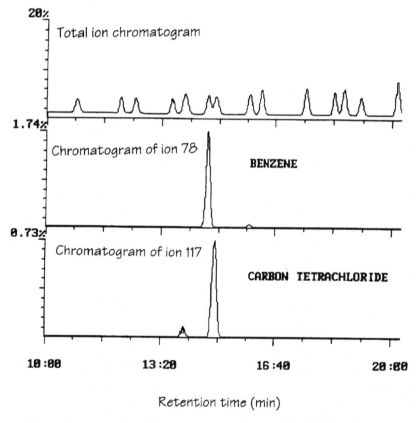

Figure 5.16 A reconstructed mass spectrum of the early part of a VOC analysis. Note that the two peaks for benzene and carbon tetrachloride are not fully separated. The individual ion chromatograms for mass 78 and mass 117 show the benzene and carbon tetrachloride peaks, respectively, free of interference from each other.

characteristic of the compound. This weighting ensures that important peaks, such as the molecular ion, which may be small in intensity, are not eliminated. The computer then takes this selected group of 15–20 peaks and searches through the specified library to find the best matches. After the list of possible matches is compiled, the computer calculates a score for each, a quantitative measure of the match, based on the entire sample and library spectra.

Often the operator may override the default number of peaks used in the presearch. A larger number of peaks will make it less likely that the correct compound will be screened out before the final matching is done. However, use of too many peaks in the presearch can slow the search process to an unacceptable degree.

When a computer search is done to determine the "best match" between a sample mass spectrum and one in a library, the definition of "match" is important. There are three usual criteria for a good match. First is the **purity** of the matched spectrum. This indicates the number of peaks which are the same in both the spectra, and how well their intensities match. A match with a high score for purity indicates that the spectra match well, with few extraneous mass peaks showing in the sample. A score of 1000 is often assigned to a perfect match. On this basis, a score above 800 indicates very similar spectra, while scores between 600 and 800 imply a significant similarity between the library compound and the sample. They may have a major molecular substructure in common.

When a mass spectrum is taken of a compound containing a significant impurity, as in a GC/MS run when peaks coelute, a search on purity may not find the compound because the extra peaks from the impurity lower the purity score. In such a case, a search based on **fit** is better. Fit indicates how well the library spectrum peaks are found in the sample. Therefore, extraneous peaks in the sample spectrum will not reduce the fit score, although if certain peaks occur in both the sample and the impurity, the score may be reduced because the relative intensities will differ from the library spectrum. A high fit score which also has a low purity result indicates that the matched compound is likely there, but that there is another compound present, or that the sample has a major substructure similar to the library compound. A low fit means that there are peaks in the library spectrum which do not appear in the sample, and implies that the compound is not present.

Finally, a **reverse fit** search can also be done. This score measures the degree to which the peaks found in the sample are also found in the library spectrum. A high fit occurs when each sample mass peak is found in the library spectrum, at the same mass and intensity. A low fit indicates that the sample has peaks which are not present in the library spectrum. This is the least useful of the search scores. Figure 5.17 shows spectral matching.

Remember, a high library matching score does not ensure that the match is correct. When the sample has been run on GC/MS, the retention time is an important piece of information which can be used to select the correct compound from the list of possible matches. Very often it is possible to determine, based on such factors as boiling point or known retention times, which of a group of compounds, all showing good fit, is the most likely.

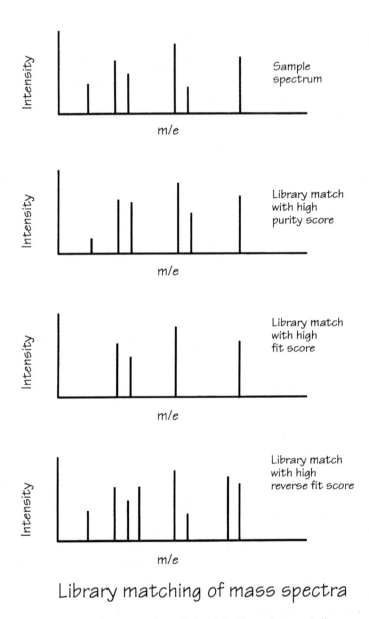

Library matching of mass spectra

Figure 5.17 When this sample spectrum is matched with the library the most similar spectrum gets a high purity score. A high fit score shows that the sample has all the major peaks of the reference spectrum, but may have more. A high reverse fit means that all the major sample peaks are present in the reference, but some of the reference peaks may be missing in the sample.

Suggested reading

Chapman, J.R. (1993) *Practical Organic Mass Spectrometry: A Guide for Chemical and Bio-chemical Analysis*, 2nd edition, J. Wiley and Sons, New York.

Date, A.R. and Gray, A.L. (eds) (1989) *Applications of Inductively Coupled Plasma Mass Spectrometry*, Chapman and Hall, New York.

Hites, R.A. (1992) *Handbook of Mass Spectra of Environmental Contaminants*, 2nd edition, Lewis Publishers, Boca Raton.

Skoog, D.A. and Leary, J.J. (1992) *Principles of Instrumental Analysis*, Saunders College Publishing, New York.

Willard, H.H., *et. al.* (1991) *Instrumental Methods of Analysis*, 7th edition, Wadsworth, Belmont, CA.

Yergey, A.L., Edmonds, C.G., Lewis, I.A.S. and Vestal, M. L. (1990) *Liquid Chromatography/Mass Spectroscopy Techniques and Applications*, Plenum Press, New York.

Study questions

1. Two ions, O^+ and NH_2^+ have the same nominal mass, 16. What is the resolution needed to separate them in a mass spectrometer? The masses of the isotopes are: $N = 14.003070$; $H = 1.007825$; $O = 15.9949$ amu.

2. What are the major parts of a mass spectrometer and what are their major functions?

3. What are the advantages of the ICP/MS over inductively coupled plasma emission spectrometry?

4. Compare the advantages and disadvantages of the ion trap and the quadrupole detectors for mass spectrometry?

5. What is the importance of background subtraction in the identification of a compound from its mass spectrum?

6. Describe the chemical ionization method in mass spectrometry, and explain its purpose. How does the information obtained differ from that given by electron impact ionization?

7. In doing GC of environmental samples, a specific detector, or a pair of detectors, can be used instead of a mass selective detector, for detection and quantitation. For example, sulfur in fuel can be measured using a FID and a sulfur specific ElCD, in parallel. Compare the methods, MS versus one or more substance-specific detectors. What are the advantages and disadvantages of each?

8. When would you use a library search for the best purity match, and when for the best fit match, for identifying a compound from its mass spectrum?

6 Sample preparation techniques

Sample preparation is the step between sampling and analysis. This is the point at which the pollutants are transferred from the environmental matrix to a form suitable for analysis. Extraction is often the first step in sample preparation. The analytes must be separated from the sample because most environmental samples cannot be directly introduced into an instrument. Extraction is done for both organic and inorganic pollutants. Some of the methods of extraction and the types of sample for which they are appropriate are shown in Figure 6.1. For example, traces of pesticides can be extracted from a large volume of water using a relatively small quantity of an organic solvent such as methylene chloride. If the pollutants are strongly bound to the matrix, the extraction may be difficult. Many metals are strongly bound to soil and require aggressive solvents to recover them quantitatively. Organics may also be strongly attached and the solvents in this case cannot be too aggressive or the organic compounds will be destroyed. Similarly, the more soluble an organic analyte is in water, the more difficult it is to extract it into an organic solvent.

Ideally, an extraction method should be fast, simple, inexpensive and should provide high extraction yield with good precision. The amount of solvent required for extraction should be kept to a minimum to produce the most

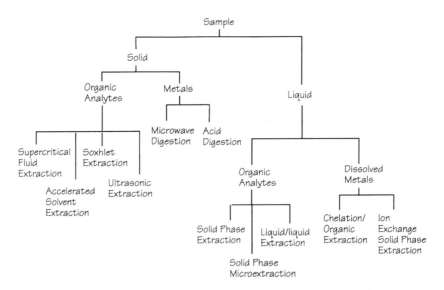

Figure 6.1 Methods of extracting the analytes of interest from various types of samples.

concentrated extract, and because the solvents themselves are a source of hazardous waste. While one would like the sample to be ready for analysis without any additional steps, this is not always the case. For instance, a small quantity of analyte may end up in a large volume of extract. The dilute extract raises the detection limit and an additional preconcentration step will be necessary for trace analysis.

Another common problem is encountered with environmental samples. The matrix itself is very complex and may contain components that interfere during analysis. This happens frequently in determining organic analytes using chromatographic techniques. In these cases, along with the concentration step, a sample clean-up step may be necessary.

Finally, one of the most difficult tasks in the preparation of samples for the determination of low concentration analytes is to carry out the sample preparation without contaminating the sample. Contaminants can lurk on surfaces which appear clean or can drift into samples with airborne dust. Some principles which will lower the chance of contamination if they are followed are:

- minimize the handling of samples
- use as few pieces of apparatus as possible
- keep the sample preparation work area as clean as possible
- use the minimum quantity of reagents

6.1 Extraction of organic analytes from liquid samples

For extraction of organic analytes from liquids, liquid-liquid extraction and solid phase extraction can be employed.

6.1.1 Liquid–liquid extraction

Liquid–liquid extraction (LLE) is a popular method for extracting organic pollutants from liquid matrices, especially water. Extraction techniques are usually employed for compounds which cannot be analyzed efficiently by purge and trap, because the analyte is not volatile enough to be purged in a reasonable time. The aqueous sample containing the pollutant is shaken with an immiscible organic solvent. The sample components distribute themselves between the aqueous and the solvent phase. When the two solutions reach equilibrium, the ratio of the concentration of the solute in the two phases is constant.

$$D = C_s/C_w \qquad (6.1)$$

where D is the **partition or distribution coefficient** and C_s and C_w are the concentration in the solvent and the aqueous phase, respectively. The partition coefficient is independent of the volume ratio of the two solvents. If a larger solvent volume is used, a larger quantity of the analyte will partition into the organic phase so that D remains constant. The fraction of the analyte extracted by the solvent is given by E.

$$E = (C_s V_s)/(C_s V_s + C_w V_w) \tag{6.2}$$

where V_s and V_w are the volume of solvent and water respectively. By dividing the numerator and denominator by $C_s V_s$,

$$E = \frac{1}{1 + C_w V_w / C_s V_s} \tag{6.3}$$

By substituting Equation 6.1

$$E = \frac{1}{1 + V_r / D} \tag{6.4}$$

where V_r is the **volume ratio**, the ratio of the volume of water to that of solvent.

For trace analysis, it is important that the maximum amount of analyte is extracted from the sample. From the above equations, it is obvious that the fraction of analyte extracted can be increased by increasing the volume of organic solvent.

6.1.1.1 *Successive extractions.* Often, a single extraction does not provide sufficiently high extraction efficiency. Better efficiency can be obtained by carrying out several successive extractions using smaller volumes of the solvent. For this reason, analytical LLE is often carried out using repeated extractions of the aqueous sample, each time using the same volume of solvent. After the first extraction, the amount of analyte unextracted is $1 - E$, or U.

The percent extracted remains constant at each step if the same solvent volume is used each time. The amount extracted in a series of n steps is shown in Figure 6.2. If C_w^0 was the initial analyte concentration of the aqueous phase, then the concentration after one extraction is C_w^1:

$$C_w^1 = C_w^0 [(V_w)/(DV_s + V_w)] \tag{6.5}$$

and after n such extractions, the concentration of analyte in the stationary phase is:

$$C_w^n = C_w^0 [(V_w)/(DV_s + V_w)]^n \tag{6.6}$$

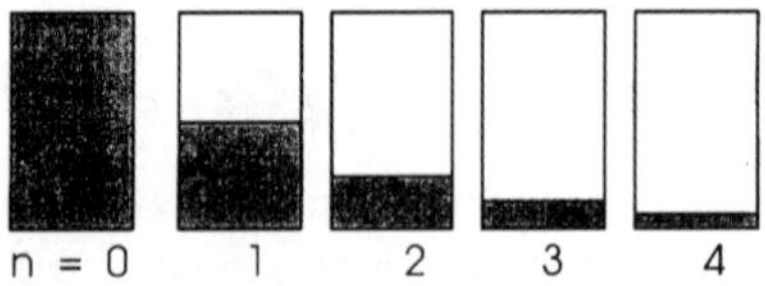

$n = 0 \quad 1 \quad 2 \quad 3 \quad 4$

Figure 6.2 Stepwise extraction. The amount of analyte remaining after each extraction is shown as a shaded area.

In the following example, the extraction efficiency increases from 0.83 to 0.92 when two successive extractions are done using the same total amount of solvent.

Example
In a certain system, the distribution coefficient for the solute is 5 and the sample volume is 100 ml. The sample may be extracted with one 100 ml portion of solvent or two 50-ml portions. How will the efficiency be affected?

For a 100-ml extraction:

$$E = \frac{1}{1 + 1/5\,(100/100)} = 0.83$$

while for a 50-ml extraction

$$E = \frac{1}{1 + 1/5\,(100/50)} = 0.71$$

A second 50-ml extraction, since the volume ratio is the same, will extract 0.71 of the remaining 0.29 of the sample, giving a total extraction efficiency of 0.92 compared with 0.83, for the same total amount of solvent used.

You can do the calculations to determine if the extra work needed to do four extractions of 25 ml each is worth the improvement in extraction efficiency. On a practical level, repeated extractions with very small amounts of solvent can become counterproductive, since the calculated extraction efficiency is not achieved in practice because the separation of the two solvents is never perfect. It is seldom beneficial to do more than three sequential extractions, because the incremental amount extracted in each successive step becomes less and less, while losses due to such effects as incomplete phase separation and adsorption on the surfaces remain constant.

6.1.1.2 *Instrumentation for LLE.*   LLE is usually carried out in a separatory funnel (Figure 6.3). The sample and a predetermined quantity of a suitable solvent are shaken together to ensure intimate contact between the two liquids. The two phases are allowed to settle and the bottom layer is drained out by opening the stopcock. The above steps are repeated with fresh solvent added for each successive extraction.

6.1.1.3 *Continuous liquid–liquid extraction.*   Continuous LLE is an alternative to multi-step batch LLE. Droplets of the extracting solvent are continuously passed through the aqueous phase. The apparatus used in this technique is shown in Figure 6.4. The extracting solvent is vaporized in the flask. The vapors condense and drip into the sample reservoir, coming into contact with the sample. If the solvent density is less than that of the aqueous phase, the solvent falls through a funnel into the bottom of the sample, rises through the sample and finally the extract overflows into the extracting solvent reservoir. If the solvent is more dense than water, it is allowed to drip directly onto the aqueous phase, sinking down and finally returning to the solvent reservoir through the tube attached at the bottom of the extraction chamber. In either case, the extracting solvent is recycled and its volume stays constant, provided the condenser functions efficiently. If the volume of the extraction chamber is too large, then glass beads are used to fill up the excess space and also to provide closer contact between the two phases.

One shortcoming of this technique is that volatile analytes may be recycled back to the extraction chamber along with the solvent, in which case the efficiency decreases. A volatile analyte also may be lost through the condenser. For these reasons, the use of continuous LLE is usually restricted to semi-volatile and non-volatile analytes. Continuous LLE can be operated unattended, and so is less labor intensive than successive extractions. While it is difficult to predict the extraction efficiency for a continuous extractor, it can be determined experimentally by extracting a sample of known concentration. Extraction efficiency increases with time, but relationship between time and efficiency is also best determined by experimentation.

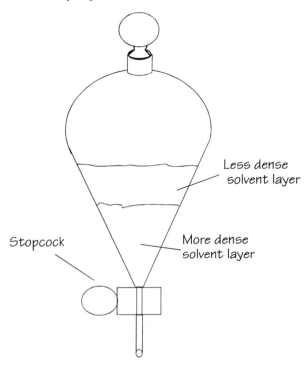

Figure 6.3 Separatory funnel for simple manual extractions

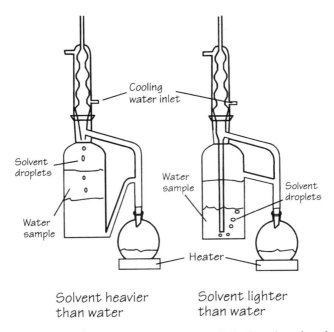

Solvent heavier
than water

Solvent lighter
than water

Figure 6.4 Apparatus for continuous extraction. Freshly distilled solvent drops through the sample or rises through it, depending on its density relative to water, then returns to the solvent reservoir.

6.1.2 *Solid phase extraction*

Solid phase extraction (SPE) is also known as liquid–solid extraction. A SPE cartridge is made by packing a few milligrams of an adsorbent in a tube. The liquid sample from which the analytes are to be extracted is passed through the SPE packing, and the analytes are retained by the sorbent. The sample is then extracted by passing a few milliliters of a suitable solvent through the cartridge. In this way, the analytes are separated from the sample matrix and transferred into a small quantity of solvent. For example, a few micrograms of a pollutant such as a pesticide or dioxin can be trapped from several hundred milliliters of a water sample, and eluted using 1 ml of a solvent such as methylene chloride. The analytes are not only extracted from the water but also concentrated in a small volume of solvent.

The extraction mechanism in SPE is the same as that for retention in HPLC. The analyte distributes itself between the solvent and the packing material, which acts as a stationary phase. The partition coefficient K is given as: $K = C_s / C_m$ where C_s and C_m are the concentrations in the stationary and mobile phases respectively. If the volume of the stationary phase (i.e., volume of the surface layer of the packing material) is V_s and the volume of the solvent phase (in the void volume of the packing) is V_m, then the fraction of analyte extracted, E is:

$$E = (C_s V_s)/(C_s V_s + C_m V_m) \qquad (6.7)$$

Dividing the numerator and denominator by $C_s V_s$, and substituting for the partition coefficient gives:

$$E = \frac{1}{1 + 1/KV_r} \qquad (6.8)$$

where V_r is the volume ratio of the two phases (i.e., V_s/V_m). Since the capacity factor k' is defined as $k' = KV_r$, E reduces to

$$E = \frac{k'}{1 + k'} \qquad (6.9)$$

Just as in HPLC, k' decreases with the strength of the mobile phase, in SPE k' decreases with the solubility of the analyte in aqueous phase. The most common type of SPE sorbent is made from porous silica particles with a chemically bonded organic coating. The bonded coatings are similar to LC stationary phases and can interact with the analytes by various mechanisms, such as hydrogen bonding, dipolar interactions, electrostatic attractions and even ion exchange. Different solid supports are used for different types of analytes.

Disposable SPE cartridges are commercially available and are made to fit onto the tip of a syringe using a Luer fitting, or are made by packing the adsorbent at the bottom of a plastic syringe barrel. The liquid is passed through it by applying a positive pressure using a syringe plunger or by pulling a vacuum on the bottom of the cartridge. A cartridge is shown in Figure 6.5.

6.1.2.1 *The SPE process.*   The SPE process is described in Figure 6.6. Four steps are involved. First, the sorbent material is conditioned or activated by passing a solvent (e.g., methanol) through the bed. The choice of conditioning solvent depends upon the packing material and the analytes for which it is to be used. The sample is then passed through the cartridge. Only the components which interact with the sorbent are retained in the packing. The packing is

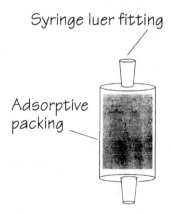

Figure 6.5 Solid phase extraction cartridge. These are packed with various materials, depending on the analyte to be extracted.

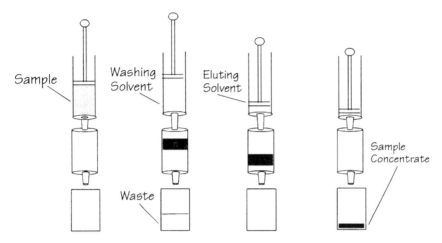

Figure 6.6 Solid phase extraction process. A syringe is used to pass the entire sample through the cartridge. Then the cartridge is washed to remove unretained contaminants. Finally, the adsorbed analyte is eluted with a small amount of solvent.

washed with a solvent, removing some of the unwanted, interfering components originally retained by the packing. To eliminate excess water, the packing is dried by drawing air through it. Finally, a small quantity (200 µl to a few ml) of suitable solvent is used to elute the analytes from the SPE cartridge. The extract is ready for further processing, if necessary, or for analysis.

Selectivity in SPE is obtained by three separate mechanisms: selective extraction, selective washing of interferents and selective elution of the analytes. To achieve maximum selectivity, it is important to choose the right combination of packing material, washing solvent and elution solvent.

6.1.2.2 *Advantages of SPE.*

SPE has several advantages over liquid–liquid extraction. This technique is rapid and does not involve extensive solvent use, and consequently generates less waste. LLE also requires a concentration step because the extract volume is usually quite large. SPE offers higher extraction yield, more selectivity and higher precision than LLE. In LLE the solvents used must be immiscible. This poses certain problems. For example, methanol may be the most favorable solvent for a certain analysis, but because it is completely soluble in water, it cannot be used for extraction. To circumvent this problem, a different solvent is used for the extraction, the extract is evaporated to near dryness to remove the extraction solvent, and then the sample is redissolved in methanol. This type of problem is not encountered in SPE. Conceptually, SPE is similar to LLE, except that the extracting solvent is bonded and does not physically mix with the matrix. SPE is also operationally simple. Systems are available which apply a controlled vacuum to many cartridges simultaneously, or use centrifugation or robotics to automate the process.

Extraction discs, containing the adsorbent particles in a filter mat are also available. These discs not only separate the components on the basis of their chemical nature, but also filter the sample at the same time. They may not have as large a capacity for analytes as the cartridges do, but are useful for sorbing analytes from large volumes of dilute sample, because the flow is much faster.

6.1.3 *Solid phase microextraction*

SPME is a recently developed system which uses a fused silica fiber, coated with a nonvolatile organic liquid or polymer to sorb organic compounds from aqueous or vapor samples. A coated fiber is held in a syringe-like device, which allows it to be extended out of the needle, and retracted, by moving the plunger. The needle is passed through a septum cap into the sample or the sample headspace, and the fiber is extended. The fiber is held there under controlled conditions of stirring and temperature, for a fixed time. It is then retracted into the hollow needle, withdrawn and inserted into the heated injection port of a gas chromatograph. The fiber is extended into the gas stream. As it heats, the adsorbed analytes are injected into the instrument for analysis. Since the fiber has a very small mass, it heats rapidly, giving a rapid injection.

Quantitation with this system is not simple, as the amount absorbed depends not only on the concentration in the sample but the exposure time, the temperature, the amount of stirring, and the distribution coefficient of the analyte between the sample and the coating on the fiber. However, it is a rapid method, and can be useful for quick, semiquantitative screening of samples. It is ideal for checking samples before injection into a high sensitivity instrument, to make sure that they do not contain excessive amounts of materials which might contaminate the instrument or require a repeat analysis on a diluted sample. Figure 6.7 shows the SPME system.

6.2 Extraction of organic analytes from solid samples

Solids often require more aggressive extraction procedures than do liquids, because the analytes can be more tightly bound into the matrix, and because the extractant has to penetrate into solid particles. A preliminary grinding or drying step may be needed before extraction, but care must always be taken not to lose analyte in these processes. Organics can be separated from solid samples by Soxhlet extraction, by accelerated solvent extraction at elevated temperature and pressure, and by supercritical fluid extraction.

6.2.1 *Soxhlet extraction*

The continuous LLE apparatus cannot be used for solids because of poor liquid–solid contact. If the condensate is allowed to drip onto the solid, it would not mix well with the solid and extraction efficiency would be poor. In order to achieve good mixing, the solid has to be immersed in the liquid and this is what happens in a Soxhlet extractor. A Soxhlet extractor is shown in Figure 6.8. The solid sample is loaded in a porous extraction thimble. Wet particles may repel a hydrophobic solvent. If the sample is wet, a drying agent such as sodium sulfate is mixed with it, to insure good solvent contact. The thimble is placed in the extraction chamber, where it holds the solids while the liquid passes through it. The solvent is distilled from the solvent reservoir, condensed, and allowed to drip into the extraction chamber. The extraction chamber gradually fills with solvent, until the level reaches the solvent return tube. This tube acts as a siphon and drains all the liquid from the extraction chamber back into the solvent reservoir.

The sample is alternately soaked in solvent and drained. This is not really a continuous process, rather a series of batch processes. Soxhlet extraction is usually best for semivolatile or nonvolatile pollutants, as volatile materials may

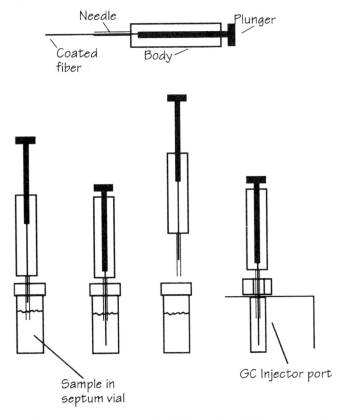

Figure 6.7 Solid phase microextraction. Useful for semiquantitative screening, the sample is allowed to adsorb on the coated fiber, then is desorbed directly into the heated GC injection port. The fiber is retracted into the hollow needle for insertion through the septa.

be lost through the condenser. It is also a slow process and normally takes several hours. Another problem is that relatively large quantities of solvents are used for extraction and a preconcentration step is necessary for trace analysis.

6.2.2 *Accelerated solvent extraction*

Just as a pressure cooker is used to cook food more rapidly than is possible by boiling in an open container, extraction of analytes from solid or semisolid samples can be speeded by carrying out the extraction under pressure and at a temperature above the normal boiling point of the solvent. Automated apparatus is available to carry out this procedure. The sample is placed in a stainless steel cartridge. Solvent is pumped into the cell and the temperature is raised to the required level. The solvent pump maintains the set pressure, and the exit valve of the cartridge opens momentarily whenever the pressure rises above the set point. The sample is held in the solvent filled cartridge for a preset time, and then the extract is purged from the cartridge into a receiving vial, first with a small amount of fresh solvent, then with a stream of inert gas. This is a static method, as the solvent only flows through the sample when the cell is being

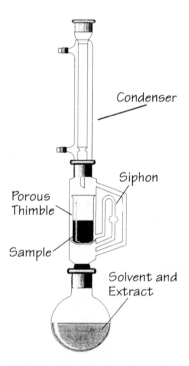

Figure 6.8 Soxhlet extractor. Distilled solvent drips onto the sample contained in a porous thimble. When the chamber containing the thimble is filled with solvent, the solution is siphoned back into the solvent reservoir.

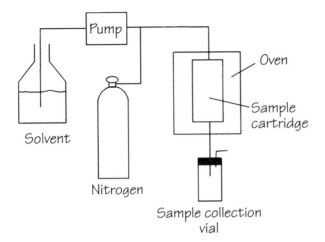

Figure 6.9 Accelerated solvent extraction apparatus. Solvent is pumped into the sample cell where it is subjected to elevated temperature and pressure. The extract is removed from the cell when the extraction is completed, by purging with nitrogen.

filled or emptied. Most of the extraction process occurs while the sample soaks in the hot, pressurized solvent. Figure 6.9 shows a schematic of the apparatus.

Samples as diverse as fish tissues, soil, and sewage sludge can be extracted by this method. Volatile analytes are well recovered, as well as the less volatile ones. Cell volumes range from 3 to 30 ml, and temperatures of 50–200°C are commonly used. The extraction takes only a short time, an equilibration time of a few minutes, and about 5 min for the actual extraction. The solvents are the same ones usually used for Soxhlet extraction, and the solvent volumes are much smaller than needed for many other methods. The high extraction efficiency of the accelerated solvent extraction method can be attributed to the better solvent contact with the sample. The high pressure forces solvent into the sample pores and the elevated temperature decreases the solvent viscosity, Also, the analyte is more soluble at the higher temperature. Relative standard deviations of less than 5% are commonly achieved, and recoveries are at least as high as those obtained by 5 or 6 h of Soxhlet extraction. The fact that the system is automated helps to keep the relative standard deviation low. Finally, the method is less expensive to operate on a day-to-day basis, since the solvent costs, both for purchase and disposal, and the operator time needed are much smaller than for Soxhlet, sonication or microwave extraction methods.

6.2.3 Ultrasonic extraction of organics

Ultrasonic extraction is another method used for extracting semivolatile and nonvolatile organic compounds from solids, sludges, and wastes. The sonication process ensures intimate contact between the extraction solvent and the solid matrix such as soil. As in other extractions with organic solvents, it is important to remove moisture from the sample to obtain high extraction efficiency. Mixing the solid sample with anhydrous Na_2SO_4 dries the solid particles, and allows them to contact the solvent. The samples are suspended in the extraction solvent, and the sample flasks or vials are set into a rack in an ultrasonic bath. The sonic waves disrupt the sample particles and agitate the solution, assisting in solublizing the analytes. For single samples, an ultrasonic disrupter probe may be inserted directly into the sample. Figure 6.10 shows an ultrasonic probe sample disrupter.

Ultrasonic extraction usually takes only a few minutes (3–10 min). Sometimes successive extraction is necessary to achieve high extraction efficiency. The sample is separated from the extract by centrifuging or vacuum filtration. The extract then is concentrated and cleaned up for analysis.

Ultrasonic extraction is more rapid than Soxhlet extraction, but its efficiency is not always as high. It requires somewhat less solvent than the Soxhlet process, but a concentration step is still often needed.

6.2.4 Supercritical fluid extraction

A supercritical fluid is a substance above its critical temperature and pressure. The properties of gas-like mass transfer coefficents and liquid-like solvent characteristics make supercritical fluids very attractive as extraction solvents. The process of using supercritical fluids as extraction solvents is known as supercritical fluid extraction (SFE). SFE has been known to the chemical engineering community for several decades. It is used in large scale, commercial processes for such purposes as decaffeination of coffee, extraction of spices, and regeneration of activated charcoal used in waste water treatment. Recently, SFE

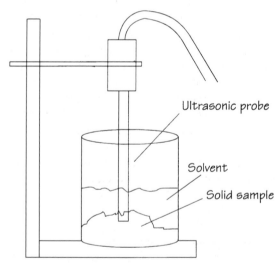

Figure 6.10 An ultrasonic probe for extraction of analytes from solid samples. The volume of sample and solvent must be appropriate to the power of the probe being used.

has gained much popularity as an analytical-scale extraction procedure. It can be used to extract volatile and semi-volatile organic pollutants from a variety of environmental matrices such as, soil, sludges, adsorbents and biological matrices such as foods, plants, and animal tissues.

6.2.4.1 *Instrumentation.*   An SFE system suitable for preparation of analytical samples is shown in Figure 6.11. A pump is used to pressurize the supercritical fluid. The compressed fluid is passed through the extraction vessel, where the sample is placed. The pollutants dissolve in the supercritical fluid and are swept out of the extraction cell into a collection device. The pressure in the extraction cell is between 80 and 400 atm (when CO_2 is used as a solvent). At or before the collection device, the pressure is dropped to atmospheric pressure using a restrictor. Once the pressure is released, the supercritical fluid turns into a gas and escapes while the analytes are precipitated in the collection container. The instrumentation used in SFE is relatively simple.

The most common type of SFE pump is a modified syringe pump, although other types of pumps are now being used. The extraction cell is made from high pressure tubing with appropriate fittings at the ends. The restrictor is a piece of tubing with a small restriction at the end. It serves as a valve and the pressure of the extraction fluid is released as it passes through the restrictor. The simplest restrictor is just a piece of 1–50 μm diameter tubing and the flow rate through the cell depends upon the restrictor orifice. The extract is usually collected in a small vial containing a few milliliters of a solvent and may be cryogenically cooled. The solution is then analyzed by GC, HPLC, or GC/MS.

The restrictor can also be used to transfer the extract directly onto the injection system of a GC (also HPLC or SFC) for on-line chromatography.

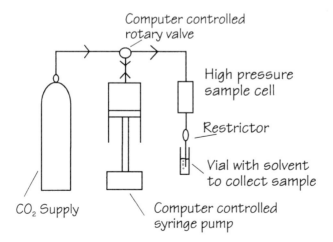

Figure 6.11   Supercritical fluid extraction apparatus. Sample is placed in the high pressure cell and the extract is collected in a small amount of solvent as the CO_2 vaporizes when the pressure is released.

Lower detection limits can be achieved by on-line chromatography because there is no dilution effect due to the solvent.

6.2.4.2 *Choosing SFE conditions.*   There are two factors that affect the extraction of an analyte. The first is solubility. The analyte must be soluble in the supercritical solvent. The second is a matrix effect which depends upon how strongly the analytes are bound to the matrix. The variables which control the extraction are temperature, pressure (or density), and the choice of solvent. The solubility of a solute in a certain solvent depends upon temperature and pressure. In fact, solubility correlates better with density (which is a function of temperature and pressure) than with pressure. Solubility has been empirically correlated with density and temperature as follows:

$$\log \text{ (solubility)} = a \text{ (density)} + b \text{ (temperature)} + c \qquad (6.10)$$

where a, b, and c are constants.

However, the matrix effect is not well understood and no methods for predicting the optimum extraction conditions are available. Some trial and error estimation is required to identify the optimum extraction parameters.

CO_2, N_2O, and NH_3 are some of the supercritical fluids that have been used for extraction of pollutants. The properties of these solvents are dramatically altered by addition of a small quantity (1–10%) of another solvent, referred to as a **modifier** or co-solvent. Organic solvents such as methanol, acetone, and acetonitrile are usually used as modifiers. Modifiers change the strength and polarity of the supercritical solvent. For example, in the extraction of PAH from an adsorbent, the extraction efficiency increased from 30% to near 90%, when 5% methanol was added to CO_2.

In SFE three variables are important, namely temperature, pressure, and solvent composition. By altering these conditions, the strength of the supercritical solvent can be changed. Some method development is necessary to optimize the conditions for each matrix and analyte. Different classes of pollutants may be selectively extracted by changing the extraction conditions, which simplifies the extraction and may eliminate the necessity for some sample clean-up.

6.2.4.3 *Advantages of SFE.*   Several studies have shown that Soxhlet extraction, sonication, and SFE yield comparable results on a variety of environmental samples. SFE offers several advantages over other extraction procedures. The extraction is rapid, usually taking 10–50 min and higher extraction yields are obtained. SFE is carried out at fairly low temperatures (32–65°C for CO_2). Therefore, thermally labile pollutants can be extracted without thermal degradation. The analyte is usually collected in a small sample volume and the sample concentration step may be eliminated. It is also the only extraction technique that can be directly interfaced with a GC, HPLC, or SFC. This makes it very useful in trace analysis. Another important consideration is that SFE may eliminate use of the large quantities of organic solvents which are required in the more traditional Soxhlet and ultrasonic extractions. These solvents are toxic in nature and require proper, often expensive, disposal.

6.3 Post-extraction procedures

6.3.1 Concentration of sample extracts

After solvent extraction, the sample analytes are usually present in a fairly large volume of solvent. When the analysis is being done at the trace level, this solution may be too dilute for an accurate measurement. Therefore, methods to concentrate the extract are often needed. If the analyte is quite nonvolatile and there is not a very large amount of solvent to be removed, the sample can be exposed to a gentle stream of nitrogen across the surface of the solution, using a tank of compressed gas. This is not efficient if a large volume reduction is required. A rotary vacuum evaporator is better suited to removal of large volumes. The sample is placed in a round bottomed flask, which is heated gently in a water bath. The flask is continually rotated to expose the maximum surface of liquid to evaporation, and a water-cooled condenser is attached to the top. The pressure inside the apparatus is reduced by use of a small pump or water aspirator. The combination of lowered pressure and mild warming removes the solvent efficiently, and the condensed solvent distills into a separate flask, eliminating the solvent vapors from the laboratory air. The evaporation should be stopped before the sample reaches dryness, and the flask must be rinsed to remove all the analyte.

For smaller volumes which must be reduced to less than a milliliter, a glass Kuderna Danish concentrator is used. This consists of an air-cooled condenser with glass floats, topping a bulb, with a narrow graduated tube at the bottom. The sample is placed in the concentrator and it is heated gently in a water bath, until the required volume is reached. Because of the narrow tube at the bottom, a small volume can be seen and measured readily. Figure 6.12 shows the Kuderna Danish concentrator.

6.3.2 Sample clean-up

Soil and many other solid matrices can contain hundreds of compounds. It is not always possible to identify the analytes of interest in the complex chromatograms produced. A clear separation of the analyte of interest is difficult, especially if it is present at much lower concentrations than some interfering compound. An additional extraction or clean-up step is then necessary to eliminate many of the interfering species prior to chromatographic analysis.

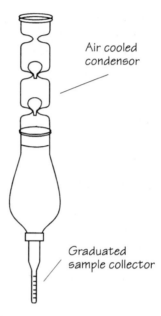

Air cooled
condensor

Graduated
sample collector

Figure 6.12 A Kuderna Danish sample concentrator. The bottom portion is heated gently, often with steam, to evaporate the excess solvent. The graduated tube allows the progress of the evaporation to be monitored.

For example, during analysis of pesticides, PCBs present at higher levels than the pesticides can interfere. The usual clean-up steps are:

- Passing the sample through a column packed with an adsorbent which selectively traps the analytes and perhaps interfering species as well.
- Using a weak solvent to elute some interfering species.
- Using a stronger solvent to elute the analyte, leaving other interfering species behind.

Columns used for clean-up are often standard low pressure, gravity flow, liquid chromatography columns packed with alumina, a porous, granular form of aluminum oxide, magnesium silicate with acidic properties, usually sold as Florisil, a trade name product of the Floridin Co., or silica gel. Reverse phase organic coated silica materials in solid phase extraction cartridges also may be used for sample clean-up and concentration. For example, clean-up of samples to be analyzed for PAH can be done using a silica gel column. The column contains about 10 g of silica gel with a 1–2 cm layer of anhydrous Na_2SO_4 on top to dry the sample. Figure 6.13 shows the clean-up column. The column is wetted with pentane and the concentrated extract in cyclohexane is added to the top of the column. An additional wash with pentane elutes many of the interfering species. Then 25 ml of a 2:3 (v/v) mixture of methylene chloride and pentane is used to elute the PAH. The cleaned up extract is then concentrated by evaporation and analyzed by GC or HPLC.

6.4 Extraction of metals from sample matrices

In the environment, metals are found in many forms, including hydrated ions, complexed ions, insoluble oxides, hydroxides or other insoluble salts. They may be lightly or strongly bound to the sample matrix. In either case, they need to be extracted from the matrix into a concentrated solution, so that the analysis can be done by atomic absorption/emission spectroscopy, ion chromatography, colorimetry, or electrochemical techniques. In a liquid matrix such as water, the metals are found dissolved in water or in the suspended particulate matter. In air, the metals are mainly found in the aerosols which can be collected on filter material.

6.4.1 *Acid digestion of samples for determination of metals*

The metals associated with particulates and solid samples are usually extracted using acids. Nitric acid is the most common acid because it can digest almost any metal and most nitrates are water soluble. HCl, H_2SO_4, $HClO_4$, HF, or H_2O_2 may also be used along with HNO_3, to obtain higher extraction yield and

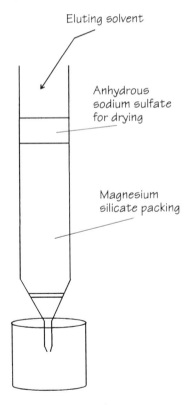

Eluting solvent

Anhydrous
sodium sulfate
for drying

Magnesium
silicate packing

Figure 6.13 Column chromatography for sample clean-up.

complete oxidation of organic compounds. Some of the other acids are more prone to cause interferences in AA analysis and may provide poorer matrices for electrothermal atomization.

Oxidation of organic matter is an important part of the digestion process. $HClO_4$, HF, and H_2O_2 are good for samples that are more difficult to oxidize. If a sample contains large amount of organic matter (e.g., sewage sludge) the sample may be dry ashed. In the dry ashing process, the sample is evaporated to dryness, then heated in a furnace at high temperature until it forms a white ash. The ash is extracted with HNO_3. Care should be taken that volatile metals are not lost during the ashing process.

6.4.2 Extraction procedures

Metals may be extracted by boiling the sample in a beaker with a few milliliters of HNO_3. The beaker is covered with a watch glass to condense some of the vapors and protect the sample from airborne dust. Alternately, the extraction may be carried out in a round bottom flask with a condenser to provide partial reflux of the vapors. The reflux minimizes sample loss in droplets formed during boiling. As mentioned before, other acids or H_2O_2 may be added to improve the extraction efficiency. In order to increase the concentration of the sample in the extract, the extract is evaporated to a final-volume of 1–5 ml.

6.4.3 Microwave digestion

Microwave digestion is often used in preference to heating in an open container. Microwave digesters are similar to the microwave ovens used for cooking, although they usually have a corrosion resistant interior surface. The digesters are equipped with a rotating turntable for homogeneous distribution of energy. The oven is programmable, so that times and power levels can be set before the extraction begins. To obtain good precision, the power delivered should be reproducible. The samples are put in Teflon vials capable of withstanding 100–150 psi of internal pressure. Each vial has a vent tube, often with a preset pressure relief valve, which is attached to a vent manifold, for removal of the acid vapors vented when the vials are heated. The vent is led to a scrubbing system to remove the acid fumes before venting the gases. The sample vials need to be thoroughly washed in acid after each extraction to prevent sample carryover.

The advantages over heating in a beaker are obvious. There is less chance of sample contamination, the samples are digested in a more reproducible, and more rapid fashion, and the acid fumes are contained, and do not pollute the laboratory air, or corrode the hood ducting.

6.4.4 Ultrasonic extraction

When a fairly mild extraction process is required for extraction of metals or salts from solid samples, ultrasonic extraction may be used. The energy transmitted through the bath water into the vials disrupts sample particles and ensures good contact between the particles and the solvent. An ultrasonic probe may also be used. This probe is inserted directly into the sample, providing somewhat more ultrasonic energy to the solution.

Since the samples cannot be heated strongly, this method is less aggressive than some others. However, it is satisfactory for the extraction of readily soluble salts, such as might be found on filters used to collect particulates from air. Metals present in solid samples in a very soluble leachable form can also be extracted by this method.

6.4.5 *Organic extraction of metals*

This type of extraction is carried out for recovery of dissolved metals from water. Ionic species, including metal ions are practically insoluble in organic solvents. When a complex of the metal ion with an organic moiety can be formed, then the metal can be extracted from the aqueous phase into an organic solvent. This can be achieved either by formation of metal chelates or ion pairs.

6.4.5.1 *Formation of metal chelates.*   Chelation is by far the most common extraction technique for metals. An organic compound which has two or more groups which can attach to a metal ion is called a chelating agent. A complex formed between a metal and a chelating agent is called a chelate. These are generally hydrophobic in nature and soluble in organic solvents. Some common chelating agents used to extract metals from water samples are shown in Figure 6.14. Methyl isobutyl ketone (MIBK) is usually used as the extracting solvent.

The partition coefficient of the metal complex into an organic solvent such as chloroform or MIBK is quite high and it can be separated by liquid–liquid extraction. The chelating agent and the organic solvent are added to the aqueous sample and shaken together. The chelate formed partitions into the organic phase. Four different equilibria are involved in this extraction, as depicted in Figure 6.15. The chelating agent, *HA*, is a weak acid which dissociates in the aqueous phase:

$$HA \Leftrightarrow H^+ + A^- \tag{6.11}$$

$$K_1 = \frac{[H^+]_{aq}[A^-]_{aq}}{[HA]_{aq}} \tag{6.12}$$

HA also forms a complex with the metal ion M^{n+}:

$$nA^- + M^{n+} \Leftrightarrow MA_n \tag{6.13}$$

$$K_2 = \frac{[MA_n]}{[M^{n+}][A^-]^n} \tag{6.14}$$

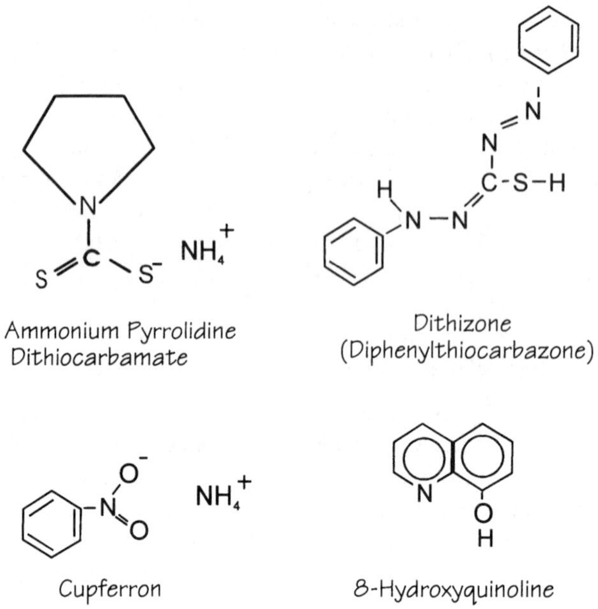

NH$_4^+$
Ammonium Pyrrolidine Dithiocarbamate
Dithizone (Diphenylthiocarbazone)
NH$_4^+$
Cupferron
8-Hydroxyquinoline

Figure 6.14 Some common complexing agents used for concentrating and extracting many metals from aqueous solution.

The complex partitions between the two phases:

$$(MA_n)_{aq} \Leftrightarrow (MA_n)_{org} \tag{6.15}$$

$$K_3 = [MA_n]_{org}/[MA_n]_{aq} \tag{6.16}$$

In addition, the undissociated chelate also partitions between the organic and aqueous phases

$$(HA)_{aq} \Leftrightarrow (HA)_{org} \tag{6.17}$$

$$K_4 = [HA]_{org}/[HA]_{aq} \tag{6.18}$$

Organic Layer
HA
MA
Aqueous Layer
K$_1$
K$_2$
HA ⇌ H$^+$ + A$^-$
+
M$^+$ ⇌ MA
K$_3$
K$_4$

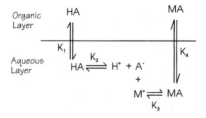

Figure 6.15 Equilibria existing when chelated metals are extracted into organic solvent. M^+ are the metal ions to be extracted and A^- the complexing anion.

The purpose of the procedure is selectively to extract MA_n into the organic phase. The distribution ratio, D, is defined as:

$$D = \frac{\text{Concentration of metal in organic phase}}{\text{Concentration of metal in aqueous phase}} \qquad (6.19)$$

$$D = \frac{[MA_n]_{org}}{[M^{+n}]_{aq} + [MA_n]_{aq}} \qquad (6.20)$$

Assuming that $[M^{+n}]_{aq} \gg [MA_n]_{aq}$ and making proper substitutions from Equations 6.11, 6.13, 6.15 and 6.17, it can be shown that

$$D = \frac{K_2 K_3 (K_1)^n [HA]_{org}^n}{(K_4)^n [H^+]_{aq}^n} \qquad (6.21)$$

If the amount of chelating agent in the organic phase is fixed, then

$$D \approx \frac{(\text{constant})}{[H^+]_{aq}^n} \qquad (6.22)$$

Therefore, D can be adjusted by controlling the pH at which the extraction is carried out. By changing the pH, not only is it possible to control the extraction efficiency, but also the selectivity. Figure 6.16 shows how the extraction efficiency of the dithizone chelates of tin and lead into organic solvent changes as the pH is varied. When the solution containing the chelated metals is extracted at a pH of 6, tin is quantitatively extracted, while the lead stays in the aqueous layer.

In a typical extraction, a measured amount of the complexing agent is added to the aqueous sample in a volumetric flask. The pH of the aqueous sample is adjusted for maximum extraction of the analyte of interest. Then 10 ml of MIBK is added (the volumetric ratio of sample to MIBK is usually less than 40) and the

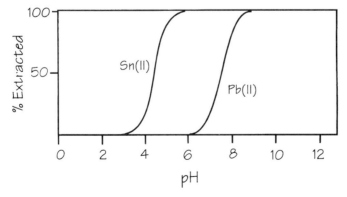

Figure 6.16 Selective extraction of dithizone chelates of two different metals at different solution pH. Extraction of tin can be done at pH6, while lead will stay in solution, as its complex does not form unless pH levels are higher.

mixture is shaken vigorously for 30 s. The metal chelate partitions into the organic phase which floats on the water. More water can be added (making sure that the pH does not change) to raise the organic level into the neck of the flask, so it can be directly aspirated into an AA for analysis.

6.5 Speciation of metals in environmental samples

Metals were among the first substances to be studied at the trace level in the environment. Large studies were done of such common pollutants as mercury and lead in water, soil, air particulates, and various living organisms. The determinations were relatively straightforward, with the sample being aggressively digested in strong acid and the analysis done by atomic absorption spectrometry. However, as further studies were undertaken, environmental scientists became aware that the impact of some metals was strongly related to their chemical form, rather than their total concentration. For instance, arsenic is generally toxic in both its As(III) arsenite, and As(V) arsenate forms, but is nontoxic in its organic forms, such as arsenocholine. Mercury, on the other hand, is toxic in all forms but is substantially more toxic as methyl mercury than it is in its inorganic compounds. Chromium in the Cr(III) oxidation state is less toxic than it is in the Cr(VI) state. Some of the many forms in which metals can exist in the environment are listed in Table 6.1.

Therefore, the total metal concentration is not always adequate to fully describe an environmental sample. The **speciation** of the metals present is required in some studies. This is a developing field, and presents difficulties to the analyst. First of all, the metal concentration present may be near the level of detection for the analysis. If this is further subdivided into several different species, greater analytical sensitivity will be required. Further, the different species are usually in equilibrium with each other in the sample. This requires less aggressive extraction processes, as the overall equilibrium should be disturbed as little as possible.

Speciation can be defined **functionally**, **operationally**, or **chemically**. A functional definition is one which specifies the type of role which the element may play in the environment. For instance, a functional definition might be "that mercury which can be taken up by plants". This definition is probably closest to

Table 6.1 Major chemical species of some metals found in the environment

Metal	Chemical forms
Mercury	Hg_2^+, Hg^{2+}, $HgOH^+$, CH_3Hg, $HgCl_4^{2-}$, $HgCl^-$
Arsenic	AsO_2^-, AsO_4^{3-}, $H_2AsO_3^-$
Chromium	$Cr(OH)^{2+}$, CrO_4^{2-}, CrO_3^{3-}
Lead	Pb^{+2}, $PbOH^+$, $Pb_4(OH)_4^{4+}$, Pb-organic complexes
Selenium	Se(IV), Se(VI)
Cobalt	Co^{2+}, Co^{3+}, $Co(OH)_3^-$

what the scientist wants to know, but is the most difficult to perform for the analytical chemist. Other than growing the plant in the contaminated water sample and analyzing the plant tissue, it is nearly impossible to carry out in practice.

An operational definition is considerably more practical. **Operationally defined species** are defined by the methods used to separate them from other forms of the same element which may be present. The physical or chemical procedure which isolates the particular set of metal species is used to define the set. "Metals extracted from soil with an acetate buffer" is an operational definition of a certain class. Finally, particular chemical species can be determined in some cases, as when arsenic content is separated into As(III), As(V), monomethyl arsonic acid, and dimethyl arsinic acid using ion exchange chromatography. If the metals present in a sample are to be separated into their different forms, the initial separations are often carried out during the sample preparation.

Sieving a soil or sediment will allow determination of lead or another metal in each particle size range, so that the distribution of the element can be determined. Species defined as "biologically active" such as free hydrated ions, may be separated from the bulk of a water sample by exposing the sample to an ion exchange resin or a chelating resin which will sorb only the species of interest. Then the sorbed species may be removed from the resin by elution with acid or may be determined by analysis of the resin. Even the distinction between the "soluble" and "insoluble" forms of an element in water can be considered a type of speciation. The separation of these species is carried out by passing the sample through a 0.45 μm pore size membrane filter. Both the filtered sample and the material retained on the filter can be analyzed.

Speciation of metal content in solids can be accomplished during the extraction process, by subjecting the sample to successive extractions with progressively more aggressive solvents, or by extracting different sub-portions of sample with the different solvents. The analysis of these subdivided samples will provide information on the most easily removed metal species, less available, and refractory metal content which is dissolved only by the strongest acid extractants. There are at least a dozen different published "speciation schemes" for metals in soils and sediments. Most include releasing metals from carbonates and hydrous oxides with acids, then an oxidation step to destroy organic materials and sulfides. However, some schemes put the oxidation step early in the scheme, on the theory that there may be an organic coating on the surface of the sample particles.

For instance, metals contained in a soil or sediment sample which are not bound in actual mineral fragments may be extracted with dilute hydrochloric acid or EDTA solutions. Metal ions sorbed on the surface of clay particles or held on ion exchange sites on the solid particles may be released by treatment with 1 M $MgCl_2$. To oxidize organics or sulfides, acidified hydrogen peroxide may be used, and to dissolve hydroxides of metals such as iron and manganese, buffers at specific pH values are used. The availability of the analytes for uptake by plants, for transport through the soil and for dissolution into water can be estimated from a well studied speciation scheme.

Suggested reading

Alegret, S. (1988) *Developments in Solvent Extraction*, Ellis Harwood, New York.
Anderson, R. (1978) *Sample Pretreatment and Separation*, Wiley, Chichester.
Luque de Castro, M.D. (1994) *Analytical Supercritical Fluid Extraction*, Springer-Verlag, New York.
McHugh, M.A. (1994) *Supercritical Extraction: Principal and Practice*, 2nd edition, Butterworth-Heinemann, Boston.

Study questions

1. Explain why two smaller volume liquid–liquid extractions may be more efficient at removing an organic compound from water than one larger volume one.
2. Why is a different apparatus needed for continuous extraction of a pesticide from water when the solvent is changed from methylene chloride to hexane?
3. What are some of the advantages of supercritical fluid extraction over liquid solvent extraction?
4. What are the two main purposes for the use of chelating agents in the preparation of aqueous samples for metal determinations?
5. A sample of water contaminated with pesticide is extracted with methylene chloride for determination of the pesticide. In experiments with a standard solution, it is found that when a 100 ml water sample is extracted with 30 ml of methylene chloride, 75% of the pesticide is removed. What fraction of the pesticide would be removed if three extractions of 10 ml of methylene chloride each were done instead?
6. Liquid–liquid extraction (LLE) has long been a standard method for removing and concentrating organic compounds such as pesticides from water. Newer methods such as solid phase extraction (SPE) are replacing LLE. Compare the two methods, discussing the advantages and disadvantages of each method.

7 Methods for air analysis

7.1 Keeping the air clean

Both natural sources and human activities put pollutants into the air. Not only do these contribute to an unhealthy atmosphere, but they can interact in the presence of ultraviolet radiation from the sun, forming ozone, oxidized organics such as formaldehyde and peroxyacetylnitrate (PAN), and acidic gases. The brownish photochemical smog which hangs over so many of our larger cities is one of the most obvious effects of air pollution. Dying evergreens in the Black Forest of Germany and on the hillsides in the eastern United States attest to the effect of acid rain, caused by the acidic components of air pollution washing down in precipitation.

Governmental regulation of air pollutants has taken many forms. Lists of pollutants are developed and specific emission standards for each of these are promulgated. Airborne contaminants in the workplace are controlled by both long term limits, to which workers may be exposed, as well as toxic short term limits which may not be exceeded.

The sources of air pollution may be divided into mobile sources, those related to transportation such as automobiles, trains, airplanes, and stationary sources such as pipelines, factories, and storage tanks. Mobile sources are controlled by automobile inspections and mandating such controls as catalytic tailpipe converters. The introduction of the catalytic converter in the US had a second environmental impact, besides the intended removal of unburned hydrocarbons from the exhaust. Since these catalysts were rapidly poisoned by lead, the addition of lead to gasoline was prohibited, and the amount of lead in airborne dust decreased dramatically.

Regulation of point sources of air pollution is complicated because industries cannot be easily compared to one another. While a coal fired power plant and a nuclear power plant both generate electricity, their emissions will be quite different. In addition, it is economically difficult to retrofit pollution controls to older plants. One tactic, taken by the US EPA, is to regulate new sources, so that each new facility must meet industry-wide standards.

Analysis is performed on many different sorts of gaseous samples and for many different reasons. Workplace air is monitored for both immediately dangerous concentrations of toxic gases and for lower concentrations of materials which may cause health effects because of chronic exposures. Outdoor air may be sampled to study the impact of local facilities such as industrial stacks, incinerators, landfills, or automobile traffic. Gaseous effluents from stacks also must be sampled to assure compliance with applicable regulations. For each of these atmospheres, sampling may include collection of solid or liquid aerosols, gases, and vapors. Sampling methods are constrained by practical matters. Electric power may be needed for running sampling pumps. If samples of air breathed by one person during a day or a work shift are to be taken, the size of the apparatus to be carried around is critical. Even the noise made by the sampler may be a limiting factor, if one expects to sample where people live or work.

Pollutants in air usually occur at very low concentration levels. Since air itself has a very low density, this means that the pollutants are present in an extremely diffuse state. If benzene is present at the level of 1 ppm (1 mg/l) in water, a 1-ml sample of the water will contain 1000 ng. One milliliter of air, also containing 1 ppm (1 ml/m^3) of benzene, holds only 3 ng of benzene. Therefore, most of the techniques specifically designed for air analysis have to do with concentrating the pollutants from a large volume into a much smaller quantity.

In air, pollutants may exist in several different forms. They are often present as gases. Compounds like nitrogen, carbon dioxide, or argon have boiling points which are much lower than the ambient temperature. Volatile compounds, normally liquid or solid, may be present in such low quantities that their partial pressures in the air are much lower than their vapor pressure. Therefore, they cannot condense, and remain as vapors.

Other materials exist in the atmosphere as fine solid particles. Particles which are smaller than 0.1 μm will not settle out, being kept in suspension by Brownian motion. Particles in the range of 0.1–10 μm will stay suspended for long periods of time. These particles contain high boiling organics, salts, particles of carbon, mineral particles, as well as materials of biological origin such as pollen grains, spores, and bacteria. High boiling liquids may also condense on the surface of such particles.

Finally, pollutants may exist in air as fine droplets of liquid. These mists are often formed when very water soluble gases such as sulfur trioxide or nitrogen dioxide dissolve in water droplets. Solid and liquid particles in air are encompassed in the term **aerosol**. Particles, both liquid and solid, of sizes below 10 μm are considered to have the largest impact on health, since they can be inhaled into the lung. Therefore, particles of this size come under air pollution regulatory control.

Some important air pollutants and their regulatory thresholds are listed in Table 7.1. These were the first group of compounds regulated by the US Government, but the list has been recently expanded to nearly 200 specific compounds including metals, acid gases, volatile and gaseous organics, especially halogenated ones, and pesticides.

7.2 Determination of gaseous species

Such gases as carbon dioxide, carbon monoxide, sulfur, and nitrogen oxides are products of combustion of various sorts, as well as emissions from other

Table 7.1 Ambient air pollutants

Compound	Regulatory concentration	Averaging period
Ozone	0.12 ppm	1 h
Carbon monoxide	35 ppm	1 h
	9 ppm	8 h
Nitrogen dioxide	0.05	1 year
Sulfur dioxide	0.14 ppm	24 h
	0.03 ppm	1 year
Particulate matter	150 $\mu g/m^3$	24 h
	50 $\mu g/m^3$	1 year
Sulfates	25 $\mu g/m^3$	24 h
Vinyl chloride	0.01 ppm	24 h
Lead	1.5 $\mu g/m^3$	3 months

processes, both anthropogenic and natural. These gases are present in the atmosphere in concentrations ranging from high ml/m^3 (ppm) to low $\mu l/m^3$ (ppb). They are considered to be pollutants, and may have serious environmental impacts. Even carbon dioxide, a naturally produced and fairly harmless gas, has been indicted as one of the most important greenhouse gases. These gases are determined using several different instrumental methods.

7.2.1 Carbon monoxide

Carbon monoxide, produced when carbon compounds are burned under conditions where a sufficient supply of oxygen is not present, is a toxic, odorless, and colorless gas. It is dangerous because it is not readily detected by the person inhaling it in an acute exposure. It has also been implicated in adverse health effects from chronic low level exposures. Its concentration can be measured by gas chromatography, by infrared spectroscopy and by the traditional volumetric adsorption method, known as the Orsat method.

7.2.1.1 *Carbon monoxide by nondispersive infrared absorption (NDIR).* When infrared radiation is passed through a sample of the gas to be analyzed, the total amount of energy absorbed can be compared with that passed through a reference gas cell. The difference in absorbance between the two cells is related directly to the concentration of CO in the sample, as described by Beer's Law. Carbon dioxide, water, methane, and ethane also absorb in the infrared, and can cause error. The reference may be a sealed cell containing a non-absorbing gas. In other cases the reference gas is the sample after passage through a catalyst which converts the CO to CO_2. This corrects for differences in the amount of interfering gases in the sample. Another method of reducing the effect of the interference is to place a cell which contains high concentrations of the interfering gases in the path of the light source. The wavelengths which would be absorbed by these species in the sample are effectively removed, so that the difference in absorption between the sample and reference paths is due only to CO. This is known as a **gas filter**. The discrimination against water vapor in a

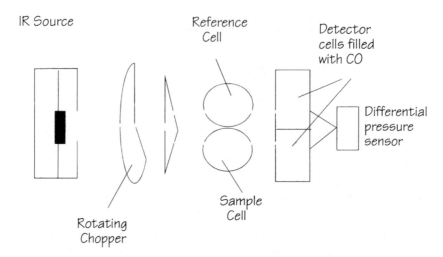

Figure 7.1 A non-dispersive IR device for measurement of CO. Transmitted radiation at the wavelength absorbed by CO raises the temperature of the fill gas in the detector cells, causing a pressure difference proportional to the relative concentrations of CO in the sample and reference cells.

gas filter NDIR analyzer is about 200 000:1, while for an analyzer using a flowing reference cell, it is about 50 000:1. Figure 7.1 shows the basic schematic of a nondispersive infrared analyzer.

The sensitivity of these instruments varies depending on the quality of the optical and electronic components. They are usually capable of detecting a concentration of CO between 0.5 and 0.1 ml/m^3. As we know from Beer's Law, an instrument which measures transmittance rather than absorbance will not give a linear response with concentration. Some instruments incorporate linearization circuitry in the electronics. Others are more accurate if calibrated against several standard gas mixtures, or if calibrated against a span gas which is about two-thirds of the expected maximum concentration.

Other factors affecting the precision and accuracy of these instruments are drift in the span and zero settings. This is instrument dependent and instruments which are more prone to drift will require more frequent recalibration. Loose dirt in the sample or on the windows of the cells can scatter light and lead to erroneous readings. Sampling at a rate higher than that which is recommended for the particular apparatus will cause an increase in pressure within the sample cell, and give high readings.

The advantage of these instruments is that they can be small and relatively uncomplicated, since they do not require a monochromator or wavelength selection. They are simple to operate, with little training needed for day-to-day operation. This will usually involve turning the power on, allowing the instrument to warm up until drift is minimized, introducing zero gas and adjusting the readout to zero. The span gas is introduced and the control is set to the corresponding reading. A more elaborate calibration, which is performed at longer intervals involves introduction of several standards of varying concentrations. The instrument readings are plotted against the concentrations of the standards and the resulting curve is used to correct the day-to-day readings.

7.2.1.2 *Carbon dioxide and carbon monoxide by gas chromatography.* Carbon dioxide may be separated from the other major components of air (O_2, N_2, CH_4, and CO) using a very polar GC column, packed with 30% hexamethylphosphoramide on 60/80 mesh Chromosorb P. The CO_2 is retained and elutes after all the other compounds which are unretained and elute together. A thermal conductivity detector is used to quantitate the sample. O_2, N_2, CH_4, and CO can be separated and determined using a molecular sieve 13-X column. This column, however, will irreversibly adsorb carbon dioxide. An apparatus using both these columns and a four element thermal conductivity detector can measure all these gases in a single analysis. A gas chromatograph which makes provision for two columns and is equipped with a thermal conductivity detector can be configured to do this analysis.

One milliliter of the air sample is injected using a gas sampling valve and the sample is dried by passage through a desiccant tube. The columns are configured so that the retention time of O_2 on the second column is sufficiently long to allow the CO_2 to elute from the first column before it. A complete analysis takes less than 10 min. The limits of detection of this method are between 250 and 500 ml/m^3, depending on the equipment.

Any compound which has a retention time near one of the components may cause an interference. Hydrocarbons heavier than methane may interfere with CO_2, but these are likely to be present at much lower concentrations than CO_2, in ambient air.

7.2.2 *Determination of sulfur species*

Oxides of sulfur are emitted to the atmosphere primarily by the burning of sulfur-containing fuels. Sulfur dioxide is the primary gaseous sulfur compound in the air, but hydrogen sulfide is also found. These can be determined by gas chromatography, with the sulfur specific, flame photometric detector being especially useful in this analysis.

These analyses can be carried out in several different ways, depending on the information desired and the time and apparatus available. A constant stream of the air to be tested can be fed to a FPD, and the response monitored continuously. The limit of detection for SO_2 in the continuous monitoring mode is about 2 μl/m^3. Because the response of this detector is not linear, and the sensitivity increases at higher concentrations, the addition of a constant amount of a stable sulfur compound to the fuel gas can improve the signal to noise ratio by as much as a factor of 10. There is the possibility of interference from hydrocarbons, arising from the CH band emission near 400 nm, which is within the bandpass of the optical filters in the FPD. Carbon dioxide is also an interference. Water vapor lowers the background current which may cause a problem in continuous monitoring. An important consideration in the analysis of these relatively reactive gases is that they can easily be lost on the surfaces of tubing or inlet systems. These should be made of silanized glass or fused silica tubing or Teflon, and be kept clean and dry.

To separate and determine individual sulfur gases, a GC column can be used with the FPD. This separates the compounds of interest from possible interference. The preferred column for this separation is constructed of specially acid washed Porapak Q packed in Teflon tubing. This will separate sulfur dioxide, hydrogen sulfide, carbonyl sulfide, and methyl mercaptan.

7.2.3 Determination of nitrogen oxides in air

Nitrogen oxides, emitted in high temperature, high pressure combustion processes where air is present, are major vehicular emissions and are important in smog and ozone formation. NO and NO_2 are measured using a chemiluminescence method. Nitric oxide (NO) and excess ozone mixed in a light-free chamber react according to the equation:

$$NO + O_3 \rightarrow NO_2^* + O_2 \qquad (7.1)$$

The excited NO_2^* species decays to the ground state, emitting light in a broad band with a peak near 1200 nm. The intensity of this light, measured by a photosensor, is proportional to the original NO concentration. To measure NO in the atmosphere, a sample of the air is mixed with ozone in the chamber.

NO_2 is measured indirectly. The air is passed through a catalytic converter at 200–400°C, and the NO_2 is converted into NO and O_2. The difference between the direct NO measurement and the measurement of both gases, using the converter, is the concentration of NO_2. As with the sulfur gases, the loss by adsorption on surfaces must be guarded against. All lines which contact the sample should be glass, Teflon, or stainless steel. Interference is rare, but any gas which can form NO in the converter can be an interference. Such compounds as peroxyacetylnitrate, amines and organic nitrates or nitrites may decompose to NO. However, these compounds are not usually present in high enough concentrations to pose a problem. Minimum detection levels of 0.01 ml/m^3 are achievable. The system is shown in Figure 7.2.

This system is calibrated against a compressed gas standard containing NO in an inert balance gas. The standard is attached to a manifold with a supply of zero air, which can be blended with the standard to produce lower concentrations. The zero air can be passed through an ozone generator, or bypass it. The NO analyzer can be calibrated by simply changing the relative flows of standard and zero air. Usually, standards are produced at the maximum expected concentration and evenly spaced over the span of concentrations to be measured. To determine the efficiency of the catalytic converter, a gas phase titration must be

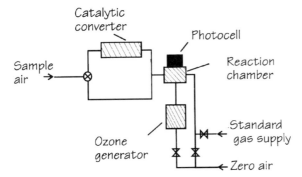

Figure 7.2 Apparatus for determination of NOx by chemiluminescence.

done. The NO concentration is set at a known, steady value by setting the flow of standard and diluent air. Then the ozone generator is turned on, and the ozone produced causes a drop in the NO concentration, with a theoretically equivalent amount of NO_2 being produced. This gas phase titration generates NO_2 in the same way as described in Equation 7.1, but the ozone is not in excess, as it is in the analyzer. The NO_x readings on the analyzer would remain constant, if the converter is 100% efficient in the conversion of NO_2 back to NO.

Example
An NO standard containing 100 ml/m^3 is to be used to calibrate an analyzer. If the flow from the standard tank is set to 200 ml/min, and the flow of zero air is 300 ml/min, what concentration of NO is produced?

$$\frac{\text{std flow} \times \text{std concn}}{\text{total flow}} = \text{concentration of mixture}$$

$$\frac{200 \text{ ml/min} \times 100 \text{ ml/m}^3}{200 \text{ ml/min} + 300 \text{ ml/min}} = 40 \text{ ml/m}^3$$

The analyzer gives a reading of 40.0 ml/m^3 for NO_x when the above standard is passed through it. When the ozone generator in the dilution system is turned on, the NO reading drops and the NO_x reading is 38.8 ml/m^3. What is the efficiency of the converter?

At 100% efficiency the NO_x readings would be the same, since the NO which reacts with the ozone is returned to its original form in the catalytic converter.

$$\% \text{ Efficiency} = \frac{[NO_x] \text{ with ozone}}{[NO_x]} \times 100$$

$$= \frac{38.8}{40.0} \times 100$$

$$\% \text{ Efficiency} = 97.0$$

(Note: an efficiency of less than 95% is an indication of instrumental malfunction and should be corrected before data collected are accepted.)

7.2.4 *Determination of ozone in air samples*

Ozone, produced in ambient air by reactions involving sunlight, nitrogen oxides, and hydrocarbons, is an indicator of other pollutants, and has deleterious effects on both health and materials. It can be measured by ultraviolet absorbance spectroscopy. Samples of air are passed through an absorbance cell, and the absorbance of light at a wavelength of 254 nm is measured. Then the sample stream is switched so that it passes through an ozone scrubber before entering the cell. The difference between the two readings is the absorbance due to ozone. The concentration is calculated from Beer's Law.

Interference is not usually a problem, but particulate matter entrained in the gas stream will scatter light and cause an error. This is avoided by passing the air through a Teflon filter before it enters the cell. The lower limit of detection is about 1 $\mu l/m^3$.

Standardization of the ozone monitor is somewhat different from other analyzers, since a span gas mixture is not available for ozone. The photometer is considered to be an absolute standard. The ozone scrubber can be tested with an ozone generator, to ensure that it is efficiently removing ozone.

7.2.5 *Determination of radon in air*

Radon, an inert radioactive gas, is a natural product of radioactive decay in rocks. It has a half life of 3.8 days and produces short lived decay products. It becomes an air pollution problem when it leaks from the soil or mineral-derived building materials such as cinder blocks or tiles, and collects inside dwellings. Radon-222 decays by emitting an alpha particle to form polonium 218, another alpha emitter, with a half-life of 3 min. A series of other daughter products are also formed, ending with a stable isotope of lead. The most lung damaging isotope, however, is the alpha emitting polonium-218, as it is usually associated with a solid particle, and therefore can deposit in the lung where it will eventually decay.

7.2.5.1 Sampling for radon. Radon is usually determined in indoor air, to determine the need for remediation. Since the emission of radon is variable, depending on the ventilation rates of the house, pressure differentials between inside and outside, and meteorological conditions, sampling methods which integrate the concentrations over a substantial period of time are preferred. Two types of passive collectors are most often used. The first is a small piece of special plastic film, which is sensitive to alpha radiation. The film is placed in a small container, which constitutes the test volume. A filter is often added across the mouth to prevent the entrance of daughter products, leaving only radon to emit alpha particles in the container. As the radon decays, the emitted alpha particles create damage tracks in the film. After exposure, the film is etched with a solution, usually NaOH or KOH, which dissolves the damaged areas, making them visible as small holes. Under a low power microscope, these are readily counted. The number of tracks is proportional to the concentration of radon, and the exposure time. These samplers are usually placed in the area to be tested for a period of 1 month to 1 year.

The second sampling method involves collection of radon on activated charcoal, for later analysis. The charcoal is contained in a canister which is opened and left in the test atmosphere for a period of 1–7 days. The radon diffuses into the charcoal and is trapped. To avoid desorption of trapped radon when the atmospheric concentration drops, a sampling time of 48 h is considered to be optimal. The sampler is then sealed and returned to the laboratory for analysis. Radon can also be trapped in an active sampling mode, by placing the charcoal in a tube and passing the air to be tested through it with a small vacuum pump.

The passive sampler can be fabricated from an 8 oz metal canister with a lid. The canister is filled with 50–100 g of activated charcoal, held in place by a metal screen. These canisters are commercially available from several sources. The active sampling system can be set up by placing 100 g of charcoal in a tube and passing the test air through it at a rate of 1–2 l/min. The sampling volume should not exceed 100 l.

7.2.5.2 *Analysis of samples for radon.*

Since radon is absorbed into pores in the charcoal grains, the alpha particles emitted from the decaying radon cannot be measured directly. Alpha particles have little penetrating power, so would not escape quantitatively from the charcoal granules. Therefore, the more penetrating gamma radiation emitted by the decay of several radon daughter products is measured. The sample is placed in a reproducible position in a gamma ray spectrometer, using a sodium iodide scintillation counter or a solid state germanium lithium detector.

In order to calculate the radon concentration in the air, the effective sampling rate of the passive sampler must be determined. This is done by exposing sampling canisters to known radon concentrations in a temperature and humidity controlled laboratory chamber. Four hours is allowed to elapse before counting the samples, to allow the adsorbed radon to equilibrate with its decay products, since the decay products are the species which are actually measured. The samples are counted and an effective rate of sampling is established. The following equation is used to establish the concentration of radon in the sampled atmosphere.

$$C_{Rn} = \frac{net\ cpm}{2.22\ E\ S_{eff}\ t_s\ \exp(-0.693t/5501)} \tag{7.2}$$

where:

- C_{Rn} is the radon concentration in the air in pCi/l
- *net cpm* is the counts/minute for the sample minus the counts/min in the background
- E is the detector counting efficiency in cpm/pCi
- S_{eff} is the effective radon absorption rate in l/min
- t_s is the sample exposure time in minutes
- t is the elapsed time between the midpoint of the sampling period and the time of sample counting

Calibration data is available for charcoal canisters, both from the EPA and from private environmental companies. The effective adsorption rate varies with humidity conditions, and can be corrected. The humidity is measured by weighing the canister to see how much water was absorbed. Then the effective sampling rate can be determined by reference to the appropriate data tables.

This method has a limit of detection of about 0.2–0.3 pCi/l, for the passively collected samples and 0.1 pCi/l for the active sampler. Water vapor, because it changes the effective sampling rate, may interfere, but this effect can be corrected for. Canisters should be dried before use, and stored airtight until used. Also, the desorption of radon during sampling may be a problem if long sampling times are used, and if the concentration of radon changes greatly during sampling.

7.3 Determination of volatile organic compounds

Volatile organic compounds in atmospheric samples include those compounds which have a sufficient vapor pressure at ambient temperatures to keep them in the vapor phase. Nonvolatile compounds have sufficiently high boiling points to condense on airborne particulate material or on surfaces, while the semivolatiles are those on the borderline of the two classes. Compounds which have boiling points well below ambient temperatures such as ethane, vinyl chloride, and acetylene are included in this class, as well as those which are volatile liquids at room temperature, such as benzene, octane, or acetone. The class ranges up to liquids which boil in the vicinity of 250°C.

These compounds reach the atmosphere through evaporation of liquids and through the escape of gases from processes. They are also frequently generated as products of incomplete combustion (PIC) from the incineration of wastes or the burning of fuels. The analysis of the VOC in urban air usually reveals many components which can be traced to gasoline, as well as those compounds being emitted by other local sources. The compounds are found in urban air at levels in the low parts-per-billion, with peaks as high as 50 μl/m^3 for an individual compound near a local source.

The analysis of this class of compounds in atmospheric samples usually involves collection of the sample, concentrating the VOC from a large enough quantity of air to obtain enough material to measure, and, finally, separating and identifying the individual compounds. The separation and identification step is usually done by gas chromatography, often coupled with mass spectrometry, but the collection and concentration steps may be troublesome and can be approached in several different ways. The sample can be seriously biased if these steps are not carefully designed and carried out.

7.3.1 *Adsorbent trap sampling*

Adsorbent trapping is one method which concentrates the desired materials as the sample air is passed through a trap or cartridge containing an appropriate adsorbent. The VOC adsorb on the surface of the packing, and are thus separated from the bulk of the air. Often, such problem compounds as water and carbon dioxide are eliminated from the analysis because they are not trapped by the adsorbent. The cartridges or traps are then returned to the laboratory, and the analyte is removed either by raising the temperature of the trap to release the adsorbed compounds, or by washing them from the trap with an appropriate solvent. There are both advantages and disadvantages to this method.

First, an adsorbent must be chosen. The compounds to be trapped must be considered, and a material chosen which will adsorb them efficiently at the temperature under which sampling will be done. However, the adsorbent cannot hold them so tenaciously that they are difficult to remove in the desorption step. Also, if the trapping material is selected to eliminate most of the water from the sample, then a material which has little affinity for polar compounds will be chosen. This material will, however, not trap such compounds as alcohols efficiently. It is important that the trapping material be selected carefully, with the properties of the compounds of interest fully in mind, to avoid both errors, that of trapping the compounds so strongly that they cannot be quantitatively recovered, or not trapping them efficiently at all.

Breakthrough and analyte recovery are the two main concerns in air sampling with traps. Figure 7.3 shows the progress of the sample in the trap. As the

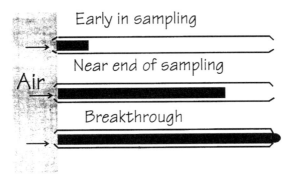

Figure 7.3 As contaminated air is sampled in a trap, the adsorbed contaminants move through the trap, and eventually break through.

sample front reaches the end of the trap, the sample begins to break through, and sampling is no longer quantitative. Since many compounds are present in the sample, there is a separation which begins to occur on the trap. The least strongly adsorbed will migrate through the trap most quickly, and will set the limit for the breakthrough volume, defined as the largest volume per gram of sorbent which can be sampled without significant loss of sample from the trap.

Adsorbent traps are available commercially, or can be packed in the laboratory. The configuration of the trap must be designed to be compatible with both the sampling pump system and the thermal desorber, if one is to be used. Traps are often designed with a backup portion of adsorbent to check for sample breakthrough. An activated charcoal trap is fabricated from glass tubing with a small amount of glass wool separating the larger front portion from the backup charcoal. The trap is sealed at both ends, and, the glass seals are snapped off each end before use. The trap is inserted into the fitting which attaches it to the pump, so that the air flow proceeds in the direction indicated by the arrow on the trap.

When no appropriate adsorbent can be found to reach the breakthrough volume needed, while still allowing the analytes to be desorbed, a layered adsorbent trap may be useful. A layered trap is shown in Figure 7.4, along with a charcoal trap. The trap contains a strong sorbent at the end, with layers of successively less retentive sorbents. The sample is drawn into the least retentive end. Readily trapped, difficult-to-desorb compounds are trapped in the front of the trap on the weaker sorbent. The compounds which are most likely to break through reach the stronger sorbent at the exit end. Table 7.2 shows some commonly used trapping materials and the compound types for which they are most suited.

To prepare adsorbent traps in the laboratory, the tubing to be used, glass or stainless steel, should be thoroughly cleaned with solvent, distilled water and purged with clean nitrogen or air while being heated. The packing is then weighed or measured out by volume to be sure that each trap contains the same amount of material. The end of the trap is plugged with a small amount of silanized glass wool, then the packing is poured into it. It is tapped or vibrated to settle the bed, and another plug of glass wool is inserted in the trap to contain the packing. Often, the traps are conditioned by heating under a gas flow. A blank can be done at this point, by desorbing and carrying one of the traps through the analytical procedure to determine if there are any compounds present which interfere with the compounds of interest.

The breakthrough volume for a variety of analytes is often supplied with adsorbent traps or packings by the manufacturer. It is a function of the type of sorbent as well as the analyte. This volume may vary from a few milliliters of sample per gram of sorbent to hundreds of liters. The breakthrough volume of methanol on a commercial carbon sorbent, Carbotrap, is 500 ml/g, while that for toluene is 650 l/g on the same sorbent. Of course, if the breakthrough volume is very high, it indicates that the analyte is tightly bound. In this case, recovery may be a problem, rather than breakthrough. Breakthrough volumes can be determined by packing a small amount of the sorbent into a small tube, and connecting this between the injection port and the detector of a GC. The analyte of interest is injected and the retention time determined. From this and the flow, the retention volume is determined, giving an approximation of the breakthrough volume.

For accurate work, the trapping adsorbent should be tested with the analytes of interest. This testing can be done by flushing a known volume of standard gas

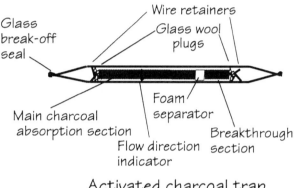

Activated charcoal trap

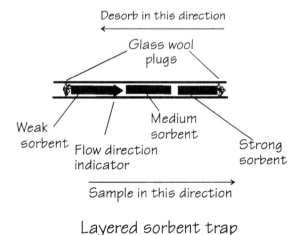

Layered sorbent trap

Figure 7.4 Sorbent traps for sampling VOCs in air. The front and back sections of the charcoal trap can be analyzed separately to check for breakthrough. For the layered traps, the proper direction of gas flow in sampling and thermal desorption must be observed.

Table 7.2 Some adsorbents for air sampling

Sorbent	Useful for	Desorption method
Tenax (polyphenylene oxide)	Nonpolar VOC with BP from approximately 40–200°C	Thermal desorption
Carbon molecular sieve	C2–C5 hydrocarbons	Thermal desorption
Activated charcoal	Low to medium boiling polar and nonpolar organics	Solvent extraction
Polyurethane foam	Polar and nonpolar semivolatile compounds	Solvent extraction
Activated silica	Amines and polar organics	Solvent extraction
Graphitized carbon	C4 to C12 hydrocarbons	Thermal desorption
	Heavy organics such as PCBs	Solvent desorption
XAD-2 resin	Semivolatiles, PAH	Solvent desorption

mixture into the trap. The trap can then be desorbed and the amount of analyte recovered compared to the initial quantity loaded into the trap. If a standard is available which has a composition similar to that expected in the sample the trap can be attached to the container of standard, with the effluent passing through a calibrated flow meter. The time and flow are used to calculate the total volume injected into the trap. However, if the samples are of low concentration, it may be expensive or difficult to obtain standards at parts per billion concentrations.

In this case, other techniques can be used. A dilution system can be purchased or assembled. This allows two streams of gas, a concentrated standard and a diluent gas, to be precisely metered into a mixing chamber. From this chamber the mixture flows into the trap to be tested. The standard dilution apparatus should be constructed so that the standard gas only passes over inert materials before it reaches the trap. For VOC, the materials of choice are stainless steel or fused silica, which have been treated to reduce surface activity. Flow meters, valves, and other parts of this apparatus should not contain organic polymers in the flow path as, these easily adsorb the VOC from the gas stream. The plumbing through which the VOC standard mixture passes should be heated to reduce surface adsorption. A temperature of 40–50°C is adequate. Too high a temperature may initiate undesirable wall reactions.

Alternatively, a small amount of a more concentrated standard may be measured into the trap and flushed into it using a carrier gas such as air or nitrogen. An apparatus for doing this can be assembled from a gas sampling valve placed in a heated enclosure. A small calibrated volume of 1–5 ml, attached to the valve, is used to measure out a known volume of standard at the temperature of the enclosure, and atmospheric pressure. When the valve is turned, the standard gas mixture is swept onto the trap. The purge gas is then allowed to flow until a volume of gas, similar to that of a sample, has been passed through the trap. The purge step is important. A small sample of gas may be readily trapped, but break through and be lost when more air is passed through the trap. Also, if a compound is strongly retained, purging it down into the trap will reveal if it cannot be readily recovered when it is adsorbed deeply in the trap, rather than at the end. Therefore, omitting the purge step will usually give unrealistically good recoveries, which cannot be matched when field samples are analyzed.

Example

A new adsorbent is being tested for collection of benzene from ambient air. Two milliliters of a 10 ml/m^3 standard is flushed into a trap containing the adsorbent. Ten liters of zero grade air are passed through the trap after the standard. When the trap is desorbed, and the chromatographic peak area of benzene is compared with that from a direct injection of 2 ml of the standard, it is found that 60% of the standard is recovered in the first trap. Repeating the experiment with a 6-l flush gives a 98% recovery.

From these experiments we conclude that the adsorbent has a break-through volume for benzene which is higher than 6 l and considerably lower than 10 l. Further experiments can be done to pinpoint an exact breakthrough volume, but, since changes in ambient temperature and humidity can affect the breakthrough volume, a safety factor should be incorporated in the sampling volume recommended.

The traps are used in the field with small low flow pumps. Usually no more than a liter a minute of air is pumped through the trap. Since both the trap and the pump can be made quite compact, these systems can be easily used for personal sampling as well as point sampling. Self contained pumps with rechargeable battery packs for power, an adjustable flow controller, and a holder for the trap are commercially available. Usual sampling volumes are in the range of 5–15 l, for most adsorbents.

In summary, the advantages of trap sampling are:

- Traps are small, and easy to transport.
- Traps can be designed to eliminate some interferents, such as water vapor, by not trapping them efficiently.
- The trap concentrates the sample in the process of obtaining the sample.
- Traps may be designed to be reused after cleaning or may be disposable.

Adsorbent traps, on the other hand, have certain disadvantages:

- They do not trap and release all species with equal efficiency, and can therefore cause the sample to be biased against very volatile or very non-volatile compounds.
- Traps can cause contamination of samples with artifacts, which may arise from the material itself or from the sample. The trap material, especially polymeric materials, may break down under desorbing conditions and release compounds which then appear to have been present in the sample. The trapped sample molecules may also decompose or react with each other, causing some species to be lost and others to be generated. Since the sample

molecules are concentrated and adsorbed on a surface, and, in addition, are being heated to high temperatures, reactions are possible which would not occur in the unconcentrated sample.

7.3.1.1 *Thermal desorption of VOC samples.* When adsorbent traps are brought to the lab for analysis, they are placed in a desorption apparatus to remove the analytes and inject them into the gas chromatograph. There are several commercially available instruments which are designed to remove the sample from the trap by heating and gas flushing, to condense the sample plug into a small volume, and to inject the plug onto the head of the column.

These instruments heat the adsorbent traps to a predetermined temperature, while passing an inert gas through the trap. The analytes are then treated in several different ways, depending on the application and the apparatus. The analytes are usually re-trapped in a small adsorbent trap, which may be cooled or not. Since this trap is considerably smaller than the original sampling trap, it can be rapidly heated, to inject a sharp plug of sample. When capillary GC columns are used for the analysis, an additional cryogenic focusing trap may be used after the sample has been desorbed and recondensed. The same effect can be achieved by starting the GC column at a reduced temperature, so that the beginning of the column serves as a focusing trap.

When desorbing samples from adsorbent traps, it is best to flush them with the flow in a reverse direction from that used for the sample collection. This allows the least mobile compounds, which were trapped at the entry end of the trap, to be desorbed without having to force them through the length of the trap. When layered traps are used, this becomes even more important, as the strongest adsorbent is located at the rear of the trap, so that only the most mobile compounds reach it. The lightest, least adsorbed compounds travel deepest into the trap, and return by the same path when they are desorbed. The most strongly adsorbed compounds are immobilized close to the entry point of the trap, and have the smallest distance to travel on desorption. A mark on the tube is essential to be sure the trap is installed with the gas flowing in the correct direction, both in the sampler and the desorption equipment. Figure 7.5 shows schematics of a commercially available trap desorber/injector system.

7.3.1.2 *Solvent desorption of VOC samples.* When sample compounds are too thermally unstable, or are too strongly adsorbed to the surface of the trap material, solvent desorption is used. This is a simpler and less expensive technique than thermal desorption, and is often used when the compounds to be measured are present in high concentrations in the sample. Activated charcoal is widely used for sampling VOC in industrial atmospheres, because it is a good trapping agent for a wide variety of sample compounds, and it is usually desorbed with solvent. The solvent used to desorb samples dilutes the sample greatly, and usually causes a large solvent peak in the analysis, which often

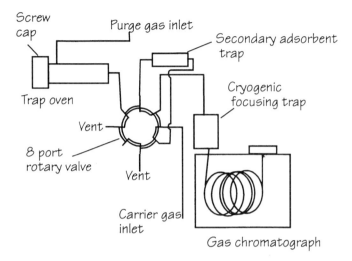

Figure 7.5 Apparatus for desorbing VOC from an adsorbent trap. The VOC desorbed from the heated trap are collected in the secondary trap, which is rapidly heated to inject the analyte into the GC. A cryogenic trap is used just before the column to focus the analyte into a sharp band for injection.

interferes with the determination of a certain range of possible analytes. Therefore, solvent desorption is usually used only for high concentrations of VOC, where the number of compounds to be determined is small.

The adsorbent is emptied from the tube into a vial containing a measured amount of the solvent. Carbon disulfide is often used to desorb charcoal tubes which will be analyzed by GC/FID. This solvent is quite toxic, and must be used in a well ventilated hood. However, the FID detector is rather unresponsive toward it, so that the interference by the solvent peak is minimized. Since there is often a considerable amount of heat generated when the charcoal is added to the solvent, the solvent is often chilled before the adsorbent is added to it. The calibration mixtures are prepared in carbon disulfide and samples and standards are injected into the GC using a syringe.

7.3.2 Whole air sampling

Samples of air can be collected in the field in containers, and these can be returned to the laboratory for analysis. The primary requirement for an air sampling container is that it must not either add or remove analytes from the sample. For low level ambient air samples, the most commonly used containers are low pressure stainless steel canisters, with an internal surface which has been electrolytically polished, or lined with silica, to deactivate it. These canisters are manufactured in many sizes, with the 6-l size being a reasonable compromise between holding enough sample so that several analyses may be done, and an

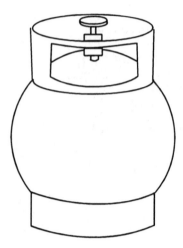

Figure 7.6 A stainless steel air sampling canister.

overall size which is not too difficult to transport. The 6-l canister has the general shape and size of a basketball, and is shown in Figure 7.6.

These canisters are filled by allowing air to flow into the evacuated sample container. Air can also be pumped into the canister, in which case it is possible to fill it to its maximum pressure rating of about 40 psig. If a pump is used, it must not contaminate the air sample or have materials which might absorb the analytes from the sample. A pump which uses stainless steel bellows to fill the canister is usually used. Sometimes, a sample which averages the contents of the air over an extended period of time is required. In this case, the flow into the canister must be controlled, otherwise it will gradually decrease as the pressure in the canister increases. The bellows pump forces the same volume of air into the canister with each stroke, as long as the canister is at a substantially lower pressure than that inside the compressed bellows. It will collect a valid sample up to at least 2 atmospheres absolute pressure.

If the sample is being collected into an evacuated canister without a pump, a small orifice or an electronic flow controller, can be incorporated in the inlet line, to allow the sample to flow in over a period of time. However, the flow decreases once the pressure difference between the canister and the atmosphere drops to the point at which the orifice or controller can no longer operate properly. The amount of sample which can be collected is less than half that which can be pumped into the canister. Finally, to analyze a sample held in a container which is filled to below ambient pressure requires specialized techniques in the laboratory. The pumpless method is better suited to grab samples rather than long-time integrated samples.

Canisters are cleaned for use by being filled with humidified air or nitrogen, heated moderately, and evacuated. The process is repeated two or three times, then the canister is filled with clean air or nitrogen, and a sample is drawn out and analyzed as a blank, to ensure that the canister is clean. The purpose of humidification is to displace contaminants from the walls of the canister with water molecules, so that this process does not occur when the canister is filled with a humid air sample, thus contaminating the sample.

Advantages of canister collected samples are that they are whole samples, with less chance of compounds being removed or added to the sample than is the case with adsorbent trapped samples. The chance of formation of artifacts is likewise less, since the sample is not heated, and the reactive materials are not concentrated while the sample is awaiting analysis. Disadvantages are those of convenience and expense. Canisters are bulky, with no more than two fitting easily into a box for mailing, or easy handling by a single person. They are also quite expensive, compared to traps, and must be cleaned after each use. Disposable stainless steel canisters are not economically practical.

There are other containers used in a similar fashion to the internally polished canisters. These are primarily polymeric bags, usually made of an aluminized film, which have the advantage of being relatively inexpensive and not bulky, at least before they are filled. However, the adsorption of VOC on their interior surface is a distinct possibility, and they should be carefully tested with the target compounds before being used for VOC sampling. The bags are usually filled by inserting them into an airtight box, with the bag open to the air. The box is sealed around the neck of the bag, and a simple pump is used to evacuate the box. This causes air to flow into the bag until it is filled. These bags are usually considered to be disposable after use, although they can be cleaned by flushing in some cases.

7.3.2.1 Concentration of analytes from whole air samples. Whole air samples must be concentrated before being injected into the column. A system for carrying out the concentration and injection is shown in Figure 7.7. The presence of moisture in the sample causes a good deal of difficulty in this step. The sample is passed through the cold trap, where an excess of water can plug the trap.

There are two approaches to dealing with the water. A dryer can be placed before the cold trap. The most commonly used dryer is formed from Nafion semipermeable membrane tubes, with the sample passing through the tube, and a counterflow of dry nitrogen in the annular space around the outside or the tube. The dryers must be maintained by heating and flushing with dry nitrogen between uses, to prevent carryover from previous samples. Alternatively, the water can be handled by passing the sample through a cooler area to condense the excess water vapor before the trap, so that the water is transferred slowly to the cryogenic trap. In that case the water freezes on the walls of the trap, without plugging it.

The amount of air from which the sample is condensed can be measured by allowing it to flow into an evacuated, calibrated ballast volume, as it exits the cold trap. This volume is equipped with a sensitive pressure gauge, and often,

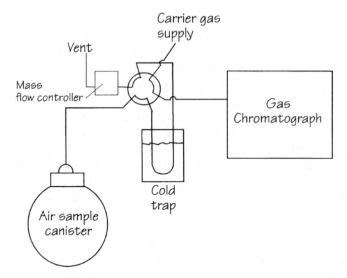

Figure 7.7 System for determining air volatiles, sampled in a stainless steel canister. The mass flow controller is used to measure the volume of sample injected, and the cryogenic trap concentrates the VOC. Then the trap is heated, injecting them into the GC column.

with a flow controller. The flow controller keeps the sample from passing too quickly through the cold trap, and the pressure gauge is used to determine the final pressure in the ballast volume. The volume of air sampled is calculated from the known volume, the temperature and the pressure read from the pressure gauge. Since the pressure is usually on the order of 2–5 psi, the ideal gas law is adequate for the calculation of the sample volume. An alternate method is to allow the effluent from the cold trap to pass through an accurate mass flow meter, and to read the volume from this meter. The first method is best for samples collected in canisters, which are filled only up to atmospheric pressure. The presence of a vacuum in the ballast volume allows samples to be drawn from canisters containing less than one atmosphere pressure. The second method requires a sample which is above ambient pressure, unless a pump is used to draw the sample through the flowmeter.

Once the sample is in the cold trap, the trap is heated, a valve is switched, and the sample is swept into the GC column with a stream of carrier gas. When a capillary column is used, a second cryogenic trap, located at the head of the column, is often used to ensure the narrowest injection plug is placed on the column head. Several commercial sample concentrators are now available.

7.3.3 GC analysis of VOC

The samples can be analyzed using either packed or, more frequently, capillary columns. EPA and other regulatory agencies have published methods, recommending particular columns for each class of VOC. Packed columns are somewhat easier to use, with larger sample capacities, but are not suitable for the analysis of ambient samples, where a very large number of compounds at similar levels are expected. Industrial hygiene samples, on the other hand, usually have a few compounds present at levels much higher than the background, so the separating power of a capillary column is less important.

For general purposes, a high resolution capillary column, coated with a nonpolar stationary phase, such as methyl silicone, is suitable. While a 0.25 mm bore gives a high resolution column, the stationary phase should be as thick as is available. The small loss of resolution caused by a thick coating is compensated by the improved column capacity, and especially by the increased retention of the early-eluting peaks. Ambient air samples usually contain many peaks which elute early, and are hard to separate.

Thermally desorbed samples are usually introduced into the GC automatically by the desorption apparatus. Solvent desorbed samples are injected with a syringe, the solvent flush method of injection being recommended. A volume of 1–5 µl is injected into a packed column, while a capillary column requires an injection splitter.

The column temperature is usually programmed from room temperature or from a subambient temperature to about 200°C. The detectors used are flame ionization, electron capture, and, most often, mass spectrometry. The effluent may be split between two or more detectors, or detectors may be used sequentially. The sample was split between FID and ECD detectors. When mass spectrometry is not being used, a combination of detectors makes it easier to identify peaks with certainty.

Example determination: VOC in ambient air by trapping and thermal desorption

The traps used are filled with the porous polymer, Tenax, and are purged with nitrogen at 200°C for several hours before use. A blank is run on at least one trap from each batch cleaned to ensure that the traps are clean before sampling. The samples are taken in an urban area, using a small suction pump with air drawn through the trap at 5–10 ml/min for 24 h. They are analyzed within 2 days of collection. The traps are heated to 250°C for 12 min, with a flow of carrier gas passing through. The analytes are refocused by trapping them on a secondary cold trap at −150°C. The focusing trap is heated rapidly to inject the compounds into the chromatograph. Separation is done on a 50 meter 0.03 mm i.d. methyl silicone coated column, with a temperature program running from 30 to 160°C in 25 min. The column effluent is split to two detectors, FID for general detection, and ECD, for specific detection of halogenated species.

Compounds are identified by comparing the retention times with standards. The response on the ECD is also used as an aid to compound identification. The sample is analyzed for a group of target compounds present in the standard gas mixture. Some samples are run by GC/MS under the same conditions to further confirm identities. The chromatogram obtained for this sample of urban ambient air is shown in Figure 7.8.

7.4 Other methods for analysis of volatiles and gases

7.4.1 *Determination of non-methane organic carbon*

To obtain a rapid and less expensive estimation of the total amount of hydrocarbon and other organic vapors in the air, a total organic carbon determination may be done. The method is designed to eliminate methane, since this gas is present in much higher concentration than the other compounds, and is not very reactive in the atmosphere. This is referred to as the measurement of non-methane organic carbon (NMOC). It is defined as the total amount of carbon regardless of the structure and functional groups in the molecules. Samples can be determined directly from the ambient atmosphere in a field installed instrument or collected in a bag or a canister and returned to the laboratory for analysis.

A NMOC analyzer is designed to produce an equal response for each carbon atom. A schematic diagram of the system is shown in Figure 7.9. The measured

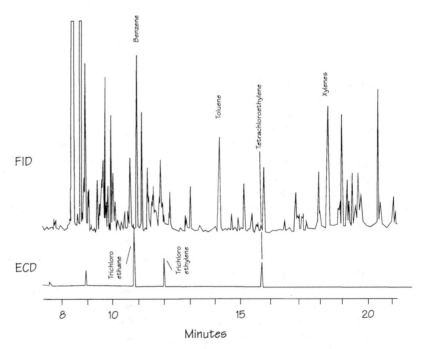

Figure 7.8 A chromatogram of VOC trapped from 1 l of urban air, separated on a methyl silicone capillary column, with ECD and FID detection.

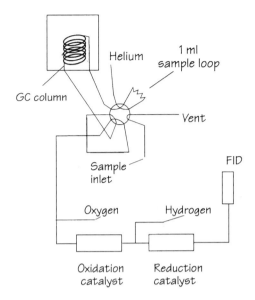

Figure 7.9 Apparatus for determination of NMOC. NMOC are separated from methane, carbon dioxide, air and other gases in the column, and are then backflushed into the detection system. To ensure that all gases give the same response, the sample is first oxidized to CO_2, then reduced to methane which is detected by a FID.

volume of sample is passed into a GC column, and the methane and carbon dioxide are allowed to pass through. Then the column is backflushed into the detection system without further separation. To eliminate the variation in response which the different organic compounds present may give, the sample is passed through an oxidation catalyst with some oxygen added to the stream. This produces a mole of carbon dioxide for each mole of carbon in the original sample. Some hydrogen is then added and the sample is passed through a reducing catalyst, producing methane. The methane is detected and quantitated by a flame ionization detector, and all the carbon present in the original sample gives the same response. The response is shown in Figure 7.10. For measurement of the higher levels of NMOC in stack samples, a preconcentration step is not usually necessary. A 1-ml sample is measured in a sampling loop and injected into the analytical system.

7.4.2 Annular denuder methods for air analysis

The annular denuder is an apparatus which uses a series of concentric glass tubes, coated on the inside surface with specific reagents. Air is passed through the series of denuders, and the acidic and basic gases and particulate matter are collected on the internal surfaces. The system was developed for the determination of reactive acidic and basic materials, including gaseous SO_2, HNO_3, NH_3, and particulate NO_3^-, SO_4^{2-}, NH_4^+, and H^+.

Sulfur dioxide and nitric acid vapors are trapped in the first denuder, coated with NaCl. The second denuder, coated with sodium carbonate solution, traps other acid gases, HCl and HNO_2, as well as any SO_2 or HNO_3 which was not absorbed in the first section. The third section, again coated with sodium carbonate, removes any remaining acid species, and the final section, coated with citric acid, removes ammonia. A final filter collects any particulate matter which was not deposited in the denuder.

Air is passed through the denuder stack, and then the system is disassembled in the laboratory. The individual sections of the denuder are rinsed and the solutions are analyzed by appropriate methods for the individual species. Ion chromatography is the most versatile and commonly used analytical tool for the analysis of denuder-collected samples. Figure 7.11 shows a denuder setup and the analysis of the first three denuder stages for nitrate, nitrite, and sulfate. The nitrate ion is found in all three of the stages, with only a trace in the third stage, while the sulfate is trapped primarily in the second stage. All three of these species are also found in the final filter stage, where they were deposited as particulate material, rather than as gases.

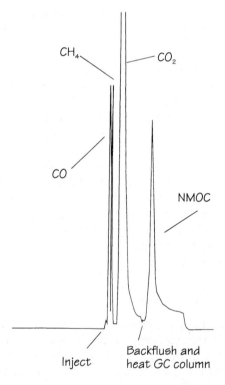

Figure 7.10 The output of a NMOC analyzer showing the gases which are separated by the column and the NMOC compounds which are backflushed from the column and converted to methane.

7.4.3 *Impinger methods for air sampling*

There are a wide variety of analyses of gaseous samples which begin by collecting a specific chemical species or class of materials in a liquid reagent solution. This solution is designed to trap the analyte efficiently. The solution is then analyzed in the laboratory. The apparatus used to collect such samples is called a bubbler or impinger. Figure 7.12 shows examples of impingers. The critical points in the design of absorbers are the bubble formation device, which may be either a glass frit or a fine orifice, and the depth of the solution through which the bubbles pass. The smaller the bubbles and the longer their path through the solution, the more efficient the contact between gas and liquid will be, giving adequate opportunity for the analyte to pass from the gas into the solution.

While glass frits tend to produce fine bubbles and good contact between gas and liquid, they are not always the best choice. Some very reactive gases, such as ozone, may be lost on the large surface area of the frit. Also, the frit may be difficult to clean and may be a source of intersample contamination. When a fine tube tip is used to produce the bubbles, the spacing from the end of the tip to the bottom of the container is critical. The bubbles should strike the bottom of the

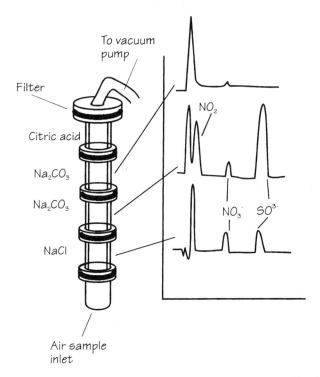

Figure 7.11 An annular denuder stack and the ion chromatograms generated by the first three levels.

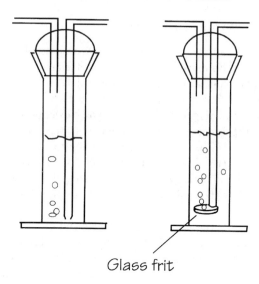

Glass frit

Figure 7.12 Impingers. One has an open tube, the other a glass frit. The fritted disk gives smaller, more efficient bubbles, but may be difficult to clean and may adsorb some of the sample in the frit.

container, so that they are further broken up before rising through the solution. It is important to control the flow rate, and the optimum rate is quite dependent on the particular analyte and the rate at which it dissolves.

Often, a train of bubblers and other absorbent devices is used for sampling. These may serve as sequential absorbers for different types of gases, or a second absorber may simply be present to check for breakthrough.

When the impingers are returned to the laboratory, the appropriate analytical techniques are used to complete the analysis. This may include chromatographic or spectrometric methods. Many of these are determined by colorimetric reactions, by adding a specific reagent and measuring a color formed. These colorimetric methods are generally quite selective to a compound or group of compounds. Table 7.3 lists some impinger-colorimetric methods.

7.4.3.1 Colorimetric indicating tubes for air pollutants. A fast and relatively simple method for measurement of many gaseous air pollutants is the indicating tube. These are marketed under the trade names of Draeger or Kittagawa tubes. They are composed of a reagent which produces a color with the particular analyte being sought, coated onto a granular support material in a small, sealed glass adsorption tube. The tube is used by snapping off the seals and installing it in a hand operated pump. A preset volume of air is drawn through the pump, and the length of the color change in the tube is related to the concentration.

The specificity and resistance to interference of these tubes is quite good, because the color reagents are usually very specific. If a suitable reagent is not available, the tube may contain a pretreatment section which screens out the interferents. If there is no appropriate reagent to form a color, the pretreatment section may react with the analyte and produce a product which can be used for the color reaction in the second section. In general, for these tubes, the sensitivity is not as high as can be obtained from more traditional instruments. However, for a rapid screening, the method is inexpensive, and requires no skilled personnel or laboratory time to achieve an estimate of the contamination. Of course, the contaminant being sought must be known beforehand, as the correct tube for each material must be used. Table 7.4 shows some of the tubes available, their range of application, and common interference.

It should be noted that most of these tubes react to a class of compounds rather than a single compound. The results must be looked at carefully, especially if several compounds of a similar nature are expected at substantial concentrations in the tested atmosphere.

7.5 Determination of pollutants in particulate material

The particulate material suspended in air contains several classes of pollutants. Metals, both free and in ionic form, and anions such as sulfate, chloride,

Table 7.3 Methods for analysis of impinger-collected samples

Compound	Absorbing solution	Analytical method	Range	Interferences
Ammonia	Dilute sulfuric acid	React with phenol to form blue indenophenol Colorimetric measurement	$20–700\ \mu m/m^3$ LOD $0.2\ \mu m/ml$ in solution	Some metal ions (EDTA prevents some interference)
Nitrogen dioxide	Triethanol amine, o-methyl-phenol, sodium metabisulfite	React with sulfanilamide and 8-anilino-1-naphthalenesulfonic acid – colorimetric measurement	$20–700\ \mu m/m^3$	HNO_2, N_2O_3
Sulfur dioxide	Sodium tetrachloromercurate	React with formaldehyde and pararosaniline Colorimetric measurement	$500\ ml/m^3$ to $10\ \mu l/m^3$	None (note high toxicity of abs. soln)
Phosgene	4-(4′-nitro-benzyl) pyridine in diethylphthalate	Colorimetric measurement	Down to $40\ \mu l/m^3$	Acid chlorides, high humidity
Chlorine	Methyl orange at pH3	Bleaching of the methyl orange is measured colorimetrically	Down to $1\ ml/m^3$ in air. $5–100\ \mu mol/$ 100 ml of soln	Free bromine, SO_2, NO_2 (used for Cl spills, in emergencies)
Ozone and oxidizers	Buffered KI soln	I_3 formed is measured colorimetrically	$0.01–10\ ml/m^3$	NO_2
Acrolein	4-hexyl-resorcinol in ethyl alcohol and trichloroacetic acid	Colorimetric measurement	Down to $0.01\ ml/m^3$ in air	Dienes (slight)

Table 7.4 Some colorimetric air sampling tubes

Compound	Range	Description	Interferences
Mercury vapor	0.1–2 mg/m³	Hg reacts with CuI reagent to give a yellow–orange complex	Cl₂ gives low readings
Carbon monoxide	5–150 ml/m³	CO reacts with iodine pentoxide giving I₂, and a preconditioning layer removes halogenated hydrocarbons, benzene, etc.	Acetylene will interfere
Ozone	0.05–1.4 ml/m³	Blue indigo dye is cleaved and bleached to white	Cl₂ and NO₂ when present above 5 ml/m³ will turn the indigo gray
Sulfur dioxide	0.5–5 ml/m³	SO₂ reacts with blue complex of I₂ and starch, changing to white	H₂S will make indicator gray
Tetrachloro-ethylene	5–50 ml/m³	First layer contains MnO₄₋ which cleaves the analyte forming Cl₂, which reacts in second layer with N,N'-diphenylbenzidine to give a gray-blue product	Free halogens, hydrogen halides and easily cleaved halocarbons

fluoride, and nitrate, comprise the major inorganic pollutants. The most important class of organic species in particulate materials are semivolatile organics, which condense on particles. Polynuclear aromatic hydrocarbons and their derivatives are an important class of these compounds, especially when the particulates arise from combustion.

Obtaining a representative sample of air with its suspended particles is not always a straightforward task. The particles have much more momentum than does the surrounding gas, if the sample is being taken from a moving air stream such as a stack or duct. When the gas stream changes direction, the particles tend to continue in a straight path, due to their greater momentum and inertia. To obtain a representative sample, the flow in the sampling inlet must have the same linear velocity as does the bulk of the gas, or the particles collected will not be representative of the size distribution in the atmosphere being sampled. **Isokinetic sampling** is the term used for sampling at the same flow velocity as the test atmosphere.

The sampling probe used for isokinetic sampling is a thin walled tube aligned so that the air flows directly into it. The streamlines of the flowing air will be bent either away from the opening, if the linear speed of the sample in the probe is slower than in the duct, or into the opening if it is faster. In the first case, particles having greater momentum than gas molecules will continue into the probe, while the gas goes around it, and in the second case, particles will pass by the sampling intake, while the gas is sucked into it. When the sample flow is slower than the system flow, the distribution of particle sizes in the sample is biased toward larger particles, while when it is faster, there are too few large particles in the sample. Figure 7.13 shows these effects. Since the smooth streamlines are disrupted by the probe itself, even a thin walled tube will cause some deviation from the ideal situation. The probe must be placed in a straight part of the duct, as far away from bends as possible, and one must also consider that the point in the duct selected for sampling may not be representative of the entire system. Often it is necessary to move the probe across the diameter of the duct, taking samples at different points to establish a particulate concentration profile. In addition to matching flow velocities, one must be careful that inlet tubes are clean, dry, and have no sharp bends. They should be as short as possible, since it is easy to lose particles on sticky surfaces or surfaces which have acquired a static charge.

In general, if samples are being collected from a relatively static atmosphere, as in ambient air sampling, the inlet is simply a hood over the filter, which eliminates the largest particles, without passing the air through any tubes or ducts at all. When particulate material suspended in air is to be collected, a high volume sampler is usually employed (Figure 7.14). This consists of a blower pump, capable of moving air through a filter at high speed. The particle size of the material to be sampled must be considered in both the selection of the type of filter and the inlet system. The samples of particulate from any atmosphere may be selectively collected, so that certain size particles may be preferentially selected. Often, inlets are designed to exclude particles too large to be inhaled. Others may be layered to collect a series of particle size fractions, or a very narrow range of particle sizes.

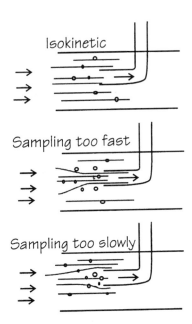

Figure 7.13 Isokinetic sampling. If the sample stream is faster than the stack flow, some particles are lost, as they do not follow the gas stream into the probe. If it is too slow, gas will flow around the probe while the particles in that gas will go into it.

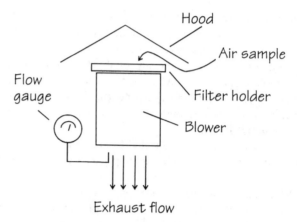

Figure 7.14 A high volume air particulate sampler. The filter is placed under the hood, and a high flow of air is drawn through it with a blower. The pressure at the exhaust is used to monitor the flow.

7.5.1 *Particle size fractionators*

The fact that changes in flow velocity and direction can discriminate toward or against certain sizes of particles can be used in a positive way to sort particles by size, in order to determine the particle size profile of a sample or to determine the differences in composition of various size fractions. By changing the diameter of the passage through which the gas stream passes, and forcing rapid changes in direction, particles of certain size ranges can be deposited. Figure 7.15 shows a cascade impactor which is used to separate particulate material into the desired size range fractions. For instance, the shelter hood over a high volume ambient air sampler is designed to exclude particles of diameters above 40 μm.

Sample inlets have been designed to mimic the human respiratory system, with each stage collecting particles which would be trapped in the nose, throat, bronchial tract, and lung. One type works by passing the air through perforated plates covered with filter paper. The holes in the perforated plates are offset, so that the air impinges on the filter paper below the hole, turns and proceeds through the next set of holes. Each plate has successively smaller holes, forcing the air stream to increase in linear velocity at each stage. Large particles are deposited readily in the first stage, with relatively mild impact, while the successively more vigorous impacts deposit smaller particles on each level. At the bottom of this fractionator, a filter removes the remaining particulates.

7.5.2 *Filters for air sampling*

Paper, glass fiber, and polymeric membrane or fiber filters are all employed in trapping particulates from gas streams. Beds of granular material and polymeric foam sponges are also used for special purposes. The most common for both ambient air and stack samples are glass or quartz fiber filters and paper filters. Glass and quartz filters have several advantages. They have good retention for particles in the diameter range from 0.035 to 1 μm, adsorb water to a lesser extent than paper filters, have good air flow permeability and will withstand high temperatures. They also has some less desirable characteristics. Glass fiber tends to have poorer mechanical properties than paper, being more likely to fray and crumble at the edges. It is also considerably more expensive than paper. Paper, however, is less uniform, making calibrations of samplers difficult, and leading to uncertainty about the size range of the collected particles.

Polymeric membrane filters have higher flow resistances, but have the advantage of a smoother surface and a very uniform pore size distribution. The surface makes them ideal for use when the particles are to be examined by microscope or when radionuclides are collected on the filter, because the sample remains on the surface, and is readily available for counting. Some of these membrane filters are also readily soluble in organic solvents, so the particulate can be separated from the filter.

Efficiency and suitability for collection of the particle size range desired is an important consideration in the choice of a filter. This includes concerns about the physical stability of the filter at the collection pressures and temperatures, as well as the behavior of the filter as it becomes loaded. Some filters rapidly

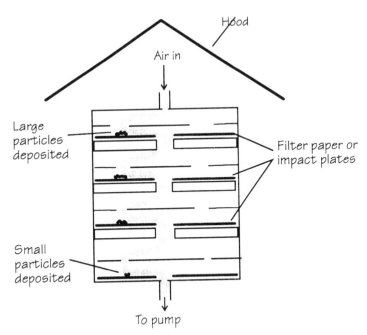

Figure 7.15 A cascade impactor to sort particles by size. As each level is passed, the gas goes through successively smaller holes, so is impacted upon the surface harder. Therefore, successively smaller particles are deposited in each level.

become resistant to flow as the particulate accumulates, while others are not as severely affected. The type of analysis to be done also must be considered. The filter material itself may contain impurities which can interfere in various ways with the analyses. These vary according to the analytical method. For instance, if a filter is to be digested in acid to dissolve metals from the collected sample, a glass fiber filter may be suitable. If, however, one was going to subject the sample, on the filter, to X-ray fluorescence analysis, the metals contained in the glass fibers themselves, which are not extractable with acid, could cause interference. In that case a paper or membrane filter, having a very low metal content, might be more suitable.

7.5.3 *Determination of total suspended particulates in air*

The change in the weight of the filter before and after sampling is used to determine the total amount of particulate matter collected. For this to be accurate, the filter must have the same moisture content during both weighings. There also must be no loss of filter material in handling or in mounting and demounting the filter in the sampler. Glass fiber filters are especially prone to fraying at the edges. The filter holder should be examined carefully to be sure that fibers are not left sticking to the filter sealing gaskets.

Moisture content before and after weighing is adjusted by holding the filters in a desiccator for 24 h before weighing, both when they are clean and after they have been returned from the field. Special analytical balances are available with extra wide cabinets, which will hold a single 8 × 10 in sheet of filter without folding. The filters can also be weighed in an ordinary analytical balance, by curling them gently into a U-shape and setting them on edge inside the pan of the analytical balance. Filters should be handled with plastic forceps or gloved hands as they are taken from the desiccator and weighed. After weighing, clean filters are placed in an envelope or wrapper, with the filter identification written on the wrapper. The filters must be weighed to the nearest 0.1 mg, since the mass of sample collected is generally in the range of milligrams.

7.5.3.1 *Division of filters.*

Before starting the analysis, the filter is usually divided into sections for replicates or to use for more than one type of determination. High volume samplers are large, rather noisy, and expensive. If several analyses must be done, it is usually easier to cut up a single filter, rather than taking several samples at the same time. Therefore, one must be careful that the sampler is well designed so that the sample is collected evenly over the entire surface. This can be checked by cutting pieces from different areas of a filter and subjecting these to the same analysis. If the variation between the resulting concentrations is greater than that obtained from, for example, replicates from a homogeneous solution, one can suspect that the sample is not being evenly distributed over the filter surface.

The division of the filter can be done most easily by using a template and a cutting wheel. A scissors with a Teflon covered blade is also often used, to avoid depositing metals from the blade on the filter. The sample-covered area of the filter must be carefully measured, as well as the area of the sample taken for analysis. The ratio of the sample portion to the whole filter is calculated, and this same fraction of the entire volume of air sampled is represented by the sample portion.

Example
A high volume sampler has a flow of 1.4 m^3/min and is allowed to run for 24 h. The area of the filter exposed to the air measures 18.0 × 23.5 cm^2. A strip 5.0 × 18.0 is cut from the filter and determined to contain 6.5 mg of sulfate. What is the sulfate concentration of the air in μg/m^3?

$$\frac{1.4 \text{ m}^3}{\text{min}} \times \frac{60 \text{ min}}{1 \text{ h}} \times \frac{24 \text{ h}}{} = 2016 \text{ m}^3$$

(air volume collected on the whole filter)

$$\frac{2016 \text{ m}^3}{} \times \frac{(18 \times 5 \text{ cm}^2)}{18 \times 23.5 \text{ cm}^2} = 428.9 \text{ m}^2$$

(volume collected on the analyzed strip)

$$\frac{6.5 \text{ mg}}{428.9 \text{ m}^3} \times \frac{1000 \text{ μg}}{1 \text{ mg}} = 15.1 \text{ μg/m}^3 \qquad \text{(sulfate concentration in the air)}$$

7.5.4 Determination of metals in airborne particulates

Most methods for analysis of particles collected from air begin by extracting the analyte from the filter with some appropriate solvent. In the case of metals, X-ray fluorescence can also be used, without removing the sample from the filter. Extraction for metals is usually done with a strongly oxidizing acid solution. A mixture of nitric and hydrochloric acids is often used, sometimes with hydrogen peroxide added, to help destroy any organic material which may be complexed with the metals. These mixtures should efficiently dissolve the metal ions and oxidize most of the organic matrix.

The extraction can be carried out using ultrasonic agitation of the filter in the acid, or heating the filter in a bath of acid (usually referred to as digestion). For ultrasonic extraction, the measured filter strip is rolled or accordion pleated and inserted into the bottom of a vial to which the acid solution is added. The filled and labeled vials are placed in a rack in the heated ultrasonic bath, usually at a temperature near 100°C. The water level in the bath should be above the filter papers, but well below the caps of the vials. After sonicating for the required time, the tubes are diluted and centrifuged to settle fibers loosened from the filters and any undissolved particulate. The clear supernatant solution is decanted for analysis.

If the sample is to be digested by heating, the filter portion is torn into small pieces and placed in a flask or beaker with the acid solution. The flask is allowed to boil gently on a hotplate, in a hood, until the extraction is complete. The solution is decanted through a filter, and the remaining filter fragments are washed a few times, with the washings added to the decanted solution.

A similar method uses a microwave oven, fitted with Teflon containers, for the digestion. This allows precise control of the heating temperature, since the containers have pressure control valves. The temperature rises until the solution vapor pressure reaches the set pressure. Then the excess steam vents and the pressure, as well as the temperature, is held at the set point. Since the extraction is done under pressure, a shorter time is usually required than would be true in an open beaker. An additional advantage of this system is that all the heat is generated within the solutions, so it is more energy efficient, and a system of vent tubes leading from the pressurized containers allows the acid fumes to be trapped and neutralized readily.

Once the metals from the air filter have been dissolved either by heating or by ultrasonic extraction, the extract is brought to a known volume and is analyzed by an appropriate technique. Atomic absorption spectroscopy is probably the most common method applied, but ion chromatography is also used.

A method which does not require digestion in acid or any removal of the particulate from the filter is X-ray fluorescence spectroscopy. Here, a section of the filter is cut and placed in a holder in the spectrometer, with the sample exposed to the exciting beam. While the sample preparation is very quick and easy in this method, the interelement interference can be substantial.

Example determination: lead and PAH in airborne particulates

The sample is taken on a filter which has been dried and weighed before being taken to the field. The high volume sampler exhaust pressure, and the time of running is recorded. The filter is removed, covered and returned to the laboratory in a cold chest. There it is again dried in a dessicator and weighed.

The filter is divided by cutting it with a Teflon coated blade, and the portion designated for PAH determination is placed in a Soxhlet extractor, where it is extracted with methylene chloride. An internal standard is added, and the extract is reduced in volume with a rotary evaporator then the extracted materials are further fractionated using a clean-up column. The final extract is evaporated to a small volume and injected into a C18 HPLC column. The separation is followed with a UV absorption and a fluorescence detector. The different response constants for the PAH on each of these detectors is an aid in identifying the compounds.

The portion of the filter for lead determination is placed in a sample container, nitric acid and hydrogen peroxide is added, and the container is placed in a microwave digester. After digestion, the sample is decanted through a glass fiber filter to remove particles. It is diluted to a known volume and run on a flame atomic absorption (AA) instrument. The instrument is calibrated against acidified aqueous solutions of lead.

7.5.5 *Determination of anions in airborne particulates*

Sulfate and nitrate salts in airborne particulates are an important end product of many acidic emissions. These and other anions can be extracted from the filter using a buffer solution, and ultrasonicating the filter as was done with the acid in the metals extraction. The analysis is done by ion chromatography, using either a suppressed or nonsuppressed column. The buffer used to extract the filter is often the same solution used as an eluent. The ions found in greatest quantity in urban air are chloride, sulfate, phosphate and nitrate.

7.5.6 *Determination of organic species in particulate material*

A portion of an air filter can be extracted to remove organic compounds for analysis. This is a much more complex mixture than the metals or anions. It consists of various high-boiling compounds such as polynuclear organic hydrocarbons (PAH) and phthalates.

These are extracted by sonication in solvent, by Soxhlet extraction, by accelerated solvent extraction, or by supercritical fluid extraction. The mixture is generally very complex and is usually subfractionated before analysis. This can be accomplished by extracting the filter with a series of solvents, beginning with a non-polar one, like octane, and continuing with methylene chloride, followed by acetone. For example, the first extraction takes off the non-polar unsubstituted PAH, while the PAH which have nitro or hydroxyl groups dissolve in the more polar solvents. Further fractionation can be done with column chromatography.

7.6 Field methods for air analysis

Since ambient air analyses require very sensitive measurements, there are few field-suitable devices. Portable gas chromatographs are useful for higher concentration analytes. These are available with FID detectors, but the requirement for flame gases limits their portability and the length of time that they can run without being refilled. Compressed gas tanks are both heavy and bulky. Gas chromatographs equipped with photoionization detectors are more useful, because the only consumable required is a battery. The columns are often designed to run at ambient temperatures, to keep power consumption down. When filtered ambient air is used as the carrier gas, only a small pump is needed to provide the pressure.

Another application of the PID is in an instrument designed to survey for higher organic compounds, ignoring the one or two carbon hydrocarbons. This is made in the form of a long wand, with the input at the tip, and a photoionization detector built into the handle. A small pump draws air from the tip through the detector, and a display on the handle indicates the concentration found. The detector is especially sensitive to aromatics. The device can be used to screen large areas, and detect spots which are emitting the largest concentrations of pollutants. The instrument is powered by a rechargeable battery. A useful application of such a device is in determining the effluent gases from a large area like a landfill. The PID is used to screen the area, and when the points of emission are found, canister or adsorbent trap samples are taken for lab analysis. The samples will have high concentrations, so that the laboratory analysis can readily identify the compounds being emitted. This saves doing laboratory analyses on many samples which would give much less information. The scans done with the PID device can be used to estimate the overall outgassing of the landfill, while the analyses of the "hot spots" will indicate the compounds being emitted.

For industrial hygiene applications, several instruments are available which will detect very high concentrations of flammable or toxic gases or levels of oxygen which are too low to support life. These are used generally to test atmospheres before workers enter confined spaces to ensure their safety. For detection of various levels of specific compounds in air, colorimetric detector tubes are useful field instruments.

7.7 Methods for stack monitoring

Special sampling trains are commonly used in stationary source or stack sampling, where several tests must be done. Stack sampling trains are usually designed to sample either particulates or gaseous emissions. A train for sampling stack gases is shown in Figure 7.16. It includes a Pitot tube for determination of the stack flow velocity, a filter to collect particulates, impingers cooled in an ice bath to reduce evaporation, a vacuum pump to draw the sample from the stack or duct, and a dry gas meter to determine the total volume. The flow must be adjusted to ensure isokinetic sampling.

This train is used for measurement of semivolatile compounds, such as PAH, and phenols. Particles are trapped in the initial filter. The sorbent trap is filled with a polymeric resin such as XAD-4. The impingers may contain water and base solutions to collect acidic gases for analysis by titration. The stack probe, inlet tube, and filter are heated to prevent the condensation of water before the impingers are reached. The empty impingers in the train prevent carryover of spray from one impinger to the next. A final impinger filled with a desiccant such as silica gel protects the pump and other apparatus from any remaining moisture. It should be remembered that stack gases are generally hot, acidic, and laden with moisture, so the train components must operate under these rather corrosive conditions.

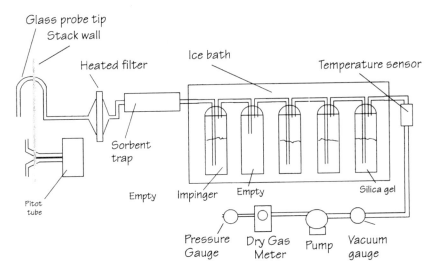

Figure 7.16 An impinger train designed to collect various species from a stack effluent. The impinger filling solutions are selected for the specific types of compounds being tested.

The filters, with the XAD-4 sorbent, are extracted using a Soxhlet apparatus. The semivolatiles in the extract are determined by GC or HPLC. For volatile organics, the trap will be filled with such sorbents as Tenax and charcoal, which will be thermally desorbed and analyzed by GC. For metals, the impingers are filled with reagents to trap the vaporized metals, and the solutions are analyzed by AA.

Suggested reading

Boleman, H.J.T. and Burn, J. (1993) *Chemistry and Analysis of Volatile Organic Compounds in the Environment*, Chapman and Hall, New York.

Lodge, J.P. Jr (ed.) (1989) *Methods of Air Sampling and Analysis*, 3rd edition, Lewis Publishers, Chelsea, MI.

Sigrist, M.W. (1994) *Air Monitoring by Spectroscopic Techniques*, Wiley, New York.

Torvela, H. (1994) *Measurements of Atmospheric Emissions*, Springer-Verlag, London.

Study questions

1. You have developed a material you think would be a good adsorbent to be used for trapping light polar vapors such as acetone and methanol from ambient air for analysis purposes. Describe the tests you would do to prove that it efficiently traps these compounds from air and releases them quantitatively, before you put it on sale.
2. A sample of particulate material is collected from air by using a filter attached to a vacuum pump. The flow through the filter is 16 cubic feet per minute. The sample is collected for 8 h, and the filter is digested in acid, and run by atomic absorbance spectroscopy to determine the lead

content. The acid from the filter is quantitatively transferred into a 50-ml volumetric flask and diluted to the mark. The standard is prepared by diluting 1.50 ml of a solution containing 200 ppm of lead to 250 ml. The AA shows the absorbance for the standard to be 0.234 and the sample to be 0.112. Calculate the amount of lead in the particulate sample in units of micrograms per cubic meter of air.

3. What are the main advantages of non-dispersive IR analyzers?

4. An analyzer gives a reading of 50 ppm NO_x when a NO standard is passed through it. What is the maximum decrease in the reading which can be accepted when ozone is added to the flow?

5. What is the effect of humidity in the determination of ambient radon, using a charcoal adsorbent. How is it corrected?

6. Compare the advantages and disadvantages of sorbent trap sampling with canister sampling.

7. Explain the importance of isokinetic sampling in collection of airborne particles.

8. An air particulate sample from a smelter facility is analyzed for cadmium by the following method. Air is drawn through a glass fiber filter. The filter is digested in strong acid and the resulting solution is diluted to a final volume of 100 ml. The sample is run on an AA spectrometer for cadmium. Data obtained are as follows:

Sampler flow	24 ft^3/h
Time of sampling	8.0 h
Concentration of standard	2.0 ppm
Absorbance reading for standard	0.234
Absorbance reading for sample	0.321

What is the concentration of cadmium in the air in $\mu g/m^3$?

9. How would you carry out sampling and analysis to determine the following?
   a. Continuous monitoring of NO_2 in a stack
   b. PAH in particulates from auto exhaust
   c. Gasoline vapors in a service station
   d. Contaminants in the air in a laboratory
   e. Ammonia in air near a fertilizer plant.

8 Methods for water analysis

Clean water is essential to human survival, and the earliest environmental regulations dealt with contamination of local water sources. Contamination of water resources has always occurred from a variety of anthropogenic activities, and, as population density increased, rivers, lakes and even oceans became contaminated. Direct discharge of industrial waste, domestic sewage, pesticide run-off from agricultural land, and acid rain all contributed to this contamination. Ground water may be polluted by landfill leachate, leaking storage tanks and pipelines, oil and gas wells, irrigation practices, waste disposal by direct underground injection, and other processes. Ground water and surface water can also contaminate each other, as they are not isolated.

Governments all over the world have regulations in place to protect water resources. Some of these laws control water quality directly, while others control the handling and disposal of chemicals and wastes. Some of the major regulations in the US are:

- **The Clean Water Act (CWA)** which controls the discharge of toxic materials into surface streams. Currently, 130 chemicals in 34 industrial categories are regulated.
- **The Safe Drinking Water Act (SDWA)** protects drinking water and ground water resources. Eighty-three contaminants, including microbiological and radiological substances, are regulated. The injection of liquid waste into underground wells is also controlled to prevent contamination of groundwater.
- **Resource Conservation and Recovery Act (RCRA)** protects ground water resources by creating a management system for hazardous waste, from the time it is generated until its proper disposal.
- **Toxic Substances Control Act (TSCA)** gives the EPA the authority to control the manufacture and disposal of toxic chemicals. For example, the manufacture of PCBs is banned.
- **Federal Insecticide, Fungicide and Rodenticide Act (FIFRA)** controls the manufacture and application of pesticides, fungicides and rodenticides. For example, insecticides such as DDT and Kepone are no longer permitted to be used.

In 1974 there was a flurry of public concern in the USA that many municipal water supplies contained a group of compounds which became known as THM, the trihalomethanes. These included chloroform, bromodichloromethane, dibromochloromethane, and bromoform, as well as a series of other light halogenated hydrocarbons. The outcry led to an extensive survey of the water supply nationwide. This study revealed the fact that the THM compounds were found more often in the finished waters than in the raw water. This led to the realization that the formation of many of these halogenated compounds took place during the chlorination procedure used for water disinfection. The study also showed that there were many other organic compounds present in the source waters, of which a substantial number obviously arose from industrial and other human activities. Some of these were not eliminated in the usual purification process. The impact of this study is still being felt, as new disinfection schemes, and more elaborate water treatments are developed to ensure safe drinking water.

Water analysis has had a long history. An early test for the suitability of drinking water was used in Istanbul in the seventeenth century. Weighed pieces of cotton were soaked in the water and dried. Those water samples which gave the least increase in weight were considered the best. Recent developments have made sophisticated analysis methods ever more important. Water is analyzed for a large number of possible substances, depending on its proposed use. Drinking water is probably the most thoroughly tested, with tests for pH, turbidity, metals, inorganic anions, and now, toxic organic materials such as pesticides and trihalomethanes, being done routinely. In addition, more intangible qualities such as odor and taste are determined. Wastewater is tested for its oxygen-consuming contents, as well as for salinity and various toxic materials. A sudden change in, for instance, the pH of the incoming water may have a serious impact on the bacterial cultures which are a vital part of the treatment process. Treated wastewater is tested for many of the same qualities, to ascertain that it is of adequate quality to be discharged into natural waterways. Testing is done on natural waters to show their suitability for various uses such as fishing and swimming, in addition to their use as feedwaters to public water systems. To give an example of the range of possible measurements, some of the standards set for regulating contaminants in drinking water in the US are listed in Table 8.1. The primary standards are for pollutants which are deleterious to public health, while secondary standards regulate such aesthetic qualities of water as color, odor, and taste.

A large number of different tests are needed to monitor such properties as color, turbidity, taste, odor, acidity, hardness, conductivity, solids, and oxygen demand as well as more specific chemical analyses for metals, inorganic non-metallic constituents, volatile hydrocarbons, trihalomethanes, volatile aromatics, base neutrals, pesticides, herbicides, and radioactivity. In addition, a range of biological examinations may be done on water samples, e.g., for algae, phyto-plankton, and various bacterial species.

8.1 Sample collection and preservation

Water samples, like other types of environmental samples, must be collected with both the purpose of the study and the requirements of the analysis method firmly in mind. It is not possible to give a set of specific directions for the collection of these samples, but some general guidelines can be set out. The specifics of a sample collection program should be determined by the analyst, taking into account all the factors which may affect the samples.

A few water characteristics are nearly impossible to preserve in a collected sample and must be measured in the field. Such characteristics include pH, temperature, and dissolved gases. These data, and all the pertinent data on sample location, collection time, and preservation methods, must be recorded and the sample clearly labeled. Samples for determination of such compounds as pesticides, PCB, or PAH are taken in large brown glass bottles to prevent photolysis of the analytes. Those designated for determination of volatiles are put into septum capped vials which are filled to the top, leaving no airspace.

A series of single point samples may be taken at varying times and locations to characterize a water system. Frequently, in order to obtain average values without doing an excessive number of analyses, samples are composited. These composites may be "time composites", where samples are collected at intervals at a single point. Large bodies of water or streams may be composited by taking samples at different depths and locations.

Table 8.1 Some US drinking water quality standards

Component	Maximum level (mg/l)	Component	Maximum level (mg/l)
Primary standards			
Metals and inorganics			
Ag	0.05	Cr	0.05
As	0.05	F⁻	4.00
Ba	1.00	Se	0.01
Hg	0.002	Nitrate-nitrogen	10.0
Pb	0.05	Cd	0.01
Turbidity	1–5 nephelometric turbidity units		
Volatile organics			
Total Trihalomethanes	0.10	1,2 Dichloroethane	0.005
1,1,1-Trichloroethane	0.20	Vinyl chloride	0.002
1,1-Dichloroethylene	0.007	Carbon tetrachloride	0.005
p-Dichlorobenzene	0.075	Trichloroethylene	0.005
Benzene	0.005		
Organics and pesticides			
Endrin	0.0002	Toxaphene	0.005
Lindane	0.004	2,4-D	0.10
Methoxychlor	0.10	2,4,5-T	0.10
Secondary standards			
pH	6.5 (min)–8.5	Chloride	250
Color	15 color unit	Cu	1.0
Corrosivity	none	Fe	0.3
Hardness as CaCO₃	50	Mn	0.05
Odor	3 odor unit	Zn	5.0
Foaming agents	0.5	SO₄⁻	250
Total dissolved solids	500		

Integrated samples may be taken by an automated system which diverts a constant small flow of the water to be sampled into a container, filling it over a day or a week. A wastewater treatment plant, for example, will monitor its inflow waters with a device which pumps water from the inlet pipe into a large bottle, held in a refrigerator. Whenever compositing is done, it must be remembered that much information is lost, since the variations with time or location cannot be determined. If a long term sampling program at a particular location is planned, it is probably best to analyze some individual samples first. When the extent of variation is known, a rational plan for compositing may be developed. Finally, one must be careful with the actual compositing process. Grab samples should not be casually poured from one container to another, since the agitation might cause loss of volatiles. A siphon below the surface of the sample should be used to transfer the water from one container to another, with as little water to air contact as possible.

The selection of sampling locations must be done with some care. In flowing streams, the most representative sample is probably in the center at mid-depth. Lakes and reservoirs also show significant variations with depth, and distance from shore, as well as distance from feeder streams. Water collected from pipes, as in water treatment plants, should be collected at a well-mixed point. Areas where the water is churning, as when it pours over a weir, however, are not good for dissolved oxygen measurements, because they will be artificially high at this point.

The handling of suspended solids must also be considered. If a sample is filtered, the results for many substances may be significantly changed. If a total value for a particular analyte is required, the solids can be separated by centrifugation or filtration, and analyzed separately. A certain amount of solids may be tolerable in some analytical methods, and can be carried through the analysis. Materials floating on the surface, such as oil slicks, also need to be treated with some care in sampling. If samples are taken below the surface, these will be ignored, which is a common way of handling them.

8.2 Potentiometry for ions and gases in aqueous solution

Electrodes which measure a specific ion in solution can provide a rapid means of determining a single species in a complex mixture. The glass electrode for pH measurements is the most common of these electrodes, giving a virtually instantaneous measurement of the activity of the H^+ ion in solution over 12 orders of magnitude. Many other specific ion electrodes are also available for measurements of free ions in solution. Specific ion electrodes for measurement of fluoride, chlorine, chloride, and hardness (Ca^{2+} and Mg^{2+}) in drinking water, as well as bromide, nitrate, sodium, sulfide, dissolved oxygen, and ammonia, among many others, are available.

8.2.1 *pH measurement*

The pH meter uses a glass electrode to determine the activity of hydrogen ions in a solution. This electrode consists of a tube, closed at one end with a thin glass membrane, filled with a buffered solution of known H^+ concentration. A silver electrode coated with silver chloride is located in the filling solution. The interior and exterior surfaces of the glass membrane are hydrated and have H^+ ions sorbed to the surfaces. The sorbed ions on each surface are in equilibrium with the H^+ ions in the solutions to which they are exposed. The difference in the concentration of ions sorbed on the inner and outer surfaces generates an electrical potential, which depends on the concentration of H^+ in the sample on the outside of the membrane, since the concentration on the inside is fixed.

The potential of the glass electrode is measured against a reference electrode, which is designed to keep a constant potential. The reference electrode is often built into the glass electrode, so that a single electrode unit is placed into the solution. Figure 8.1 shows the glass pH electrode. Because the current carrying circuit includes the glass membrane, the electrode has a very high impedance and the measurement has to be made using a voltmeter designed to measure such a high impedance source.

The response of the pH electrode to changes in the $[H^+]$ is governed by the Nernst equation, and the potential is related to the pH by the following equation:

$$E_{electrode} = \varepsilon + 0.059 \, pH$$

where $E_{electrode}$ is the voltage measured by the pH meter, ε is a constant which depends on the particular electrode, and pH is that of the solution being measured. ε is set for each measurement by placing the electrode sequentially into two different pH buffers, and adjusting the meter to read the proper values. These adjustments are setting the meter for the intercept (ε) of the straight line relating voltage and pH. The constant 0.059, the slope of this response line, is valid for a temperature of 25°C, and should be adjusted if the temperature of the solution is significantly different. Sophisticated pH meters include a temperature sensor in the electrode, and an electronic circuit which adjusts the meter automatically for the solution temperature. Others require a manual adjustment to change the slope in order to compensate for a sample temperature which is different from 25°C.

Interference in measurements made with the pH electrode is not usually a problem, except at extremely high or low pH values. When the pH of the solution is below 1 or above about 12, erroneous values may be read. Fortunately, few environmental measurements have to be done at these extreme levels. In cases where the H^+ or OH^- must be measured at these high concentrations, titration may be more suitable than use of the pH meter.

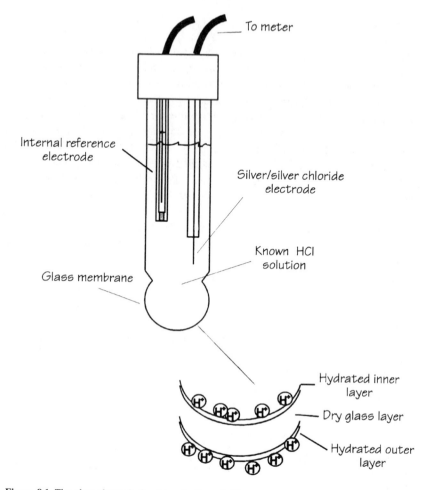

Figure 8.1 The glass electrode for determination of pH.

8.2.2 *Other specific ion electrodes*

Specific ion electrodes can be a valuable tool in the speciation of many elements in water. The electrode measures only free hydrated ions, as opposed to that portion of the element tied up in complexes or present as organometallic species. In addition, the electrode does not change the concentration of the ion being measured, so the act of doing the measurement does not disturb any equilibria within the sample.

Several different types of electrodes, similar in configuration to the pH electrode, are used for specific ion measurements. These are categorized by the type of membrane used. A solid state electrode uses a solid homogeneous substance as the sensing membrane. This material has an affinity for the analyte ions, and tends to adsorb them. For example, an electrode with a membrane made of a pellet of silver sulfide can be used for either silver or sulfide measurements, as silver sulfide will sorb either of these ions, in preference to others. For fluoride measurements, a lanthanum fluoride membrane is used. These membrane materials must have a very low solubility constant, to avoid contamination of the solution and to remain stable over a long period of time.

A liquid membrane electrode has an ion exchange material within a solid polymeric membrane. This material exchanges the analyte ion across the membrane, which creates the potential. Calcium and nitrate ions, among others, are measured by liquid membrane electrodes. Other electrodes of similar construction use a neutral molecule which has an affinity for the ion to be determined. For instance valinomycin has an affinity for potassium ions, and is incorporated in a suitable solvent behind a polymeric membrane to form a potassium selective electrode.

Gas sensing electrodes are used to measure gases dissolved in aqueous solutions. The membrane is a permeable polymeric material, such as silicone rubber, which allows the dissolved gas to permeate through it. The membrane is backed with a small volume of solution specific for each analyte. The dissolved gas concentration in the internal filling solution is sensed by an appropriate electrode couple within the electrode body. Gases such as ammonia, oxygen, and carbon dioxide can be measured using these electrodes.

Each ion selective electrode requires a reference electrode which provides a constant known voltage against which to measure the changes in voltage of the indicating electrode. There are different configurations of reference electrodes available, since the reference electrode filling solutions may interfere with some analyses. Each reference electrode has a porous plug which serves as a contact between the solution and the internal electrode filling solution. While a very small quantity of solution diffuses through this plug, it can be enough to change the measured concentration in the sample. In addition, a precipitate of insoluble salt, formed when the sample and the filling solution come into contact, can plug the channel between them, so that the electrode does not function properly. A silver/silver chloride electrode is often used as a reference electrode. This electrode may contain a filling solution of potassium chloride saturated with silver chloride, which can cause a problem if the sample is being analyzed for chloride. Enough chloride ion may pass into the solution to interfere with an accurate measurement. A double junction reference electrode has a second filling solution in an outer electrode chamber, which contacts the inner silver/silver chloride electrode through a porous plug, and the sample through a second plug. The outer chamber filling solution provides a non-reacting solution in contact with the sample. For instance, in the measurement of chloride, a filling solution of potassium nitrate can be used in the outer electrode jacket.

The electrodes measure the activity of the analyte, which depends on the total ionic strength of the solution. Ionic strength adjustment buffers (ISAB) are often

Table 8.2 Some specific ion electrodes

Electrode	Type	Concentration range	Interferences
Ammonia	Gas sensing	$1-5 \times 10^{-7}$ M	Volatile amines
Chloride	Solid state	$1-5 \times 10^{-5}$ M	OH^-, S^{-2}, Br^-, I^-, CN^-
Fluoride	Solid state	Saturated to 0.02 ppm	OH^-
Lead (Pb^{+2})	Solid state	$0.1-10^{-6}$ M	Ag^+, Hg^{+2}, Cu^{+2}, high Cd^{+2} or Fe^{+2}
Oxygen	Gas sensing	0–14 ppm	
Silver or sulfide	Solid state	$1-10^{-7}$ M (Ag^+ or S^{-2})	Hg^{+2}
Water hardness (M^{+2})	Liquid membrane	$1-6 \times 10^{-6}$ M	Na^+, Cu^{+2}, Zn^{+2}, Fe^{+2}, Ni^{+2}, Sr^{+2}, Ba^{+2}, K^+
Calcium (Ca^{+2})	Liquid membrane	$1.0-5 \times 10^{-7}$ M	Pb^{+2}, $Na+$, Hg^{+2}, H^+, Fe^{+2}, NH_4, Mg^{+2}

added to both the samples and the standards to "swamp out" differences in ionic strength between the solutions, and give more accurate potential readings. The manufacturers will supply information for each electrode on the specific ISAB recommended, the concentration range, special storage conditions, and interferences. Table 8.2 lists information on some commercially available specific ion electrodes and interferences which affect each of these.

These electrodes often respond to other materials in the sample solution, although with a smaller response than to the primary analyte. The interferences specified for each electrode must be noted. Selectivity factors are usually provided with each electrode, so that the tolerable concentration of the interferent can be calculated. The selectivity factor is the ratio of the response to the analyte and that of the same quantity of interferent. The higher the selectivity factor, the less serious the interference. For instance, the selectivity factor for a calcium ion electrode for interference by sodium, defined as (response to Ca^{2+})/(response to Na^+), is 630. This indicates that the signal is 630 times greater for a specific Ca^{2+} concentration than for the same concentration of Na^+.

Example 8-1

A solution containing about 0.001 M Ca^{2+} is being measured using a calcium specific electrode. If an interference from Na^+ of 2% is acceptable, what is the maximum permissible concentration of Na^+?

The amount of Na^+ which will give a signal equal to 2% of the Ca^{2+} concentration is

$$0.001 \text{ M } Ca^{2+} \times 0.02 \times \frac{630 \text{ mol } Na^+}{1 \text{ mol } Ca^{2+}} = 0.013 \text{ M } Na^+$$

Therefore Na^+ concentrations of below 0.013 M will not interfere with this determination, above the acceptable 2% level.

8.3 Metals in water samples

Metals are most often determined spectroscopically. Atomic absorption and inductively coupled plasma emission are most suited to determination of total content of a particular metal, without regard to its oxidation state. Both flame and electrothermal methods are used in AA, with flame methods being preferred for moderate to high metal levels. Visible or ultraviolet spectroscopy is also used for colorimetric measurements of specific metal ions. These methods can be more specific and are useful for determination of a particular metal-containing species, even in the presence of other forms of the same metal, although the methods may be less sensitive.

Metals present in water samples can be arbitrarily divided into the two types: **dissolved metals** or **suspended metals**. These are defined as the metals which remain after passage through a 0.45-μm filter and those which are filtered out, respectively. If these are to be determined separately, the sample should be filtered immediately after sample collection and before acid preservation, because acid will begin to dissolve some of the suspended metal content. Acid extractable metals are defined as those which are found in solution after the unfiltered sample is treated with hot dilute mineral acid. Finally, total metals may be determined either by summing the analyses of both the filtered water and the solids, or by aggressively digesting an unfiltered sample with acid before analysis.

Sampling of water for metals determination is best done in quartz or TFE containers. Since, these are expensive, polypropylene or polyethylene bottles are usually used. They are cleaned with an acid rinse before use. The samples are acidified with high purity concentrated nitric acid immediately upon collection, unless the dissolved and suspended metals are to be analyzed separately. In that case, the samples are filtered immediately, and the filtrate is then acidified. Samples should be stored on ice or refrigerated until analysis. They are usually stable for several months, although samples containing very low levels should be analyzed as soon as possible because any loss or contamination will lead to seriously flawed data.

An additional concern in sampling is the cleaning of any pumps, tubing, or siphons used in the sampling process, as well as laboratory glassware, pipettes, and other apparatus. These must not be contaminated from previous samples. Other sources of contamination include some plastic materials which may contain metals such as zinc or cadmium, and airborne dust which almost always contains lead in urban areas. Blanks are prepared using the storage bottles which have been subjected to the same treatment procedure as the sample bottles. These are filled with high purity distilled water, acidified in the same way as the samples and carried through the entire analytical procedure. Properly done blanks will show up many sources of contamination, and they should not be neglected. Finally, one must be very careful about the storage of mercury samples. If mercury vapors are present in the storage area, as they are in many laboratories, the vapors may diffuse through plastic bottles and contaminate the samples. Samples should be analyzed for mercury within 5 weeks of collection.

8.3.1 Sample filtration

Polycarbonate or cellulose acetate filters with 0.45-μm pores are used for initial separation of suspended metals from dissolved metals, if this is desired. The filter should be rinsed with distilled water before use, and the rinse water can be tested as a filter blank. If the filters prove to contain metallic contaminants, they may be soaked in acid and rinsed before use. In very precise trace analyses, it may be necessary to check that the filter is not removing analytes by sorbing metal ions from the sample. This can be done by passing a standard through the filter and checking the metal content before and after. If suspended metals on the filter are to be determined, the volume of water filtered must be recorded and the filter stored in a clean container for analysis. If samples are very turbid, it may be better to centrifuge them before trying to filter them, since the loading will be less and the filters will not be as easily fouled.

8.3.2 Digestion of metal samples

Metals associated with particulates, whether the particles are organic or inorganic in nature, will need to be brought into solution before AA or ICP analysis. This is done with an acid digestion. The appropriate acid mixture differs depending on the matrix and the metals to be determined. In general, the mildest conditions compatible with the analytical method which will give reproducibly complete recoveries, is best. The more acid used, the more the trace metal impurities present in the acid become a problem.

Nitric acid is generally the most widely used acid for digestion, since nitrate salts are generally soluble, and nitric acid makes a good matrix for analysis. Other acids, including perchloric, hydrochloric, hydrofluoric, or sulfuric may be added, to extract more difficult metals, but these will interfere with certain determinations. For instance, the addition of HCl will help in the extraction of antimony and tin, but may cause precipitation of lead.

Digestion can be done in an Erlenmeyer flask or a beaker, on a hotplate. Fifty to 100 ml of the sample is mixed with acid and boiled down to the lowest volume possible, without allowing it to dry. Portions of concentrated acid are added, and heating continues until the sample is clear and light in color, at which point digestion is considered complete. The sample is then cooled, diluted to volume, and is ready for analysis. Use of microwave digestion gives more reproducible results, because temperature and digestion times are automatically controlled by a user defined program. Microwave digestion has the additional advantage that it takes place in a sealed container, with the vented fumes being directed to a scrubber. This eliminates corrosive acid fumes in the laboratory hood, and prevents contamination of the samples by airborne dust.

8.3.3 *Preconcentrating the sample*

For very low concentration samples, a preconcentration step may be necessary. The metals are complexed with a reagent such as ammonium pyrrolidine dithiocarbamate (APDC), and the complex is extracted into methyl isobutyl ketone (MIBK). Each of the metals is complexed at an optimum pH, which varies from metal to metal. A 100-ml aliquot of the sample is brought to the correct pH with HNO_3 or NaOH. Then 1 ml APDC solution is added to form a complex. This is extracted with 10 ml of MIBK. The extraction is done in a 200-ml volumetric flask, and at the end, sufficient distilled water acidified to the correct pH, is added gently down the side of the flask to bring the organic layer into the neck of the flask. The organic layer is aspirated directly into the flame of the AA. Air–acetylene flame AA may be used for cadmium, chromium, cobalt, copper, iron, lead, manganese, nickel, silver, and zinc, and nitrous oxide–acetylene flame AA for aluminum, barium, beryllium, molybdenum, osmium, rhenium, silicon, thorium, titanium, and vanadium. The instrument should be zeroed using water-saturated MIBK. Blanks and samples are done in the same way.

8.3.4 *Separating "labile" metal species*

Since the form in which the metal is found is very important in determining the significance of its impact on the environment, whether as a toxic material or as a nutrient, the most "labile" metal species may be concentrated from the sample before analysis. Labile species are loosely defined as the free hydrated ions and those ions which are easily exchanged from various complexed forms, as opposed to metals securely bound to particles or involved in strong complexes from which they cannot be readily released. These relatively free ions may be separated from the sample by sorption on a chelating resin such as Chelex 100®. The resin is either packed in a column through which the sample is passed, or is added to the water, stirred for a fixed time and then filtered out. This is an operationally defined measure of speciation, since the amount of metal ion which is sorbed by the resin will depend heavily on such parameters as the time of exposure to the resin, the temperature, the presence and concentration of competing ligands, the pH, and the ionic strength. However, it is generally assumed that the resin will sorb the free hydrated ions and readily exchanged complexed ones, which are likely to be the most bioavailable species. The resin is then eluted with acid to remove the metal for analysis, or the resin beads are analyzed directly, using electrothermal atomization.

8.3.5 *Atomic absorption methods*

8.3.5.1 *Direct flame analysis of water samples.* After the sample has been prefiltered if necessary, digested according to the method being followed, and filtered again if any particulate is present which might clog the nebulizer, it is ready for analysis.

Interference has been discussed in the chapter on spectroscopy. When the water to be analyzed has a high salt content, the MIBK extraction may give better results, even if the metal concentration does not require preconcentration. Removing the metal from the salty matrix makes interferences less likely. Otherwise, brines and sea water should be analyzed using background correction, and should be diluted before analysis if possible. Quantitation by standard addition is also recommended. Frequent recovery checks should be done when working with these difficult samples.

8.3.5.2 *Cold vapor atomic absorption method for mercury.*   Mercury is difficult to determine by flame at low levels. The cold vapor method is much more sensitive because the flame background is eliminated. The mercury is freed from any organic matter by digesting the sample with potassium permanganate and potassium persulfate in acid solution. This also oxidizes any chloride ion to free chlorine, eliminating it from the sample. The excess permanganate is removed with hydroxylamine hydrochloride. Then the mercury salts are converted to free mercury by reaction with stannous chloride.

$$Hg^{2+} + Sn^{2+} \rightarrow Hg^{\circ} + Sn^{4+}$$

The free mercury has sufficient volatility so that it can be sparged out of solution. The sparged gas is passed through a quartz-windowed tube placed in the light path of the atomic absorbance spectrometer. The response is a peak in the absorption, whose area is proportional to the amount of mercury.

For even more sensitive determination of mercury, the sparged gas may be passed through a gold plated mesh. This collects the mercury as it amalgamates with the gold. When the sparging is complete, the gold mesh is heated rapidly, releasing the mercury in a sharp peak. Detection limits of ng/l have been achieved with this method.

8.3.5.3 *Hydride generation methods for arsenic and selenium.*   Arsenic and selenium are also difficult to measure in the AA flame because they absorb in portions of the UV spectrum which are especially subject to molecular interferences. They are analyzed by first reducing them to the gaseous hydrides. The samples are acid digested in a mixture of nitric and sulfuric acids, to oxidize any organic matter and bring all the arsenic to the As(V) state. This is reduced to As(III) with KI and sodium borohydride ($NaBH_4$) is added to react with the As(III) and form the volatile hydride, arsine (AsH_3). The borohydride reacts much more slowly with arsenic in other oxidation states, so the initial oxidation and reduction steps are important. Selenium undergoes similar reactions. The hydrides produced are swept into a quartz windowed tube in the light path. The tube is heated to 800–900°C and the hydrides thermally decompose to the free atoms. Figure 8.2 shows the setup for this analysis. The generated hydrides may also be passed into an air-entrained hydrogen flame, but the quartz tube gives a less noisy signal and better detection limits. The hydride generation method can detect less than 0.5 µg of arsenic or selenium per liter.

8.3.5.4 *Electrothermal atomic absorption methods.*   Electrothermal atomization or graphite furnace atomic absorption gives better detection limits for most metals than flame AA, generally 20–1000 times lower. This may allow the analysis to be done without preconcentration steps, eliminating much sample handling. However, the precision is often higher in flame methods. Graphite furnace methods are subject to many more interferences and matrix effects, and, because of the high sensitivity, contamination may be a serious problem. The sample matrix may cause molecular absorption. A good background correction method is usually essential for all but the simplest matrices. Chemical interferences and matrix effects are caused by the sample volatilizing unevenly, and by formation of refractory compounds before atomization. Matrix modifiers may be added to overcome these problems, but each metal and each type of sample matrix must be studied to determine the best modifier. Standard addition is often used to correct for matrix effects in electrothermal atomization methods.

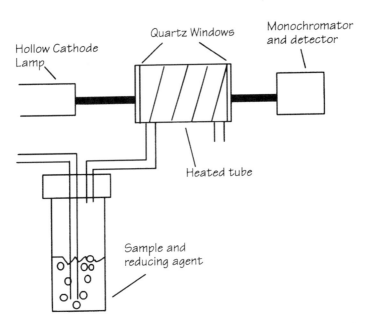

Figure 8.2 Hydride generation system for AA determination of arsenic and selenium. The hydride is formed in the sample and is swept into the heated tube. There it decomposes to form atomic arsenic or selenium, which is detected by its absorbance. The cold vapor system for mercury determination is similar, but the sample cell tube is not heated.

The electrothermal atomizer is recommended for aluminum, antimony, arsenic, barium, beryllium, cadmium, chromium, cobalt, copper, iron, lead, manganese, molybdenum, nickel, selenium, silver, and tin. Samples of 100 ml are digested as above, with 2 ml of 30% hydrogen peroxide added if arsenic or selenium are to be done. A matrix modifier is added if needed. The samples are transferred to a 100-ml volumetric flask and brought to volume. A blank is done in the same way. Standards are prepared, and should be matched as closely as possible to samples, especially to the acid concentration.

The system is optimized by selection of the best drying, charring, and atomization temperatures. Unless a test sample shows an accuracy between 85 and 115%, samples should be analyzed by the method of standard additions. A sample is run, then a measured amount of standard is added to a new portion of sample and this is analyzed. The addition is repeated with fresh portions of sample and increasing amounts of standard. If a sample is much higher in concentration than the standards, it must be diluted. If large dilutions are required to bring the sample into the linear range, the method may be too sensitive, and flame AA may be more appropriate. Large dilutions are undesirable because they magnify any error in the analysis.

Since this is an extremely sensitive method, low detection levels are expected. However, several replicates should be done at the start of each batch of samples, to ensure that the reproducibility is adequate.

8.3.6 Inductively coupled plasma methods

ICP is well suited to water analyses, because all the elements desired can be analyzed in one run, saving analysis time and money. Standards are prepared which contain all the target elements. Samples are digested in the same way as for AA, and standards again should be matched to the sample matrix as much as possible. Standard addition may be used if matrix problems arise. One should be careful when high-solids samples and standards are analyzed, because the salts may build up at the nebulizer tip. Chemical interferences are much less likely to cause problems in ICP than in flame methods, because of the much higher temperatures involved.

A method blank is prepared in the usual way, by carrying distilled water through the entire digestion and analysis process. The instrument is zeroed and rinsed between samples with a solution of dilute HCl and HNO_3. A method quality control sample should be analyzed with each set of samples. New types of sample should be checked for matrix interferences by running a series of samples at different dilutions, using the calibration blank solution for the dilutions. These samples, corrected for dilution, should agree within 5%. If the matrix appears to be causing a problem, the sample can be diluted to the point at which the problem disappears, if this leaves a high enough concentration to analyze. Otherwise, the standard addition method is the best solution to the problem.

8.3.7 Determination of metals by voltammetry

The electrochemical method known as **voltammetry** has been applied to analysis of water for metal content, and is especially suited to rapid on-site determinations. Voltammetry depends upon the oxidation or reduction of a chemical species at a very small electrode. The amount of current passed through the electrode during the process is indicative of the concentration of the species being determined, and each species has a particular potential at which it will be oxidized or reduced under a given set of conditions.

Let us examine an electrode placed in a solution containing Pb^{2+} ion. When a small current is passed through the electrode at a voltage sufficient to reduce Pb^{2+} to Pb, the electrode becomes polarized. This means that a concentration gradient is formed around the electrode. Figure 8.3 shows this gradient. The Pb^{2+} can be transported to the electrode by three mechanisms: diffusion, convection, or ion migration. The method depends on the ions being transported reproducibly by diffusion only. In order to prevent migration of the Pb^{2+} ions through the solution under the electrical field around the electrode, a supporting electrolyte which will carry the major portion of the current is usually added, unless the sample already contains a substantial salt concentration. Convection is kept to a minimum by avoiding temperature gradients and vibrations, or is kept constant by careful stirring. The concentration of the Pb^{2+} at the surface of the electrode can be assumed to be zero, since every ion which reaches that point will be immediately reduced. The concentration in the bulk of the solution will remain at the original concentration, C_o, as the electrode is very small. The current, therefore, is determined only by the diffusion of Pb^{2+} to the electrode surface. This **limiting current** depends on the concentration gradient of Pb^{2+} between the electrode surface and the bulk solution, which in turn, depends only on the value of C_o, the value we wish to determine.

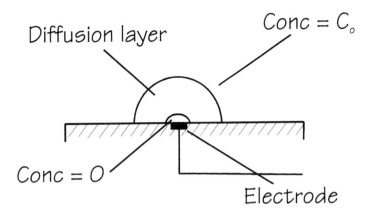

Figure 8.3 Diffusion layer at a microelectrode. Current is limited by the transport of analyte from the bulk solution to the electrode layer, and this in turn is governed by the concentration in the bulk, C_o.

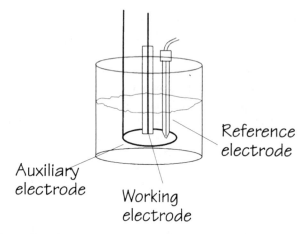

Figure 8.4 System for voltammetry, and for anodic stripping voltammetry.

The experiment is performed by use of a three-electrode system (Figure 8.4). One electrode is the micro electrode at which the reaction takes place. A reference electrode, designed to produce a constant potential under a wide variety of conditions, is used as a standard against which to measure the potential being applied to the working electrode. A larger auxiliary electrode is used to carry current, so that no current is drawn through the reference electrode, to prevent that electrode from becoming polarized and changing its potential. The potential of the working electrode is gradually made more negative, until the potential is reached at which Pb^{2+} is reduced. At that point, a current will begin to flow, polarizing the working electrode. The current will rise as the reaction begins to occur, and it will stabilize at a **limiting current** which is governed by the original concentration of lead, C_o.

However, in many cases, especially for drinking water, direct voltammetry is not sufficiently sensitive. **Anodic stripping voltammetry** has been developed to increase the method's sensitivity to the nanomolar range or lower. In this case, the Pb^{2+} or other analyte ion is preconcentrated by electrodeposition at a constant potential, using a mercury drop or mercury coated microelectrode. The reduced lead dissolves in the mercury. After a definite period of time, during which the solution may be stirred in a reproducible manner or kept undisturbed, a reproducible fraction of the metal present in the sample will be deposited into the mercury. Then the electrode potential is ramped toward a more oxidizing potential, and the collected metal is oxidized. The current will increase as the oxidation proceeds, then drop as the metal deposited in the drop is depleted. The area under the current/voltage curve is proportional to the amount of lead deposited upon the electrode, which will be determined by the original concentration, if all the other parameters such as deposition time and stirring are kept constant. The system can be calibrated by standard addition, or by use of a surrogate standard (called a pilot ion in electrochemistry), an added ion which is known not to be in the sample. The pilot ion must be oxidized at a potential sufficiently different from the analyte so that the two can be distinguished. It is preferable to calibrate in the same run as the analysis, since transporting the sample solution to the electrode in a very reproducible way by stirring is difficult.

Stripping voltammetry can be used for simultaneous determination of several ions if their oxidation potentials are sufficiently different, and if the stripping potential is changed slowly enough that one ion is completely oxidized before the potential for the next one is reached. Several factors have joined to bring this method out of the research laboratory, and make it into a fieldable instrument. Because stripping voltammetry requires measurement of very small electrical currents and modest voltage supplies, with close control of time and voltage, it is ideally suited for development of computer controlled instruments. In addition, methods used in production of solid state electronics can be used to make very small, very reproducible, electrodes or arrays of electrodes, at much lower cost than mechanically built ones.

Example determination: chromium in fresh water

Chromium, in several different oxidation states, exists in species such as Cr^{3+} and CrO_4^{2-} in water. Cr(VI) is of particular interest, because it is the most soluble and most toxic. The water sample can be analyzed for several different chromium species. First the sample is filtered through a 0.45-μm membrane filter. The filter containing the removed particulate matter is digested in a microwave digester, in a solution of nitric acid, with added hydrogen peroxide. The digest is filtered and the total chromium present is determined by flame or graphite furnace atomic absorbance at 357.9 nm, using standards made up in nitric acid for calibration. This gives the total chromium content of the filterable matter, the insoluble chromium, in the sample.

A strong anion exchange resin, Dowex 1 X 8, 100–200 mesh, is used to concentrate soluble Cr(VI) from the filtered water sample. This resin has trimethylbenzyl ammonium active groups, with an ion-exchange capacity of about 1.2 meq/ml and is used in the chloride form. To sorb Cr(VI) on the anion exchange resin beads, a 500-ml water sample, containing 0.45 g of the anion-exchanger is mixed in a beaker and stirred for 30 min at low speed with a magnetic stirrer. After the sorption process is completed. the beads are filtered from the water sample. Species containing chromium in other oxidation states are cations and therefore are not sorbed. The sample-laden beads are mixed into 5 ml of 5% nitric acid solution, and an aliquot of the slurry formed is pipetted into the graphite furnace. The sample is dried at 150°C, pyrolyzed at 900°C and atomized at 2300°C. The absorbance is monitored at 357.9 nm, and background correction is used. Calibration curves are prepared using 5,10, 15, and 50 μg/l Cr(VI) solutions in dilute nitric acid. A linear calibration plot is obtained over this concentration range. This will detect chromium(VI) to concentrations below 0.01 µg/l.

Finally, the total chromium in another aliquot of the unfiltered sample can be determined. The sample is acidified with high purity nitric acid, and heated in a beaker to near dryness. Another small portion of nitric acid is added to the residue, and the beaker is covered with a watchglass. The sample is digested on a hotplate until the residue is completely digested, usually until no further color change is seen. The sample is again evaporated to near dryness, the sides of the beaker are washed with distilled water, and 1:1 nitric acid to dissolve any residue present. This is then filtered to remove insoluble silica. The filtrate is again evaporated to near dryness and redissolved in 5% nitric acid. The sample is transferred to a 5-ml volumetric flask and diluted to volume with 5% nitric acid. Then the total chromium is determined by flame or graphite furnace atomic absorption. Standards are made in 5% nitric acid, and blanks of pure deionized water are taken through the entire process, to detect any contamination during the analysis.

8.4 Inorganic anions in water

Ions containing nitrogen, phosphorous, sulfur, and the halogens have significant environmental roles, both as essential nutrients for living systems and as pollutants. Nitrates and ammonia, for instance, are valuable in plant growth, while the same quality causes eutrophication of lakes, choking them with excessive weed growth. The concentrations of these materials in water samples can be measured by several colorimetric methods as well as by ion chromatography or with specific ionic electrodes.

8.4.1 Ion chromatographic analysis for common anions

While titration or colorimetric methods are available for most of these, ion chromatography is the only method which can give a quantitative determination of chloride, nitrate, nitrite, phosphate, and sulfate in a single analysis. A suppressed or non-suppressed ion chromatography system can be used, with a conductivity detector.

Samples are prepared by filtering through a 0.2-µm filter, removing particulates. Standards of each individual analyte, or a mixture containing all the desired anions, are injected at a minimum of three different concentrations. The standard calibration curve for each ion should be plotted and checked for linearity. If linearity has been established for a particular set of system conditions, then a single point calibration may be adequate. A sample of 10–100 µl is usually injected.

There is often an interference with the fluoride and chloride peaks due to the presence of low molecular weight organic acids, as well as a negative-going peak due to the elution of the matrix water. The water peak can be reduced by adding enough concentrated eluent to the sample to match the final eluent composition, if possible. The minimum detection limits are around 0.1 mg/l for most of the ions, except for fluoride. The detection limit can be further lowered by preconcentration or using a larger injection volume. Fluoride determination is more difficult, and a careful study of the precision and accuracy in the particular matrix should be undertaken when fluoride is to be done.

8.5 Organic compounds in water

Most natural waters contain naturally occurring organic substances in addition to those of anthropogenic origin. While each compound present may be determined, the presence of hundreds of organic compounds makes this impractical for any routine testing. Groups of substances may be assembled according to the effects they have in common, or to some common property. For most purposes, the determination of the concentration of each of these groups is sufficient to characterize the water. In cases where a single pollutant is of interest, for instance in tracking a source of pollution, more detailed analyses must be done.

The determinations of the classes of total carbon are described as follows:

- **Biochemical Oxygen Demand (BOD)**: This test measures the amount of organic material which can be oxidized by bacterial action. It is an empirical test, and allows judgments to be made on the amount of treatment required before water can be released to streams, or the potential of streams to support aquatic life. High BOD values cause the oxygen levels of the water to be lowered, often leading to fish kills and other undesirable results.
- **Chemical Oxygen Demand (COD)**: This measures the total amount of oxidizable material in the sample. It includes the BOD material and any other substances susceptible to chemical oxidation.
- **Total Inorganic Carbon (IC)**: This category comprises the carbonates and bicarbonates, and carbonic acid in a sample. It is of interest primarily because the CO_2 which can be generated from these species will interfere with the determination of total organic carbon. Therefore this is measured by acidifying the sample to a pH below 2 and purging out the CO_2 produced.
- **Total Organic Carbon (TOC)**: TOC may be further subdivided into purgeable volatile organic compounds (VOC) and non-purgeable organics (NPOC). Another distinction can be made between dissolved organics and particulate organics. These are defined as the organic material which passes through a 0.45 μm filter and that which does not, respectively. TOC usually must be determined in several steps. If IC procedure is carried out to remove and measure the carbonate-related materials, this will also remove purgeable volatile organics. A total carbon analysis done at this point would actually measure the NPOC. A determination of the VOC done on a fresh sample can

be added to the NPOC to obtain a TOC value. An alternative route to the TOC is to carry out a total carbon determination, and an inorganic carbon determination. The difference between these will be the TOC.

While the TOC and the COD and BOD may seem to be measuring the same thing, they are not equivalent. TOC measures all organic carbon species, regardless of their oxidation state, while the BOD and COD only measure the capacity of the sample to be oxidized either biologically or chemically. This may include non-carbon compounds, such as nitrites. When a particular organic matrix is studied over a period of time, it may be determined that these parameters have a rather fixed relationship to each other. A single one of these measurements may prove to be adequate to predict the others, depending on the purpose of the analysis and the degree of accuracy needed. One would expect that when a large change in the measured property occurred, a check of the other tests might be indicated, to assure that the relationship was still valid.

Example

The total carbon in a sample is determined to be 5.0 mg/l. When the sample is purged with nitrogen for a period of time, the total carbon drops to 4.3 mg/l. If a fresh sample of the same water is acidified and then purged, the total carbon becomes 2.2 mg/l.

From these analyses, estimate the purgeable organics (VOC), total inorganic carbon, and TOC.

Total carbon comprises all the forms of carbon. When this is purged, the VOC are removed. Therefore $5.0 - 4.3 = 0.7$ mg/l carbon is present as VOC. When the sample is acidified before being purged, the VOC and the carbonates and bicarbonates are all removed. This group of substances is present at $5.0 - 2.2 = 2.8$ mg/l. Subtracting the VOC (0.7 mg/l), leaves 2.1 mg/l of carbon in carbonate and bicarbonate. The 2.2 mg/l of carbon left at the end is primarily non-purgeable organics. We may add this to the VOC determined above, giving TOC of 2.9 mg/l.

Other groups of compounds commonly determined are total organic halogens or trihalomethanes. This class of compounds, arising often from chlorination of organic-bearing waters, includes several suspect carcinogens. Oil and grease are determined in wastewaters before and after treatment to assess the amount of treatment required and the suitability of the water for release.

8.5.1 *Biochemical oxygen demand*

The BOD is an empirical test in which the amount of oxygen required to decompose the digestible components of the sample at a specific temperature and time. It is most often used to determine the loading of waste treatment plants and evaluating their efficiency in removal of oxidizable material. The dissolved oxygen (DO) is measured on the diluted samples, and then they are incubated at 20°C for 5 days. The DO is remeasured, and the loss of oxygen is calculated.

Samples are taken in 250- to 300-ml bottles, filled to overflowing and sealed. These may be kept at 4°C for a maximum of 24 h, but should be analyzed as soon as possible. Some wastes, such as industrial waste, waste with extreme values of pH or high temperature waste, might not contain a native population of microorganisms. To these a bacterial "seed" should be added. The seed can be material from the aerobic digester tanks at a water treatment plant. If the waste contains unusual materials which may be toxic to ordinary bacterial cultures, a specially cultured seed may be used.

Samples should have enough dissolved oxygen present so that the oxygen is not the limiting factor in the analysis. Therefore, samples will require dilution to bring them into the acceptable range of oxygen to organics. The best results are obtained when the residual dissolved oxygen level is no lower than 1 mg/l and the uptake is at least 2 mg/l after 5 days of incubation. Generally, several different dilutions are made, to ensure that at least one will be in the correct range.

The dilution water contains phosphate buffer, magnesium sulfate, calcium chloride, and ferric chloride, and is saturated with air by shaking in a partly filled bottle. It can be checked by measuring its dissolved oxygen and incubating it at 20°C for 5 days. Its BOD should not be higher than 0.2 mg/l. There are many variables in this method, especially the bacterial population of the "seed" and the presence of compounds toxic to the bacteria. The method should be checked for accuracy by determining the BOD of a known solution of pure organic compounds. A mixture of glutamic acid and glucose is convenient for such a check. Matrix problems can be detected by adding the same mixture to a sample as a spike and determining the increase in BOD. Dilution water can be used as a blank.

The dissolved oxygen is usually determined using a membrane electrode. This electrode is actually an electrochemical cell, covered by an oxygen permeable membrane, usually made of polyethylene or TFE. It contains a gold or platinum working cathode, a reference anode, which keeps a constant potential, and an electrolyte solution. Silver coated with silver chloride is usually used as a compact micro reference electrode. A constant potential sufficient to reduce oxygen is applied across the two electrodes, and as oxygen diffuses in through the membrane, it is reduced at the cathode. The current which flows through the system is monitored, and is directly related to the amount of oxygen which reaches the cathode. This, in turn, is related to the concentration of dissolved oxygen in the solution.

8.5.2 Chemical oxygen demand

The chemical oxygen demand is determined by aggressively oxidizing the sample with a measured amount of potassium dichromate in sulfuric acid solution. The excess dichromate is then determined colorimetrically or by titration with ferrous ammonium sulfate, using ferroin indicator. Samples are collected and preserved by acidification with H_2SO_4, if they cannot be analyzed immediately.

Samples are measured into the digestion vessel, which may be an Erlenmeyer flask fitted with a condenser, for samples of several hundred milliliters, or a small tube, sealed with TFE lined cap. Use of the smaller sample size requires care that the sample is representative. If there is particulate material present, the material should be homogenized before the sample aliquot is taken. When a substantial part of the organic material in a sample is volatile, the closed tube method is preferred, because the volatiles are kept in better contact with the oxidizing reagent.

After the sample is placed in the sample vial, a measured amount of potassium dichromate solution, sulfuric acid and silver sulfate are added. The silver ion is present as a catalyst to improve the oxidation of straight chain hydrocarbons. The sample tube is digested at 150°C for 2 h. Any particulate is allowed to settle and the tube is placed in the spectrometer, measuring the absorbance of the remaining dichromate at 600 nm. Calibrations are carried out by digesting solutions of potassium acid phthalate (KHP) using the same reagents and measuring the absorbance in the same way. KHP has a theoretical COD of 1.176 mg O_2/mg KHP.

The method measures all oxidizable organics and also reduced inorganic species such as nitrites and ferrous iron. These can be determined separately, and subtracted from the final value. Nitrates can be eliminated by the addition of sulfamic acid before the dichromate is added.

8.5.3 *Total organic carbon*

Total organic carbon is determined by the oxidation–infrared method. A small volume of the sample is injected into a heated reaction chamber packed with an oxidative catalyst such as cobalt oxide. The water is vaporized and the organic carbon is converted into CO_2 and H_2O. The CO_2 is carried by a carrier gas and is measured using a non-dispersive infrared analyzer. Automated instruments based on this principle are commercially available. The detection limit of this method is around 1.0 mg/l of total organic. A schematic diagram of a TOC analyzer is shown in Figure 8.5.

Alternatively, the sample can be oxidized with persulfate in the presence of ultraviolet light. The entire carbon content of the sample is converted to carbon dioxide and swept into a cell where the infrared absorption is measured at the CO_2 wavelengths. The detection limit of this method is as low as 0.05 mg/l. This method is used for low concentration samples.

When the sample contains carbonates and bicarbonates, these also produce CO_2. The inorganic carbon interference is dealt with by injecting a second sample into a reactor containing silica beads coated with phosphoric acid. This

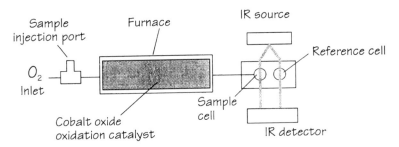

Figure 8.5 A total carbon analyzer. The sample is combusted in oxygen atmosphere, forming CO_2 which is detected and measured using a nondispersive IR detector.

will convert the inorganic carbon into CO_2, while not affecting the organic carbon. The value obtained from this is subtracted from that obtained by thermal oxidation of the entire sample to obtain the total organic carbon.

8.6 Volatile organic compounds in water

Water samples for determination of VOCs are collected and stored in TFE-faced septum capped bottles, 25 or 40 ml in size. They are filled at the sampling site without leaving any headspace. Samples are often preserved by acidification with HCl, and with ascorbic acid added if the sample contains residual chlorine. Acidification and chilling stops the bacterial action which may be digesting the analytes. A reducing agent like ascorbic acid removes any free chlorine which may remain from the chlorination of raw sewage, effluent waters, or drinking water. If not removed, the chlorine will continue to react with organics, forming various chlorinated compounds. The samples should be kept chilled and ana-lyzed as quickly as possible.

Analysis of VOCs utilizes the high volatility of these compounds. The normal analytical approach is to transfer the VOCs into a vapor phase or head space and then analyze the vapor phase for these compounds. The transfer can be done in a static system where the sample is allowed to equilibrate with its head space, or in a dynamic system where a gas is used to purge the analytes. The latter is more common.

Total VOCs are determined by purging the volatile compounds from the water, trapping them and injecting them into a flame ionization detector. If the individual compounds present are to be determined, the sample may be injected into a gas chromatographic column, for separation and quantitation.

8.6.1 *Measurement of VOCs using purge and trap*

Purge and trap or dynamic headspace analysis is done by purging the VOCs from water using an inert gas such as N_2. The purged VOCs are trapped and concentrated in an adsorbent trap. Subsequently these compounds are injected into a GC or a GC/MS for analysis.

Figure 8.6 shows the schematic of a purge and trap analytical apparatus. The sample is transferred into a glass purge chamber using a syringe. When one syringeful of water is removed from the sample container, a head space is left. If a second sample is anticipated, the second syringe should be filled immediately after the first, and reserved for later analysis.

The purge chamber is brought to a known temperature, and it may be heated to increase the vapor pressure of compounds of lower volatility. Purge gas is introduced into the bottom of the purge chamber containing the sample. The gas is dispersed into fine bubbles to ensure good gas-water contact. This may be accomplished with a fritted disperser or a fine tube. The frit style tube is more efficient, but some samples tend to foam excessively, and frits are not suitable for these. The purge gas passes into a cooled trap packed with Tenax or a multi-sorbent trap such as Tenax, silica gel and charcoal. Specially designed carbon molecular sieve materials and graphitized carbon based adsorbents are more recently developed packings for these traps. The trap is usually about 4 mm i.d. and about 25 cm long. When the purging is completed, the trap is backflushed with carrier gas, as the trap is heated rapidly to 200–350°C. The sample is then passed into the GC. Because the complete desorption of the trapped analytes may take several minutes, the sample may need to be refocused before the

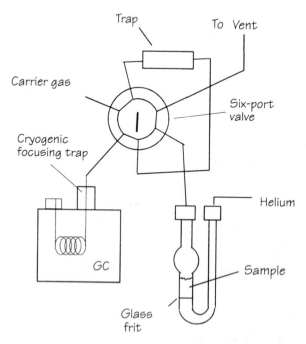

Figure 8.6 Purge and trap system for determination of VOC in water samples.

injection into the GC, to produce a sharp band of analyte. This is accomplished by trapping it in a second, smaller trap, by condensing it on a cold spot, or by chilling the GC column so that it serves as a trap. The analytes are then released by heating, and the chromatographic run commences.

Commercial instruments are automated, and are usually set up so that the purging time, purge temperature, trap temperature, and trap desorption parameters are all programmable. Systems which analyze a series of samples with no operator attention are available. Either packed or capillary column systems can be used. Mass selective detectors, electron capture detectors, electrolytic conductivity detectors, photoionization detectors and flame ionization detectors are all suitable for use with purge and trap analyzers, depending on the species sought.

Several parameters can have a significant influence on the precision and accuracy of purge and trap analyses. Purge time and temperature both should be optimized. If the time is too short and temperature too low, purging is inefficient, while if these are too high, a large amount of water may be transferred to the trap. Water in the trap may cause breakthrough, ice plugging or poor peak shape when the sample is injected into the GC.

Response factors are determined by spiking clean water with standards made as concentrates in methanol, then diluted in water to the desired concentration. Calibration plots are made using standards which cover the range of expected sample levels. Compounds are identified by their retention times, by their mass spectra, where available, or by their relative responses on different detectors, if the column effluent is split to two or three detectors. Blanks are done with distilled water.

Because the compounds measured in these analyses are often present at low trace levels, a great deal of care must be taken in cleaning all glassware and equipment. Especially, syringes used for handling high concentration standards or samples should not be used for trace level samples. Carryover of compounds between high and low samples is also possible. System blanks done after high samples or standards will show if this is a problem. Glassware and syringes used to handle samples and standards should be carefully cleaned with detergent, and heated to remove adsorbed organics. Storing this equipment in a warm vacuum oven will sometimes prevent high blanks due to adsorption of volatiles from the laboratory air. Quality control samples consisting of mixed standards should be run regularly to ensure good work. In general, it is better for these QC standards to be made up by someone other than the analyst, to avoid the problem of unconscious bias in the data.

Purge and trap is an excellent qualitative and quantitative tool. Table 8.3 shows some of the compounds for which this method is suited. Using a long (30–60 m) capillary column, 100 or more individual compounds can be separated and identified. This includes a wide range of compounds such as aliphatic and aromatic hydrocarbons, chlorinated organics, oxygenated and nitrogenated organics. Detection limits can be below 1 µg/l. Water soluble

Table 8.3 Some compounds determined by purge and trap and gas chromatography

Benzene	1,2-Trichloroethene
Bromodichloromethane	1,2-Dichloropropane
Bromoform	1,3-Dichloropropene
Carbon tetrachloride	Ethyl benzene
Methylene chloride	Chlorobenzene
Chloroethane	1,1,2,2-Tetrachloroethane
2-Chloroethylvinylether	Tetrachloroethene
Chloroform	Toluene
Chloromethane	1,1,1-Trichloroethane
Dibromochloromethane	1,1,2-Trichloroethane
Dichlorobenzenes	Tichloroethene
Dichloroethanes	Trichlorotrifluoroethane
1,1-Dichloroethene	Vinyl chloride
Styrene	Xylenes
Saturated and unsaturated C4 to C12 hydrocarbons	

compounds such as alcohols and ketones cannot be purged as effectively from water, and consequently have detection limits that are an order of magnitude higher.

In some cases, if volatile organics are present in high concentrations, the instrument used for purge and trap may become seriously contaminated. If high concentrations are suspected, a screening analysis of the gas in equilibrium with the sample, a head space sample, can be a good first step, to prevent contamination.

8.6.2 Head space screening for VOCs

Static head space analysis does not involve purging the water with a gas. The sample is sealed into a septum-topped vial. The VOCs equilibrate between the liquid and the vapor space according to Henry's Law. The concentration of an analyte in the head space gas is a function of the concentration in the liquid phase :

$$\text{Conc}_{gas} = K \times \text{Conc}_{liq} \qquad (8.1)$$

where K is a constant which is a function of temperature and the matrix. The vial has a flexible polymeric septum on top which can be punctured by a syringe. A gas tight syringe is used to take a few microliters to a 1-ml sample of the head space gas. This is then injected into a GC or a GC/MS for analysis. It is obvious that the highest values of K will give the best sensitivity, since more analyte will be present in the vapor phase. K can be increased by raising the temperature, and by adding salt, which helps to reduce the solubility of organic compounds in the water.

Automated head space analyzers are also available. When a vent is opened at the far side of a sampling loop, the pressure in the vial forces the sample into the loop. A pressure control at the vent allows the pressure in the loop to be regulated. The contents of the loop are then injected into the gas chromatograph, as a gas sampling valve is rotated.

8.6.3 *Screening for VOCs by solid phase microextraction*

Solid phase microextraction, SPME, is also very well suited to rapid screening of water samples for volatiles and semivolatiles. The coated fiber is inserted into the sample, through a septum, and allowed to sorb the organics present. The fiber is retracted into the carrier needle and then inserted into the heated injection port on a gas chromatograph. The desorbed analytes are swept into the column for analysis.

8.7 Semivolatile organics in water

The designation of semivolatile organics includes a wide range of pollutants: PAH, PCBs, phenols, pesticides, etc. These are low vapor pressure compounds which cannot be purged from water and so must be extracted. The two most common extraction methods are liquid-liquid and solid phase extraction. The extract may require clean-up and concentration before analysis by GC, GC/MS, or HPLC.

Liquid-liquid extraction can be done using a continuous extractor, or extraction can be carried out manually, using a separatory funnel. Commonly used solvents are methylene chloride and hexane. To obtain the best efficiency, two or three successive extractions are performed. An extraction efficiency of at least 70% is usually required.

Solid phase extraction is fast becoming the method of choice because it is rapid and uses very little solvent. A variety of SPE stationary phases are available. The most popular phases are reverse phase HPLC packing materials. C8 and C18 hydrophobic phases are used as well as more polar phases incorporating CN, NO_2, or NH_2 groups. In solid phase extraction, some trial and error method development is required to determine the best washing and eluent solvents. Proper selection of these solvents provides high extraction efficiency and can enhance the selectivity of extraction. It may even eliminate the clean-up step required for complex samples.

8.7.1 *Extractable base/neutrals and acids*

It helps to simplify the extracts, if the organics present in water are separated into classes based on their acidity. When the solution is made basic, acidic compounds, such as phenols, are ionized, and become more water soluble. These are not extracted into the organic solvent. When the water sample is made acidic, their ionization is repressed, and the molecular acids formed are soluble in the organic layer. A dual extraction gives an initial separation, and produces two fractions which can then be analyzed on the columns which are most appropriate for each.

To determine the base/neutrals and acids, a liter of sample is brought to pH 11 using NaOH, and extracted three times with 60 ml of methylene chloride in a separatory funnel, combining the extracts. The extract is dried by pouring it through anhydrous sodium sulfate and is evaporated to a final volume of 1 ml. This is the base/neutral fraction, containing extracted PAH and phthalates. 2,3,7,8-Tetrachlorodibenzo-p-dioxin (TCDD) and other dioxins are also extracted in this fraction. The original aqueous sample is acidified to a pH of 2 and is again extracted with three more 60-ml portions of methylene chloride. This extract is concentrated in the same manner as the previous one, yielding the acid fraction which contains mostly substituted phenols.

The two fractions are analyzed by GC with mass selective detector or by GC with electron capture detector. It is recommended that a surrogate compound, one with similar properties to the analytes, but which is not present in the samples, be added to the sample. Addition of a surrogate before extraction allows correction for losses in extraction.

8.7.2 Pesticides

Organochlorine pesticides show up in water in areas where there is substantial agricultural use of these chemicals. These may also occur in suburban areas where there is runoff from treated lawns. The water samples are extracted in a separatory funnel, using similar volumes to those used for extraction of the base/neutrals, with a mixed solvent of 15% methylene chloride in hexane or 15% diethylether in hexane. Three extractions are made, dried by pouring through sodium sulfate, and evaporated in a Kuderna Danish concentrator to a final volume of 10-ml. A sample of this extract can be injected to give an initial scan of the presence or absence of pesticides. If peaks are present, the extract may be concentrated or diluted to bring them into the range for which the chromatograph is calibrated. Since an ECD may be used for detection, no methylene chloride should be added for dilution at this stage. The initial methylene chloride will have boiled off during the concentrating procedure. The analysis of samples for pesticides requires great care in avoiding contamination, because the analytes are present at such low levels. As with all trace analyses, one must be especially careful to segregate the glassware used with high concentration standards. Also, only the highest purity solvents may be used, because any low volatility material present in the solvent will be concentrated by a large factor when the sample extract is evaporated. Pesticide grade solvents are available which have been tested and certified not to contain unacceptable levels of high boiling compounds.

Several possible interferents, particularly phthalates and PCB, may be present. If these are not well separated from the pesticide peaks, a column clean-up of

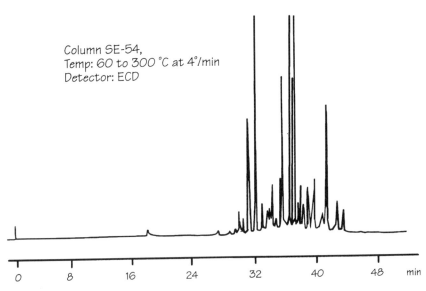

Column SE-54,
Temp: 60 to 300 °C at 4°/min
Detector: ECD

0   8   16   24   32   40   48   min

Figure 8.7 A chromatogram of the pesticide chlordane, separated by GC and detected by ECD.

the extract will be necessary to remove them. This is usually accomplished using a column packed with magnesia and silica gel. The extract is eluted through the column using successively higher concentrations of diethylether in petroleum ether, which will eliminate PCB and separate the pesticides into several fractions, simplifying the chromatography.

To determine extraction efficiency, water may be spiked with a pesticide standard, extracted and concentrated in the same way as samples, and chromatographed. The peak areas can be compared with those generated from more concentrated standards which are injected directly, and the extraction efficiency determined.

The samples may be separated using capillary or packed column GC, but a capillary column will give much better separation, and is less likely to cause irreversible adsorption of the analytes. Figure 8.7 shows a chromatogram of the pesticide, chlordane, run on an ECD GC. The detection limits using the electron capture detector are in the low nanogram per liter range for most of the pesticides. With GC/MS analysis the detection limits are in the low microgram per liter range. Therefore, most pesticide work is done by GC/ECD, although new technology is making the MS detection of pesticides more practical.

Example determination: haloacetic acid and trichlorophenol in drinking water

During disinfection of drinking water with chlorine, monochloro- or dichloroacetic acids can form as natural organic acids, humic and fulvic acids, react with the chlorine. Similarly, phenols can be chlorinated to form chlorophenols. Some of these compounds are known to be carcinogenic. These polar acids are not easily separated by GC, so they must be derivatized to esters. These are more volatile and can be separated on a GC column and detected by a specific detector such as the ECD.

Sampling: Water samples collected from a tap are collected after flushing long enough to eliminate stagnant water in the tap and pipes. The water is allowed to run directly into 40 or 60 ml vials containing 65 mg NH_4Cl to tie up any remaining free chlorine. The samples are taken with no head space to eliminate interaction with air. Samples may be stored at 4°C for a few days, but the analysis should be done as soon as possible.

Analytical procedure: The measured volume of sample is acidified by addition of 1.5 ml of concentrated H_2SO_4, 3 g $CuSO_4$, and 12 g Na_2SO_4. The sample is then extracted with 3 ml methyl tertiary butyl ether (MTBE) to extract the undissociated acidic compounds. The salts are added to increase the extraction efficiency. The sample is shaken with the MTBE on a mechanical shaker until the salts dissolve. The vial is allowed to stand till the phases separate. A 2-ml aliquot of the MTBE extract is pipetted from the top of the vial and is passed through a tube containing anhydrous Na_2SO_4 to dry it. Care should be taken that none of the aqueous phase is taken up with the extract.

The extracted compounds are methylated with diazomethane to produce methyl ester or ether derivatives which can be separated by gas chromatography. For example, dichloroacetic acid is converted to methyl dichloroacetate. The methylation is carried out by adding from 250 μl to 1 ml of diazomethane solution to the sample, until a yellow color persists, indicating the presence of excess diazomethane for esterification.

The extract is analyzed using GC-ECD. Typically, a 30-m, 0.25-mm i.d. capillary column is used, and is temperature programmed up to 250°C. An internal standard such as 1,2,3 trichloropropane or 1,2 dibromopropane can be used to correct for any deviation in injection volume. The detection limits should be in the range of 0. 25–15 μg/l.

8.8 Field methods for water

Many water analyses can be done in the field. Kits are available for colorimetric analyses, which can be done with a hand held, battery powered spectrometer. The kits are designed to avoid the need for much laboratory glassware, with most of the reagents packaged in one-use packets, or dropper bottles. The analysis often requires measuring out a volume of sample, adding a specified reagent packet or several drops of reagent solution, and reading the absorbance of the sample in the spectrometer. Even simpler analyses are done by matching the colors produced with a color chart.

Other water measurements usually done in the field are pH, temperature, and conductivity. For these, pocket sized, battery powered instruments are available. These are dipped into the water, and read directly. Immunoassay kits for petroleum hydrocarbons, PCB, and other pollutant classes are also usable in the field.

Portable gas chromatographs can be used to carry out some analyses for organics in water in the field. Direct injections of small quantities of water can be used for analyses of fairly high level contaminants, at μg/ml levels. Techniques such as SPME and head space analysis can be done in the field, and even small scale purge and trap systems are being made portable.

Suggested reading

APHA-AWHA-WPCF (1989) *Standard Methods for the Examination of Water and Waste Water*, 17th edition.
Boleman, H.J.T. and Burn, J. (1993) *Chemistry and Analysis of Volatile Organic Compounds in the Environment*, Chapman and Hall, New York.
Shipgun, O.A. and Zolotov, Y.A. (1988) *Ion Chromatography in Water Analysis*, Ellis Horwood Series in Analytical Chemistry, Ellis Horwood, Chichester.
Zuane, J.D. (1990) *Drinking Water Quality Standards and Controls*, Van Nostrand Reinhold, New York.

Study questions

1. How would you ensure that there was no interference when performing an analysis of a seawater sample for chromium, using flame atomic absorption?
2. If they are not the same, which would most likely be higher, the BOD or the COD? Explain how this can happen.
3. A sample shows a total carbon of 3.6 mg/l. After it is purged with nitrogen for an hour, the new total carbon measurement is 2.9 mg/l. When another aliquot is acidified, purged with nitrogen and then analyzed, the total carbon is 2.3 mg/l. What is the concentration of purgeable organics, non-purgeable organics and of inorganic carbon?
4. What is the purpose of adding sulfamic acid to a sample before conducting the COD test?
5. A water sample is analyzed for chloroform by using a purge and trap system. A 5-ml water sample is purged for 10 min with nitrogen. The trap is dried for a minute, then the trap is heated and the chloroform is desorbed into the GC column. The chloroform peak area from the sample is 23 823. A 10 μl injection of a 20 μg/ml chloroform standard in methanol gives a peak area of 44 523. What is the concentration of chloroform in the water sample in μg/l?
6. A water sample is tested for lead by adding a complexing agent to a 100-ml sample of the water. The complexed lead is then extracted into hexane, and the hexane volume adjusted to 5-ml. A standard is prepared by dissolving 0.0123 g of dried Pb(NO$_3$)$_2$ in 100 ml of water and diluting 1 ml of this solution to 100 ml. The standard gave a reading of 0.493 on the atomic absorbance spectrometer, while the sample gave 0.346. Extraction of 100 ml of distilled water with complexing agent and hexane gave a blank which read 0.012. Calculate the lead content of the water sample in mg/l.
7. What factors must be considered in the optimization of the parameters (flow, purge time and temperature) for a purge and trap method for organics in water?
8. Why does the extract have to be concentrated prior to analysis for low concentration samples?

9. Why is sample clean-up necessary?
10. How would you carry out the following analysis? Give all the necessary steps required and also mention your choice of instruments.
    a. Hg contamination in a river
    b. Parts per trillion level of Pb in drinking water.
    c. PAH in river water.
    d. Gasoline contamination in ground water.

9 Methods for analysis of solid samples

Contamination of soil is now recognized as being as important as air and water pollution. Soil contaminants leach into ground water as rain water percolates through it. Food crops may absorb and concentrate contaminants. Even children may be exposed when they play in contaminated soil.

Most land contamination occurs in the disposal of mining and industrial wastes, or from land disposal of domestic and municipal waste. The mining and industrial waste can contain high concentrations of specific chemicals which may be very toxic, while domestic waste usually is very heterogeneous and contains fewer toxic chemicals. With the explosion of consumer goods, our society currently produces an enormous quantity of domestic waste which is generally buried in landfills. Before 1970s, landfills were often chosen haphazardly. As a result, contaminants have leached from many of them and contaminated soil and ground water. Today, sites are selected for their geological characteristics and may be lined with clay and graded to reduce percolation of leachate.

Another source of soil contamination is leakage from underground storage tanks, especially gasoline tanks. Over the years, these tanks can leak millions of liters and can contaminate large areas of soil as well as ground water. Many old industrial sites are found to be contaminated because, in the past, chemicals were used for variety of activities and little attention was paid to their proper containment. For example, solvents used for degreasing machine parts often were drained directly onto the ground surface. Over the years, this could contaminate large areas of land. More dangerous chemicals were injected into underground disposal wells. This type of disposal is now strictly regulated. Military installations, where ordnance was tested, have now been found to be contaminated with residues associated with explosives.

Airborne pollutants also contribute to land contamination, from both wet and dry deposition. For example, lead contamination from auto exhaust still exists widely in urban areas. Airborne pollutants from coal burning utilities, petroleum refineries, and smelters can be deposited over large areas of land.

A variety of regulations have been put forward and implemented to reduce the contamination of land. Some of these address the management of hazardous waste. Regulations are being enacted by governments all over the world to protect land from such contamination and to manage hazardous wastes. In 1983, an Act was passed in the US to fix responsibility for clean up of chemical contamination due to past activities. This landmark Act was called Comprehensive Environmental Response, Compensation and Liability Act (CERCLA), commonly known as the Superfund Act. This act uses taxes and other means to generate funds to clean up contaminated land sites and ground water.

Some common strategies for preventing contamination of the land and ground water are:

- Control of the location of industrial facilities.
- Clean up of hazardous waste sites.
- Regulation of solid waste disposal sites, such as landfills.
- Regulation of the discharge of wastes.
- Implementation of recycling to reduce the overall waste burden.
- Regulation of underground waste disposal wells.
- Testing and monitoring of underground storage tanks.
- Control of the use of agricultural chemicals.

Solid environmental samples may consist of soil samples, perhaps from a contaminated site or a landfill. They may be parts of animals or plants which require analysis to determine the amount of certain pollutants they have absorbed from their surroundings or from their food. Agricultural products, ranging from watermelons to wheat, may require analysis for pesticide residues. The fact that these materials are solid immediately makes the process of obtaining a representative sample even more difficult than it is for air or water, where one can assume reasonable mixing, at least on a local scale. For solids, a substantial difference in the concentration of a specific compound may exist even from one grain of wheat to the next. Analytical methods usually require small samples, only a few grams or even milligrams of sample, which requires even more exacting sampling. Consider the problem involved in obtaining representative 1-g samples of soil from a contaminated landfill, which may cover an area of thousands of square meters.

It is not possible to discuss all the analyses of solids in detail. However, analyses of solids have certain common components. Some of the most widely analyzed solid matrices are soil, sludge, and particulates from air and water. Methods for air and water particulates were described in previous chapters. Here much of the discussion will center around soil, but the same principles are applicable to other solid matrices.

Soil is a complex matrix just as many other environmental materials are. Many components occur naturally in soil, and others are introduced into it by human activities. Table 9.1 lists the concentration of some species naturally found in soil and plant materials. One can see the wide concentration ranges encountered in environmental samples.

9.1 Sampling

The basic rule of sampling for solids is that the variability due to sampling becomes greater as the particle size of the material becomes greater, as the sample size becomes smaller, and as the concentration of the analyte becomes more variable from particle to particle. By applying statistics to the sampling process, the variability due to sampling can be determined and controlled. In environmental samples, the statistical information required for such a treatment may not be readily available. However, even if the exact statistics of a given sampling task cannot be calculated, knowledge of basic sampling theory will assist us in obtaining the most representative samples possible.

As always, the purpose of the study is important. In an area, for example, where the soil is contaminated by chromium, we may decide to try to determine the maximum concentration of chromium in the area. Or, we might try to determine the integrated concentration over the entire area. Either of these goals can be justified in terms of effects on the health of organisms in the environment. A small spot with a high concentration can have serious health effects if, for instance, it is in an area where children might play and dig, even if the integrated concentration over a larger area appears to be reasonably safe. On the other hand, a wide area overall concentration may better show the concentrations to which people moving about the area are exposed.

The important consideration is that the plan be made, and the goals set, before sampling is attempted. A sampling plan which attempts to obtain a representative overall picture of the concentration of the chromium in the area might involve laying out a grid over the area, and taking a specified size sample at each

Table 9.1 Concentrations of some naturally occurring species

Substance	Concentration
Usual concentrations in soil	
Potassium	1–2%
Exchangeable potassium	100–200 µg/g
Manganese	20–3000 µg/g
Lead	15–25 µg/g
Mercury	0.03–0.08 µg/g
Arsenic	0.2–40 µg/g
Bromine	0.3–40 µg/g
Fluorine	around 300 µg/g
Molybdenum	0.1–40 µg/g
Cobalt	0.2–30 µg/g
Sulfur	100–500 µg/g
Aluminum	0.5–5.0%
Iron	3–10%
Selenium	0.1–100 µg/g
Usual concentrations in plants	
Chromium	0.003–1 µg/g
Fluorine	around 3 µg/g
Lead	0.2–12 µg/m
Manganese	20–250 µg/g
Bromine	5–120 µg/g
Selenium	0.002–1 µg/g
Cl^-	around 35 µg/g

grid intersection. If a randomized sampling plan is chosen instead, the plan should be made up before the sampling begins, perhaps by selecting grid squares to be sampled, using a random number generator. If the sampling personnel are allowed to select what seem to them to be random, representative samples, the chances are that the sample will be biased, since a "good sampling spot" will differ between different operators, and may depend, consciously or unconsciously, on what they expect or hope to find.

The search for a high concentration "hot spot" is quite different. In this case, the operator's judgment and powers of observation must come into play. Knowledge of where the particular pollutant would be likely to collect, and how it might migrate from the point of discharge would be useful. However, even in this case, a properly representative sample of the chosen area, even if this area is small, will still be required. In the chapter on sampling, apparatus for taking some samples is discussed.

9.1.1 Preparation of solid samples

Solid samples often require reduction of particle size before analysis. As the size of the particles in a sample is lowered by milling, crushing, or grinding, the amount of the sample which must be taken for a representative analysis becomes smaller. An extraction procedure will probably be done to remove the analytes of interest from the sample. This also requires small particles, to improve the mass transport from the sample into the extract. Therefore, field samples which consist of large chunks of material usually require a process to reduce their size before analysis. This step must be carried out with great care to prevent both contamination and loss of analyte. For instance, a fish liver to be analyzed for PCB may be ground in a blender before being extracted, but if the same sample was to be analyzed for trace metals, a metal blade in the blender might cause contamination. Finally, if the sample was to be analyzed for volatile compounds, such as chlorinated organics, a blending step would allow the volatiles to escape.

Results are usually reported on a dry basis, because moisture content can vary from day to day. Drying of samples is important for many analyses, as moisture can interfere. For example, solvent extraction of organics from solids with hydrophobic solvents is difficult with wet samples. A sample can be dried by heating it to a temperature of 100°C in an oven. However, this might result in a loss of volatile metals and organics. Another method of drying is by mixing the soil with anhydrous Na_2SO_4 or $MgSO_4$, which take up moisture to form hydrated salts. Alternately, samples may be air dried at 20°C or 30°C. For very fragile samples such as biological tissues, which might be subject to decomposition if air dried, freeze drying under a vacuum is possible. Samples are often analyzed without drying, to avoid baking the sample into a hard mass and to prevent the more refractory metals from being converted into less soluble oxides.

If an analysis is done on a wet sample, the results may still be reported on a dry basis. A separate portion of the wet sample is used to determine the total solids, by drying a weighed amount to constant weight. Then the analyte concentration can be calculated on a dry basis.

Example

A 1-g wet sludge sample is extracted and found to contain a total of 0.023 g of petroleum oil. A 2-g sample of the same sludge is dried and found to contain 1.259 g of solids. What is the percentage of oil on a dry weight basis?

$$\frac{0.023 \text{ g oil}}{1 \text{ g wet sample}} \times \frac{2.00 \text{ g wet sample}}{1.259 \text{ g dry sample}} \times 100 = 3.65\% \text{ oil}$$

The amount of sample which must be carried through the analytical procedure depends both on the concentration in the sample and the sensitivity of the instrument. The sample may need to be concentrated or diluted to bring the amount of analyte into the optimal range for the instrument to be used. It is important to decide whether one wishes to determine major components of the sample, minor components or traces. A major component is one which is present at levels above 1%. Traces are present below 1000 mg/kg, or 0.1%, while minor components fall between these limits. There are several techniques which may be applied to bring the sample into a state suitable for analysis, depending on the analytical method to be used.

The analytes are also divided into general classes of substances which are determined by similar methods. These classes include metals, volatile and semivolatile organics, and inorganic ions and elements. The method is chosen based on the class or classes of analytes to be determined. The detection limits of the method must also be considered. If the analysis is being done to meet a legal regulation, the detection limit must meet the requirements of the regulation.

In most cases, samples will not be homogeneous. The division of the field sample into test portions for analysis must be done carefully to avoid bias. For instance, a wet sludge sample is composed of both a solid phase and a liquid phase. The analyst has to decide whether to dry the sample and convert it all to solids, or to blend the sample to distribute the solids and moisture evenly before dividing it.

9.2 Measurement of soil pH

Soil pH, one of its most important properties, is a measure of the activity of H^+ in the soil solution. A soil with a pH below 4 may contain free acids from processes such as the oxidation of sulfides. A pH below 5.5 may indicate the presence of exchangeable aluminum, and a pH in the range of 7.7–8.2 often indicates the presence of $CaCO_3$. Based on pH measurements, corrective actions may be taken. For example, lime may be added to acidic soil to raise the pH. The soil pH is influenced by the nature of the organic and inorganic species present in soil. These may be residues of fertilizer, may have been transported by irrigation water, or formed as a result of microbial action. The carbon dioxide content of soil also affects the pH. It is usually present in the dissolved form as H_2CO_3.

To measure soil pH, 5 g of air dried soil is suspended in 5 ml of distilled water. The mixture is shaken together and allowed to stand for a few minutes. The pH is then measured with a pH meter. The pH can also be measured using a colored indicator or pH paper. Field test kits based on indicators are commercially available.

9.3 Analysis of metals in soil and solids

Metals occur naturally in soils, biological samples, and other environmental solids. Wastes generated by mining and manufacturing may contain many metals. Metals present in waste water end up as solid waste in water treatment sludge. Sometimes sludge is recycled onto land as an agricultural soil conditioner. This practice may raise the metal concentration of soil. Metal is deposited from air as particles settle. For example, auto exhaust is a major source of lead in urban dust wherever leaded gasoline is used. Metals present in soil can be taken up by plants and passed into the food chain.

The preparation of soil or sludge samples for metals analysis is similar to that used for extraction of metals from air aerosol filters. The representative sample is digested in nitric acid and hydrogen peroxide, and then refluxed with either nitric or hydrochloric acid. If the extract is very dilute, it can be concentrated by chelation and extraction with an organic solvent. The extract can also be evaporated, but there is always a possibility of sample loss by spattering or by entrainment of particles in the vapor.

Because soils and sludges can be complex and unpredictable, the analyst must carry out appropriate tests to be sure that the analyte is being efficiently extracted. Spikes of analyte into the same matrix should be made, stored for a sufficient time to ensure thorough equilibration, and extracted. If a standard reference material of similar composition is available, that material should also be carried through the analytical scheme, to ensure the accuracy of the analysis.

After extraction, the metal content of the extract must be determined. Many methods are available to detect and quantify the concentration of the metals in the extract. Spectroscopy, electrochemical techniques, or even colorimetric reactions may be suitable. Most soil or biological samples are complex and contain many metals and other components. While all methods are subject to some possibility of interference, atomic spectroscopy is relatively free from interference and is by far the most popular technique for metal analysis. Its detection limits are generally in the parts per billion range. However, these instruments are more expensive than electrochemical and colorimetric tests.

Which atomic spectroscopic method is chosen depends to some extent on the detection limits necessary. Atomic absorption spectroscopy, either flame or electrothermal, inductively coupled plasma emission spectroscopy or ICP/mass spectrometry, are the methods most often used. Flame emission and atomic fluorescence, are used less commonly. Cold vapor absorption is used for mercury determination, and hydride generation for metals such as arsenic and selenium, which are subject to serious interference in flame AA. ICP-AES and ICP-MS are by far the most sensitive instruments, providing simultaneous multi-element analysis. However, they are also more expensive. In some applications, graphite furnace atomization may be convenient because it may eliminate the need for sample preparation. For example, to determine lead levels in blood, a few microliters of sample can be injected directly into the graphite furnace.

Example

A sludge sample was homogenized in a blender and a 1-g sample was weighed out and dried to constant weight. Its final weight was 0.668 g. A second portion of wet sludge, weighing 1.283 g was mixed in a beaker with 10 ml of 1:1 nitric acid, covered with a watch glass and heated below boiling for 10 min. An additional 5 ml of concentrated HNO_3 was added and the sample heated for 30 min. A second 5-ml portion of HNO_3 was added and the digestion repeated.

The sample was heated until almost all the liquid was evaporated, and 3 ml of 30% hydrogen peroxide with 2 ml of water was added. The sample was warmed gently until all the effervescence subsided, then it was cooled. Further additions of H_2O_2 were made until no more reaction was observed. A 5-ml portion of concentrated HCl was added and the sample was refluxed for 15 min. The sample was then diluted to 100 ml and allowed to settle until clear liquid was present at the top of the flask. The clear liquid sample was aspirated into the AA, as were several lead standards. The concentration of lead in the analyzed solution was determined to be 2.9 mg/l, by comparison with the calibration standards. What is the concentration of lead in the original sludge, on a dry basis?

First calculate the amount of solid sample on a dry basis present in the portion taken for analysis.

$$\frac{0.668 \text{ g solid}}{1 \text{ g sludge (wet)}} \times 1.283 \text{ g sludge (wet)} = 0.857 \text{ g dry solid}$$

Then calculate the amount of lead in the final 100 ml of solution upon which the analysis was done.

$$100 \text{ ml solution} \times \frac{2.9 \text{ g lead}}{10^6 \text{ ml solution}} = 2.90 \times 10^{-4} \text{ g Pb}$$

Finally, calculate the mg of lead which would be present in 1 kg of dry sample.

$$\frac{2.9 \times 10^{-4} \text{ g Pb}}{0.857 \text{ g dry solid}} \times \frac{10^3 \text{ g}}{\text{kg}} \times \frac{10^3 \text{ mg}}{1 \text{ g}}$$

$$= 338 \text{ mg/kg lead in sample on dry basis}$$

9.3.1 Determination of mercury

Background levels of mercury in soil can be between 30 and 80 ng/g, but mercury contamination is quite extensive all over the world. Recycling of sewage wastes, and contamination from landfills, have made it an important pollutant. Mercury exists in organic as well as inorganic forms in soils, sludges, and other solids. Inorganic mercury may be transformed into highly toxic methyl mercury in soil by chemical or microbiological processes. Since volatile mercury species can be lost during drying, the analysis is often carried out on a wet subsample and the results are corrected for the moisture content. If sample is dried, it should be done carefully at a low temperature.

Speciation of mercury into the different forms in which it exists is not a simple task. Total mercury is measured by converting all forms of mercury into Hg(II) in solution by a digestion and oxidation process. This is converted to mercury vapor and is determined by the cold vapor atomic absorption method. The sample is digested with a mixture of strong oxidizing acid such as HCl, HNO_3, or $HClO_4$, in the presence of a strong oxidizing agent such as $KMnO_4$ or $K_2Cr_2O_7$. This ensures complete oxidation of any organic matter present. Digestion is carried out at a temperature below 60°C to reduce vaporization losses. An oxidizing environment needs to be maintained so that free mercury is not formed.

The digest is treated with a reducing agent, $SnCl_2$, to produce free metallic mercury. The mercury is then purged from the solution into the absorption cell for measurement.

9.3.2 Determination of arsenic and selenium

Arsenic and selenium determinations are carried out using the hydride generation method. The soil is digested in acid. The extract is treated with $NaBH_4$, generating arsine. A flow of argon is used to sweep the arsine into a heated quartz cell for analysis by atomic absorption. The hydride generation method has been described in the previous chapter.

Example determination: simultaneous measurement of metals such as Pb, Cr, Fe, Cd and Ba in dry sewage sludge

Since sewage sludge has a high organic content, HNO_3–$HClO_4$ digestion is preferred over the more common HNO_3 extraction. This strongly oxidizing digestion decomposes organics quickly and efficiently. Care must be taken because $HClO_4$ can react violently with organic material, especially when the acid is added to a hot solution. The sample is first digested with HNO_3 before $HClO_4$ is added. A few milliliters of concentrated HNO_3 is added to 1 g of the sludge in a flask or a beaker and heated on a hot plate. After visible reaction ceases, the flask is cooled and 10 ml each of HNO_3 and $HClO_4$ are added, and the flask is reheated. The beaker is covered with a ribbed watch glass to provide a reflux, and heating is continued till white fumes of $HClO_4$ are generated. The sample is diluted with water, and is boiled gently until a clear solution is obtained, allowing any chlorine or oxides of nitrogen to be expelled. The solution is cooled and filtered.

For simultaneous multi element analysis, ICP-AES is the method of choice because it is a rapid and sensitive method and has a long linear dynamic range. Attention should be paid to interference if the concentrations are above 1500 mg/l. The sample digest is nebulized directly into the ICP torch. Suggested wavelengths for Pb, Cr, Fe, Cd and Ba are 220.35 nm, 267.72 nm, 259.94, 226.5 and 455.4 nm, respectively. However, if spectral interference occurs, other wavelengths can be used. The method of standard addition can be used for calibration, because it reduces matrix interference. The amount of added standard should be between 50 and 100% of the analyte concentration. This ensures that precision is not sacrificed, and spectral interferences which depend upon element/interferent ratios are similar for both the sample and the standard. Standard addition can be used for all the elements in the sample if the added standards do not cause serious spectral interference.

9.3.3 Analysis of solids with X-ray fluorescence

A method which does not require extraction of the analyte from its matrix, X-ray fluorescence, is useful for screening solid samples, sometimes in a non-destructive manner. Portable X-ray fluorescence is becoming popular for on-site analysis. A hand held X-ray fluorimeter is available for screening for lead-based paints in older homes.

For quantitative and semiquantitative X-ray fluorescence analysis, soil samples are first ground to a particle size smaller than 50 µm. Care must be taken that the grinder does not contaminate the sample. Biological tissues, such as plant materials, are first oven dried and then are chopped finely. These particles can then be ground to 50 µm particle size using a mechanical grinder. The ground sample is poured into a pellet press. Borax or cellulose is added around the edges and on the top of the sample to form a backing for the pellet. Pressure is applied to form the pellet. Approximately 2 g of sample is usually adequate to produce a pellet suitable to determine most metals. The pellet is then placed into the X-ray fluorescence spectrometer and the measurement is done. Standards are made up in similar pellets.

9.4 Analysis of soil for total nitrogen

The Dumas method is an automated method for analysis of total nitrogen in solids. A few milligrams of fine dry soil is placed in a boat and is introduced into a furnace at a temperature above 1000°C. The sample is rapidly heated in a flow of oxygen. Combustion produces products such as CO_2, H_2O, N_2 and NO_x. These gases flow into a furnace containing copper at 650°C. Oxygen is removed, and NO_x is reduced to N_2. Then the effluent passes through traps to eliminate CO_2 and H_2O. Finally the gas passes into an on-line gas chromatograph which separates and detects the N_2. The whole analysis takes about 3 min.

9.5 Colorimetric tests for soil and sludge analysis

Colorimetric tests may be used for measuring a variety of components in soil. These tests can be used to measure ions (e.g., NO_3^-, SO_4^{2-}), elements (e.g., N, S, P, C), metals (e.g., Pb, Cr, Zn, Cd) and organic compounds in soil. These tests usually involve transferring the component of interest into a solution by extraction, or fusion with Na_2CO_3. A colorimetric reagent is added to the extract to form a species which is determined quantitatively by UV–Vis spectrophotometry. These methods do not usually have the sensitivity of other instrumental methods, are usually prone to interference, but they may be simple and inexpensive. They are often suitable for use as portable field methods. A few examples will be given to demonstrate their applicability.

9.5.1 Determination of phosphorous

Phosphorous is an important component in soil because it is an essential nutrient for plants. For total phosphorous analysis, it is important to convert all the insoluble analyte into a soluble form. About 1 g of soil is fused with 4–5 g of Na_2CO_3 in a platinum crucible at around 1000°C. The fused sample is dissolved in sulfuric acid and filtered. The pH of the solution is adjusted with Na_2CO_3, and ammonium molybdate is added. A blue phosphomolybdate complex is formed, and its absorbance at 380 nm is measured. The detection limit is between 1 and 3 µg/ml of total phosphorous in the extract.

9.5.2 *Determination of aluminum*

Aluminum occurs in several forms in soil, depending upon pH and the geo-logical composition of the soil system. It plays an important role in soil acidity. Aluminum can be toxic to plants in acid soils if the concentration is high. It can be determined by X-ray techniques, and atomic spectroscopy, as well as by colorimetry.

The colorimetric determination of aluminum is prone to interference from other metals, such as Fe, Ca, and Mg. These are removed prior to analysis. Aluminum can be extracted from soil or sludge by Na_2CO_3 fusion as described above. The fusion residue is boiled with $HClO_4$ to dehydrate the silica which is removed by centrifuging. Aluminum, iron, and titanium are then precipitated by adding NH_4OH. This leaves calcium, magnesium, and manganese in solution. The precipitate of aluminum, iron and titanium is dissolved in hot, dilute HCl. NaOH is added to precipitate iron as $Fe(OH)_3$, leaving the aluminum in solution to be determined colorimetrically.

Several colorimetric tests are available for aluminum. A mixture of 1% 8-hydroxyquinoline and hydroxylamine hydrochloride in acetic acid is added to the extract. Then 1 N sodium acetate containing 0.2% phenanthroline at pH 5 is added. Any remaining iron is complexed with the phenanthroline. An aluminum complex with 8-hydroxyquinoline is formed which absorbs at 395 nm.

9.6 Measurement of total organic carbon in soil

Total organic carbon is measured by heating the soil in a furnace in presence of oxygen at 950°C (Figure 9.1). All carbon in the sample is converted into carbon dioxide at these temperatures. The evolved carbon dioxide is measured using infrared absorption. A correction for carbonate is required, since carbonate also produces CO_2 when heated. About 0.5–1.0 g of soil is ground to pass through a 0.5-mm sieve. If carbonate is present, the sample is heated to 80°C with 0.5 ml of 2 M HCl. This converts any carbonate to CO_2, which volatilizes. The sample is then placed in the furnace, heated, and the effluent gas is passed into an IR

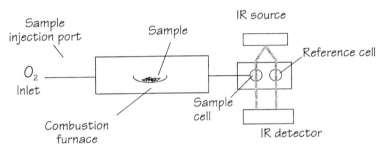

Figure 9.1 Apparatus for determination of total organic carbon in solid samples.

cell where the CO_2 absorption band is measured. Automated, computer controlled instruments are available for this analysis.

9.7 Volatile organics in soils or solids

The determination of organic volatiles in soil requires careful sampling, with the samples being sealed into containers with minimal exposure to air. Methanol is often added to the soil samples to help retain the VOCs. It is difficult to divide the gross samples after collection, because the usual grinding, sieving, blending, and dividing processes cannot be carried out without losing the analytes of interest. The determination of VOCs in soil, sludges, and solids takes advantage of the high vapor pressure of VOCs. The determination is similar to that for VOCs in water and the same methods are used, namely purge and trap and head space analysis. Solid phase microextraction has also been used.

9.7.1 Measurement of VOCs using purge and trap

The apparatus used for purging and analyzing volatiles from aqueous samples can be adapted for soil analyses. Solids are commonly suspended in a liquid before purging. Usually water is used, but other liquids such as methanol and polyethylene glycol can be added to the water in some applications. The purge chamber is designed so that the solids do not impede the flow of gas. An apparatus for solids purging is shown in Figure 9.2.

Purge and trap gives better precision and accuracy than some of the other methods for VOCs in solids. It is usually satisfactory for compounds with

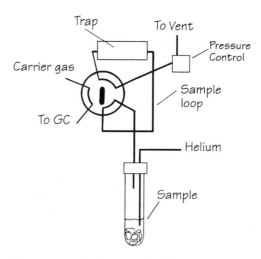

Figure 9.2 Purge and trap apparatus for the removal of VOC from solids.

boiling points lower than 200°C. This includes aliphatic and aromatic hydro-carbons, halogenated hydrocarbons, alcohols, ketones, acrylates, nitriles, ace-tates, and ethers. The limit of quantitation is approximately 10 times higher for water-soluble polar compounds such as alcohols and ketones, because these are not as efficiently purged. Sometimes, salt is added to the purge matrix to decrease the solubility of polar compounds.

Standards can be made by spiking a similar solid matrix (for example, precleaned soil) with known quantities of analyte, and analyzing by the same method as the sample. Purging of volatiles is more difficult from soil than from water. Often a known quantity of a spike or a surrogate compound is added to each sample. For example, a deuterated benzene, which will not be present in the soil, can be added and detected by GC/MS. A recovery of at least 70% of the surrogate ensures good purging efficiency. Purge and trap instruments are prone to sample carryover. It is advisable to run deionized water blanks between samples to ensure that nothing is left over from a previous run.

9.7.2 Head space analysis

Head space analysis described for water samples can also be used for measuring VOCs in solids. This technique suffers from low precision, so is mainly used for screening large numbers of samples rather for than quantitative determinations. If a preliminary analysis using this method shows contamination, then the more elaborate purge and trap method may be used for quantitation. Detection limits vary widely among samples depending upon matrix, and the type of sample.

The weighed sample is placed in a container of known volume and allowed to equilibrate at a constant, known temperature, usually between 70 and 90°C. Figure 9.3 shows a solid sample placed in a vial with a septum closure for head space analysis. Water or other liquid is sometimes added. The vials can be placed to equilibrate in a water bath at elevated temperature. An ultrasonic bath may be used to agitate the vials containing soil and water, so that the particles are broken up, and the VOCs equilibrate with the water.

A sample of the air above the sample is withdrawn through the septum with a gas-tight syringe, and injected into a GC or GC/MS for analysis. Quantitation, however, is not always straightforward for these types of samples. Clean soil spiked with a known quantity of analyte may be used as a standard.

9.7.3 Determination of VOC using solid phase microextraction

Solid phase microextraction (SPME) can also be used for concentrating and injecting volatiles from the head space. A coated fiber is exposed to the head space. The fiber is placed into the head space through the septum cap on the sample container. After adsorption, the fiber is inserted into the heated injection port of the gas chromatograph where the analytes desorb and are swept onto the column for separation and quantitation. Control of both temperature and sample

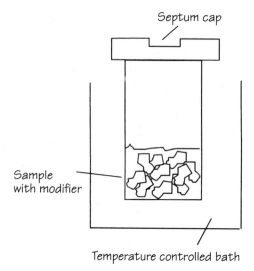

Figure 9.3 A sample being prepared for a head space analysis. A modifier, a salt which decreases the solubility of the organic in the liquid, may be used to increase the head space.

equilibration time is important. Calibration is usually done by preparing standards in a matrix similar to that of the sample, and using SPME to concentrate and inject the analyte from the standard under the same conditions as the samples. Because accurate calibration is difficult, this is used only as a screening technique.

9.8 Total petroleum hydrocarbon in soil and sludge

To determine total petroleum hydrocarbons, the soil or sludge is extracted using trichlorotrifluoroethane. The C–H stretching band for the hydrocarbons in the extract is measured using an infrared spectrophotometer. Because the chlorofluorocarbon solvent contains no C–H bonds, it does not interfere with the IR measurement. The analyte is usually removed from the soil by Soxhlet extraction. If a wet soil or sludge is to be analyzed, it is not dried by heating, as volatile hydrocarbons will be lost. The soil or sludge is mixed with $MgSO_4.H_2O$, which can absorb 75% of its own weight in water, forming $MgSO_4.7H_2O$. After drying, the sample is extracted using the chlorofluorocarbon solvent. This method is being modified to avoid the use of CFC solvents. Supercritical extraction with CO_2 has been developed as a less environmentally damaging method for extraction of petroleum hydrocarbons. A major advantage of supercritical extraction is that it takes only about 10–30 min.

The extract is analyzed by infrared spectrometry using a 1-cm path length cell. Standards and samples are scanned between 3200 cm^{-1} and 2700 cm^{-1}.

Total petroleum hydrocarbons are determined in water by a similar procedure, using a liquid–liquid extraction with CFC, followed by IR quantitation.

9.9 Semi-volatile organics in soil and sludge

Semi-volatile compounds such as PAH, PCBs, and pesticides may be strongly bound to the solid matrix and must be extracted before analysis. The major approaches are Soxhlet extraction, sonication and supercritical extraction. In addition to extraction, several clean-up steps may be required to remove interference from the samples before analysis by GC, GC/MS or HPLC.

For extraction with organic solvents, it is necessary to mix the soil thoroughly with anhydrous Na_2SO_4, to absorb moisture. Presence of moisture prevents the hydrophobic solvents from efficiently contacting the surface of the particles to remove the organic pollutants. The extraction efficiency is checked by spiking the matrix with a known quantity of an analyte and extracting it under the same conditions as the sample. Usually, a minimum extraction of 70% is expected.

9.9.1 Dioxin in a herbicide

Let us examine in detail a rather complex analysis of a herbicide preparation, 2,4-D, to determine its dioxin content. The dioxin in these materials is present in the low parts per billion range, and careful technique is required to be able to measure it at all. In addition, because of the high toxicity of dioxins, suitable precautions must be taken throughout the analysis to protect the analyst and to avoid having waste materials become a hazard themselves. In this method, the initial extraction removes not only the analyte but also a series of compounds which will make the final analysis more difficult. Since the analyte is present in much lower concentrations than the interference, this must be reduced to acceptable levels before analysis is attempted. In addition, the analyte must be concentrated to obtain enough to be measured. A flow diagram of the analysis is shown in Figure 9.4.

The sample of herbicide powder is dissolved in a mixture of acetonitrile, water and methanol. The solution is extracted three times with 100 ml of hexane. The combined hexane extracts are extracted with methanol/water to remove polar interference, and the methanol wash is discarded. The extract is then extracted with sodium hydroxide solution, hydrochloric acid solution and with water, in succession with the aqueous phases each being discarded. The acid and basic washes remove any compounds which are either acidic or basic in nature, leaving the neutral dioxin. This washed hexane extract is filtered through sodium sulfate to dry it, and then evaporated to 25 ml. It is washed with concentrated sulfuric acid and with water and is dried again with sodium sulfate and further evaporated to about 5 ml.

At this point the extract is placed on an alumina column and first hexane, then 2% methylene chloride in hexane, is passed through the column. Finally the analyte is eluted with 200 ml of 30% methylene chloride in hexane. The initial wash with hexane takes off the most non-polar compounds, with the wash containing 2% methylene chloride removing the slightly more strongly held compounds. The final elution with 30% methylene chloride removes the dioxin.

The proper mixture of solvents to use for elution can be determined by placing the desired compound on the column and applying successively stronger washes, to determine which mixture of solvents is just strong enough to move the compound of interest. Then, weaker mixtures may be used to separate off the less well retained compounds, and those compounds even more retained than the desired analyte will be left behind. In essence, this is programmed column chromatography, but done in batches, rather than with a continuously changing solvent flow. The experience gained by doing HPLC separations of the initial extracts may be very useful in setting up a column clean-up method.

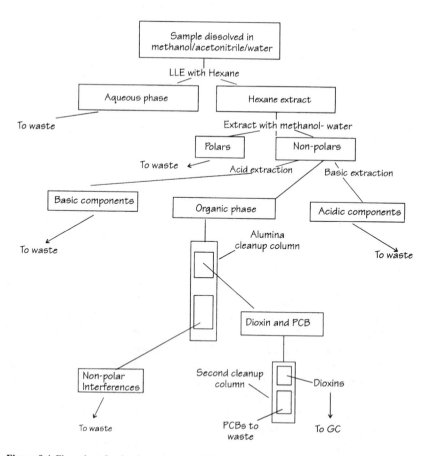

Figure 9.4 Flow chart for the determination of dioxin in a herbicide preparation.

If there are any PCBs in the sample, these will interfere, because their molecular properties are not much different from those of dioxins. Therefore, PCBs will go through the separation scheme in the same way as the dioxins. A second packed column separation will have to be performed to eliminate these, before the GC/MS analysis is done. An alumina column is prewetted with hexane and 1.0 ml of the sample is applied. Then the column is washed with carbon tetrachloride to remove PCBs. Finally, the dioxins are removed with methylene chloride, which is evaporated to near dryness, and taken up in 1.0 ml of methylene chloride for analysis.

The solution recovered from the column which contains the target analyte is evaporated again to 1.0 ml, and is injected into the GC/MS for analysis. A methyl silicone column is used, and detection is done in the specific ion monitoring mode, since the amounts to be detected are very low. The peaks to be monitored are the dioxin peaks containing 1 to 8 chlorine atoms, m/e ratios of 218, 252, 288, 322, 356, 391, 426, and 460 amu.

In a complex analysis such as this one, it is important to keep track of all the dilution steps for the final calculation of the concentration. Sometimes, standards are taken through all the steps, using the same dilutions and therefore the final results are directly proportional, between sample and standard. However, in many cases the GC/MS is calibrated against a prepared solution of the desired compound in a suitable solvent, injected directly.

Example

Using the above method and the following data, let us calculate the concentration in the sample.

Initial sample weight	0.5098 g
Standard solution of TCDD	0.500 ng/ml
Area of 288 peak (standard)	28 946 units 0.10 µl injected
Area of 288 peak (sample)	34 804 units 0.20 µl injected

All volumes are as indicated in the method description.

$$\frac{0.500 \text{ ng/ml in standard}}{28\,946 \text{ (standard)}} \times 34\,804 \text{ (sample)} \times \frac{0.1 \text{ µl (std)}}{0.2 \text{ µl (sample)}}$$

$$= 0.301 \text{ ng/ml in sample solution}$$

Looking through the analysis method, we see that all the dioxin extracted from the original sample was carried through all the steps and was finally diluted to a volume of 1 ml. Therefore, there is 0.301 ng of dioxin in the whole sample, which weighed 0.5098 g.

$$\frac{0.301 \text{ ng dioxin}}{0.5098 \text{ g sample}} \times \frac{1 \text{ mg}}{10^3 \text{ ng}} \times \frac{10^3 \text{ g}}{\text{kg}}$$

$$= 0.59 \text{ µg/kg or ppb in original sample}$$

9.9.2 *Organochlorine pesticides and polychlorinated biphenyls in solids*

The method for organochlorine pesticides (EPA Method 8080A) provides another example of a determination of traces of pollutants in a solid sample, including soil, plant material or animal tissue. The method will determine these compounds at low levels, but it is difficult to specify what the detection levels are, because these depend more on the concentration of interferents than on the instrumental method. The method detection limits for pesticides which can be determined by this method, assuming no serious interferences are present, are listed in Table 9.2.

When interference is present, a Florisil column clean-up may be needed to reduce the interference. Commonly, phthalate esters are found to interfere. These substances, used in the manufacture of many types of plastic, are nearly ubiquitous in the laboratory as well as in the environment. The electron capture detector, usually used in these analyses, responds well to phthalates, which show up as large peaks, late in the chromatogram. Because the phthalates are easily leached from many flexible plastics, these materials should be avoided as much as possible when working with trace analyses using the ECD. Special glassware clean-up may also be required. Finally, the presence of phthalates in the sample may well be real, as these materials are widely distributed. If the

Table 9.2 Pesticides which may be determined by EPA method 8080A

Analyte	Chemical name	Method detection limit (mg/kg)
Aldrin	1,2,3,4,10,10-hexachloro-1,4,4a,5,8,8a-hexahydro-1,4,5,8-endo-exo-dimethanonaphthalene	0.004
Lindane	Hexachlorocyclohexane	0.004
Chlordane	1,2,4,5,6,7,8,8-octachloro-4,7-methano-3a,4,7,7a-tetrahydroindane	0.014
4,4'-DDD	2,2-Bis (4-chlorophenyl)-2,2,2-trichloroethane	0.011
4,4'-DDE	2,2-Bis (4-chlorophenyl)-1,1-dichloroethylene	0.004
4,4'-DDT	1,1-Bis (4-chlorophenyl)-2,2,2-trichloroethane	0.012
Dieldrin	1,2,3,4,10,10-hexachloro-6,7-epoxy-1,4,4a,5,6,7,8,8a-octahydro-1,4,-endo-exo-5,8-dimethanonaphthalene	0.002
Endosulfan	6,7,8,9,10,10-hexachloro-1,5,5a,6,9,9a-hexahydro-2,9-methano-2,4,3-benzodioxanthiepin-3-oxide	0.014
Endrin	1,2,3,4,10,10-hexachloro-6,7-epoxy-1,4,4a,5,6,7,8,8a-octahydro-1,4-endo-endo-5,8-dimethanonaphthalene	0.006
Heptachlor	1,4,5,6,7,8,8-heptachloro-3a,4,7,7a-tetrahydro-4,7-methanoindene	0.003
Heptachlor epoxide		0.083
Methoxychlor	2.2-Bis(para-methoxyphenyl)-1,1,1-trichloroethane	0.176
Toxaphene	Chlorinated camphene (approximate formula: $C_{10}H_{10}Cl_8$)	0.24
PCB-1242	Polychlorinated biphenyl (technical mixture)	0.065

phthalate contamination is an unwanted interference, it can be totally eliminated by using an electrolytic conductivity detector (ElCD) instead of the ECD.

The GC is calibrated by injection of standards of each component of interest at a minimum of five concentrations in isooctane. These calibration mixtures may be prepared in the lab, and are also commercially available. Internal standard calibration may be used, in which case the analyst must select one or more internal standard compounds which are similar in their analytical behavior of the analytes of interest. This means that they must elute well on the selected GC column, but not interfere with any of the components present in the samples. The internal standard must also not be contained in the sample, and must have a similar volatility to the analytes. There is no single internal standard which is recommended for this method because of these limitations. The analyst must use good chemical judgment here.

In addition, the extraction and clean-up must be monitored for efficiency. This is done by spiking samples, standards, and organic free water blanks with pesticide surrogates. The compounds most often used as surrogates are tetrachloro-*m*-xylene and decachlorobiphenyl. The samples, standards, and blanks are then taken through the extraction and clean-up processes and the recovery of the spike is used to determine if the preparation of the sample yielded an efficiency which was within tolerable limits. Two surrogate compounds are used, because in the event that there is an interference with one, the other can be used.

The samples are extracted with methylene chloride, either in a Soxhlet extractor or in an ultrasonic bath. Because methylene chloride would cause a high background in the ECD or ElCD detectors, a solvent exchange must be performed before analysis. The methylene chloride is evaporated using a Kuderna–Danish concentrator, leaving only about 1 ml. Hexane is then added and the sample is again concentrated to 1 ml. Since the methylene chloride remaining has a lower boiling point than the hexane, it will be distilled off with the excess hexane. The sample is then diluted to a known volume with hexane and is ready for further clean-up, if necessary, or for addition of the internal standard and injection into the GC.

In such trace analyses, there are several ways that analytes can be lost in the analysis, besides the obvious ones of incomplete extraction and losses in sample clean-up. If the injection port is dirty, the carbon deposits, combined with the high injector temperature, can cause some of the pesticides to break down before they enter the column. DDT and endrin are especially susceptible to such breakdown. The problem will be detected if a standard containing only these two pesticides is injected and their degradation products, DDD, DDE, endrin ketone, and endrin aldehyde are detected in the chromatogram. The relative extent of breakdown can be determined by dividing the sum of the breakdown product peaks by the sum of the parent compound peak and its breakdown peaks.

Example

A pure DDT standard is injected and the following peaks are noted:

Peak	Area
DDT	23 487
DDD	663
DDE	8 482

What is the percent degradation of DDT in this system?
The total area due to DDT and its breakdown products is

$$23\ 487 + 663 + 8482 = 32\ 632$$

while the area due to breakdown products is

$$663 + 8482 = 9145$$

Therefore, the percent degraded can be expressed as:

$$\frac{9145}{32\ 632} \times 100 = 28.0\%$$

Since this is greater then 20%, the system should be disassembled and cleaned before samples are run.

Pesticides are not always single compounds. For example, chlordane is a technical mixture containing more than ten major components and about 30 minor ones. PCBs are even more complex because there are many isomers, and these were marketed as different blends of these isomers, in formulations designed to have different physical properties. For example, Figure 9.5 shows capillary chromatograms of different commercial PCB formulations, sold under the Arochlor trade name. One can see that the distribution of the separate isomers is quite different, and the overall mixture volatility, as shown by retention times, also differs. A chromatogram of chlordane, shown in the chapter on water analysis, also showed a large number of components. The largest peaks can be used for quantitation, but the analyst must be aware that the composition of technical grade formulations are usually not completely known, and change from batch to batch.

Adding to the complexity of the problem are the differences in the breakdown products which will be produced depending on the history of the sample. Animals, plants, air, and sunlight all produce breakdown products differently. For quantitation, the analyst must judge which standard best matches the sample and use that for quantitation. In some cases, the standard will be a technical mixture with a peak pattern similar to that of the sample, while in other cases single chlordane components, several of which are commercially available, will be used to quantitate individual peaks in the sample. PCB samples are done in a similar way. Technical mixtures, which were marketed for many years, are used to produce chromatograms and the one which best matches the sample is used

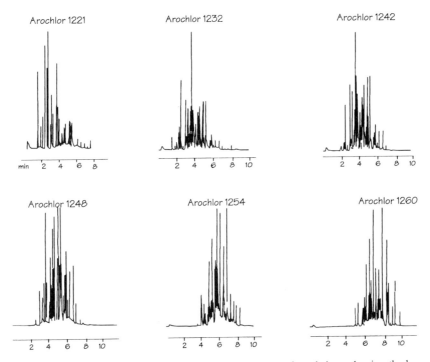

Figure 9.5 Chromatograms of a series of technical grade PCB formulations, showing the large differences among them.

for quantitation. Since these are multi-peak chromatograms, the total area under the PCB "hump" is used for quantitation.

Since these compounds are present at very low levels, GC/MS may not always provide sufficient sensitivity to use for confirmation of peak identities. The injection of the samples into two different GC columns, and comparison of the areas found for each corresponding peak is a good method for confirmation.

9.10 Leaching tests for wastes

Some wastes pose a hazard because toxic substances may leach from them when they are exposed to rainwater in a landfill. If a waste is deposited in a landfill, the leachates from the landfill can eventually contaminate the ground water. In the early 1980s, tests were designed to monitor the leachability of heavy metals. Recently, the US EPA has published protocols referred to as Toxicity Characteristics of Leaching Procedure (TCLP) for studying the leachability of metals as well as organic components of waste materials.

For very dilute wastes which contain less than 0.5% solids, the waste is filtered through a 0.6–0.6 μm glass fiber filter and is defined as the TCLP

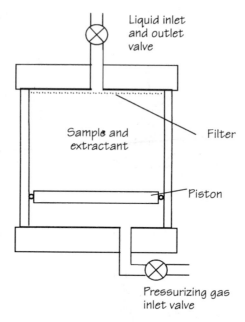

Figure 9.6 Vessel for performing leaching tests on samples which contain volatiles. The sample is put in the cell, leaching solution added, and the head space removed by moving the piston up. After extraction, the leachate is removed for sampling by raising the piston and forcing leachate out the top valve.

extract. For wastes which contain above 0.5% solids, the liquid phase is separated from the solid phase. The solids may have to be ground to reduce their particle size. The solid phase is extracted with a weakly acidic solution such as acetic acid at a pH between 2.88 and 5. The extract is then analyzed for metals using AA or ICP. Similarly, the extract is analyzed for volatile organics using purge and trap, and for semivolatile organics using an extraction procedure followed by GC or GC/MS.

When volatiles may be present, the extraction is carried out in a zero head space extraction vessel which precludes the exposure of the extract to the atmosphere, thus eliminating loss of volatiles. A schematic diagram of such a vessel is shown in Figure 9.6. The vessel has an air-tight piston at the bottom and a filter at the top. At the end of the extraction, the extract can be pushed through the outlet valve by applying gas pressure behind the piston, without opening the vessel.

Table 9.3 gives some of the compounds which are regulated by TCLP tests, along with the levels of pollutant in the extract which would trigger classification of the material as a hazardous waste. The material then would not be allowed in an ordinary landfill, but would have to be placed in a disposal facility which did not allow the leachate to flow into the groundwater.

Table 9.3 Some TCLP contaminants

Metals	Regulatory level (mg/l)	Pesticides	Regulatory level (mg/l)	Other organics	Regulatory level (mg/l)
Arsenic	5	Chlordane	0.03	Benzene	0.07
Barium	100	2,4-D	1.4	Chloroform	0.07
Cadmium	1	Endrin	0.003	Cresol	10
Chromium	5	Lindane	0.06	1,4-Dichlorobenzene	10.8
Mercury	0.2	2,4,5-TP (Silvex)	0.14	Pentachlorophenol	3.6
Lead	5	Heptachlor	0.001	Trichloroethylene	0.07
Silver	5	Methoxychlor	1.4	Toluene	14.4
				Vinyl chloride	0.05

9.11 Immunoassay tests

A method for rapid screening for some compounds, PCB, and petroleum hydrocarbons, for instance, is immunoassay. These tests use the very specific interactions between compounds and **antibodies** which are produced when a living organism is exposed to the specific chemical. Each antibody is quite specific to the target substance, called the **antigen**. Antibodies may be **monoclonal**, which means that they target a specific antigen, or **polyclonal**, which bind to a class of antigens. To monitor the reaction between the sample antigen and the reagent antibody, the antibody or the antigen is usually tagged with a chemical label which allows easy detection. Early applications of immunoassay involved radioactive tags. Radioimmunoassays were simple in concept and provided great sensitivity and selectivity. However, the use of radioisotopes has inherent risks, and the reagents have a short shelf life. Fluorescent tags are also used, but these are limited by the presence of materials in the sample which may quench fluorescence. Enzymes have proven to be more useful labels. These are conveniently assayed by incubating them with a substrate, producing a colored product.

Non-competitive immunoassay is easiest to understand. A tagged antibody is mixed in excess with the antigen in the sample. The complex formed between the antibody and antigen is insoluble and is removed by centrifugation. The remaining tagged antibody in solution is monitored. The amount of the antibody removed during the test is related to the amount of analyte. However, this system is not readily automated, nor is it suitable for field use.

Competitive immunoassay is much more frequently applied. In this system, some labeled antigen is mixed with the antibody, along with the sample. The total antigen (that added and that from the sample) is in excess, so the labeled and unlabeled molecules compete for binding to the antibody. The amount of tagged antigen in the precipitated complex is measured and related to the total amount of sample antigen present. To eliminate the centrifugation or other separation step, environmental immunoassays are usually performed using a kit in which the antibody is immobilized on a surface. The test, which can be used on an extract of the soil to be tested, or on a water sample, generally consists of a test tube or a plastic strip coated with the antibody. A reagent containing the enzyme-tagged antigen and the sample are mixed and allowed to react with the antigen coating for a specific time. The solution is then removed and the antibody-coated surface is rinsed. A substrate which will react with the bound enzyme is added and the color developed by reaction between the bound enzyme and the substrate is compared with a color chart.

The test works by competition between the reagent and the analyte compound from the sample. The concentration of the reagent is kept constant and the number of binding sites is likewise constant. As more sites are occupied by analyte, fewer will hold the color-forming reagent. Therefore, a more concentrated sample will yield a less intense color in the test. However, competition between reagent and analyte is necessary for the test, so the range of concentrations over which good results can be obtained is limited. Some samples may require dilution before running. Frequently, several dilutions of a sample will be run simultaneously, to ensure that at least one will give useable results. Figure 9.7 shows how these test kits work.

Since these kits use biological compounds, there are a number of problems which may occur in their application. The reagents have a limited shelf life and can be denatured at high temperatures. They are also pH and temperature sensitive, and the timing of the various steps of washing and incubation need to be optimized and controlled.

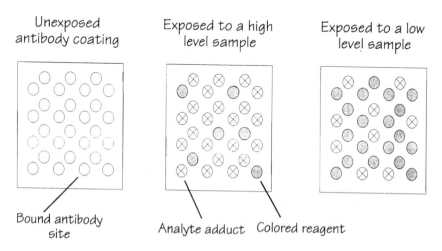

Figure 9.7 Immunoassay kit; as analyte binds to the specific antibody sites, represented by circles on the diagram, the sites become unavailable for the reagent to bind. The color developed is proportional to the amount of reagent bound, and therefore is inversely related to the amount of analyte in the sample, within the range of the test.

9.12 Field methods for solids

Many of the instruments used for water analysis can be adapted to use in soil and solids analysis. If soil is suspended in water or a potassium chloride solution, a pH meter can be used to determine soil acidity. Other simple extractions of this type can be used for immunoassay measurements of pesticides or petroleum hydrocarbons. However, solid samples are inherently harder to analyze without prior sample preparation, and fewer reliable field methods are available.

Suggested reading

Bohn, H.L., McNeil, B.L. and O'Connor, G.A. (1985) *Soil Chemistry*, 2nd edition, John Wiley & Sons, New York.

Boleman, H.J.T. and Burn, J. (1993) *Chemistry and Analysis of Volatile Organic Compounds in the Environment*, Chapman and Hall, New York.

Carter, M.R. (1993) *Soil Sampling and Methods of Analysis*, Lewis Publisher, Ann Arbor, MI.

Dragun, J. (1988) *Soil Chemistry of Hazardous Materials*, Hazardous Materials Control Research Institute, Silver Spring, MD.

Study questions

1. Tree leaves are collected from town streets and are composted. The resulting leaf mold is used in gardens to improve the soil. However, there is a concern that the material may contain an excessive amount of lead from automobile exhaust emissions remaining in the road dust. To test the material a representative sample of the leaf mold is dried and a 2.2453-g sample is weighed out. This sample is digested in acid and hydrogen peroxide, and the extract is diluted to a final volume of 100 ml. A 10-ppm lead standard is prepared and the two solutions are run on a flame atomic absorbance spectrometer. The results are:

   - Absorbance of standard $= 0.746$
   - Absorbance of sample extract $= 0.623$
   - Absorbance of blank $= 0.012$

   What is the concentration of lead in the leaf mold?

2. It is suggested that it would require less sample preparation to do the analysis of leaf mold by graphite furnace AA, just by putting a weighed sample of the leaf mold into the furnace and running the analysis. What problems can you foresee if the analysis was done this way? If you were the analyst, how would you test this method to be sure that it was giving correct results?

3. Think of some applications where analysis of a head space sample would be useful.

4. A sample of wet sludge is analyzed and found to contain 20 μg of Cd/g. A 3.457-g sample of the same sludge is dried and reweighed. The dry sample weighs 1.778 g. What is the Cd concentration on a dry sample basis?

5. What are some of the practical uses for SPME?

6. For a DDT analysis, a dry soil sample is spiked with decachlorobiphenyl (DCB). Then the sample is extracted with methylene chloride in an ultrasonic bath. The solvent is evaporated and the sample taken up in hexane. It is further cleaned up by passing it through a column, and the extracted DDT and DCB internal standard are finally obtained in a hexane solution. In a GC/MS

analysis, the total ion current for the two compounds is monitored, and the total mass spectrum of the DDT and DCB peaks are checked to ensure that they are not contaminated by coeluting substances. From the following data determine the amount of DDT in the sample in $\mu g/g$.

Weight of dry soil sample	1.2345 g
Weight of DCB added as I.S. to soil sample	3.4 μg
Volume of final hexane extract solution	1 ml
Standard solution contains:	5 $\mu g/ml$ DDT
	5 $\mu g/ml$ DCB
Volume of standard injected into GC/MS	2 μl
Volume of sample extract injected into GC/MS	2 μl
Total ion current peak area for DDT in standard	5221
Total ion current peak area for DCB in standard	6227
Total ion current peak area for DDT in sample	4411
Total ion current peak area for DCB in sample	3894

7. How would you carry out the following analyses? Describe the necessary sample preparation steps and your choice of appropriate analytical instrumentation:
   a. Simultaneous determination of Pb, Zn, Cr, and Cd in crops grown in soil treated with sewage sludge.
   b. Determination of pesticide residues in apples
   c. Determination of Pb in human plasma.
   d. Determination of SO_4^{-2}, NO_3^- and PO_4^{-3} in compost to be used for lawn fertilizer.
   e. Screening for gasoline contamination around underground storage tanks.
   f. A simple field test for detection of PCBs at a contaminated land site.
8. Explain how an immunoassay test kit works. What are its advantages over conventional analytical methods?

10 Quality assurance and quality control

An environmental measurement involves planning, sampling, analysis and finally reporting. Systematic as well as random errors are encountered at each step and the purpose of quality assurance (QA) and quality control (QC) is to identify, measure, and keep these errors under control. QA/QC measures are necessary during field sampling as well as in laboratory procedures.

Quality assurance refers to the activities which allow us to demonstrate that a certain quality standard has been met at a stated confidence level. Quality control refers to procedures that lead to statistical control of the measurement process and provide the desired accuracy of the measurement. So, QC consists of specific technical procedures such as the running of blanks or spiked samples to assess and control the measurement process, and QA refers to the management process which implements and documents effective QC.

Some basic requirements of a QC system are shown in Figure 10.1. Competent personnel and suitable facilities are the most essential requirements of QC. These factors are becoming increasingly important as more sophisticated instruments are employed in environmental measurements. Good laboratory practice (GLP) refers to the general practices and procedures involved in running a laboratory, such as efficient sample handling and management, record keeping, and facility maintenance. GLPs are independent of analytical techniques, while good measurement practices (GMP) refer to the specific analytical techniques being used. A good laboratory will have formally documented GLPs and GMPs, which are carefully followed.

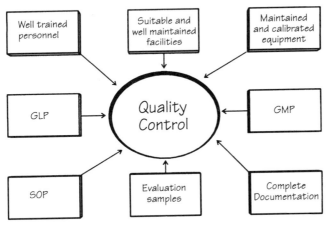

Figure 10.1 Components of a quality control program.

Standard operating procedures (SOPs) are written descriptions for specific procedures. SOPs are useful for repetitive operations such as calibration and sample preparation. Precision can be significantly improved by following SOPs strictly. SOPs may be published standard methods obtained from the literature, or they may be developed in house. Most methods taken from the literature will require additional detail to be formalized, to be a proper SOP. Guidance in developing SOP protocols can be obtained from such organizations as the American Society for Testing and Materials, The United Kingdom Accreditation Service, and the United States Environmental Protection Agency. Last but not the least, proper documentation is very important. This should comprise a written record of every aspect of the measurement process. The sample information, analytical methodology, calibration procedure, instrument outputs, and data analysis, should all be readily available for each sample. The routine QC procedures, such as calibrations, blanks, and spikes are part of the record for each sample done during that day. It should be possible to find the calibration and blank data which apply to each sample. Good documentation is vital in case the measurement must be repeated in the future, or if the data have to be defended in a court of law. Poorly documented laboratory work is simply a waste of time and money.

The major parameters which should be known and kept in control in order to generate quality analytical data are presented in Table 10.1. The first is the **accuracy of the determination**. Accuracy is determined by the analysis of evaluation samples. The objective is to prove that the methods used are regularly producing correct values by running samples which have known concentrations in a similar matrix to that of the samples. Then the **precision** of the measurement must be defined and checked to be sure it is remaining stable. This is done by measuring replicate samples. Possible **contamination of samples** also must be controlled. The running of various types of blanks ensures that each step of the analysis is free of significant contamination. Finally, extraction steps must be checked for extraction efficiency. A matrix spike or standard reference material is used to ensure that the analyte is being properly extracted. Each of these areas of control must be checked on a regular basis. The areas most likely to become "out of control" must be checked more frequently. For instance, if a particular sample type has been thoroughly investigated and a set of conditions for extraction has been found to be efficient, it may not be necessary to repeat

Table 10.1 Important procedures in quality control

Measured parameter	Procedure
Accuracy	Analysis of reference materials or samples of known concentration
Precision	Analysis of replicate samples
Extraction efficiency	Analysis of matrix spikes
Contamination	Analysis of blanks

matrix spikes frequently. If, however, the matrix varies from day to day, more frequent analysis of spiked samples may be needed.

10.1 Determination of accuracy and precision

Determination of accuracy and precision is the most important step in generating high quality data. Without an understanding of these, the data are useless. Accuracy is determined by analyzing samples of known concentration and comparing the measured value to the known, true value. Standard reference samples are available from regulatory agencies and from commercial companies. A sample of known concentration may also be made up in the laboratory, for this purpose.

Determination of precision involves making replicate measurements. The precision should be measured at different concentration levels which are within the range expected in the samples. The precision of the analytical method can be determined by splitting a sample into several aliquots and analyzing them to obtain a measure of relative standard deviation. This is also a measure of intra-sample variance. For example, in analysis of VOCs in air using canisters, several samples from the same canister can be analyzed to determine the precision.

Inter-sample variance can be measured by analyzing several samples from the same source. For instance, several canister samples can be collected at a certain site at the same time and under the same conditions. They are all analyzed by the same method and the relative standard deviation calculated.

10.2 Statistical control

Statistical evidence that the precision of the process is within a certain specified limit is referred to as statistical control. Statistical control does not take into account the accuracy or bias of the measurement. However, the precision of the measurement should be established and statistical control be achieved before accuracy can be estimated.

10.2.1 Control charts

Control charts provide a graphical display of statistical control and can also be used to monitor the variability of repetitive measurements. A standard, a reference material, or a sample prepared with a known concentration, is run at specified intervals. The result should be the same each time, as the samples run are replicates. Of course, they will show some variation, because of random error. These results are plotted on a control chart to be sure that the random error is not increasing or that a bias is not creeping into the analysis. One type of control chart is shown in Figure 10.2, where replicate measurements are plotted as a function of time. The center line is the average or expected measurement.

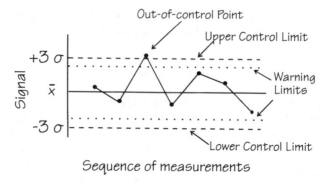

Figure 10.2 Control chart. The upper and lower control lines are usually set at 3 σ from the mean. The warning limits are closer to the mean, and mark the points at which some action is scheduled to be taken to prevent out-of-control occurrences.

Upper (UCL) and lower (LCL) control limits are the values within which the measurement must lie. Normally the control limits are ±3σ, within which 99.7% of the data should lie. For example, the same calibration standard may be run on a gas chromatograph each day, before the day's work begins. The peak area obtained is plotted on the control chart. When the plotted value falls outside the control limit, a specified process of readjustment of the instrument is carried out and the standard is analyzed again.

Control limits are established based on the analyst's judgment and the experimental results. For example, in a manufacturing process, the control limits are often based on product quality. The common procedure is to use the mean of several measurements as the center line and the standard deviation to set the control limits. Control charts are often used to plot the regularly scheduled analysis of a standard reference material, spiked sample, or audit sample. These are then tracked to see if the measurements show a trend or systematic departure from the center line, and to monitor the precision.

Another type of control chart is called the cumulative sum or CUSUM chart. In this method, a running total of the deviation of each control sample from the average is plotted against the sequential number of the measurement. If the control samples show a normal random scatter, the sum will consist of roughly equal numbers of positive and negative deviations from the average, and the sum will remain near the zero line. If, however, a series of errors in the same direction begin to occur, the sum will rapidly deviate from the normal line. This type of control chart will identify a trend more quickly than the control chart described above. To determine the "out of control" limits, a clear plastic overlay is used. The marked point on the overlay is placed on the point corresponding to the last sample. When data points begin to fall outside the limit lines marked on the overlay, the system is deemed out of control. The angle θ between the limit lines and the distance of the mark from the vertex of the two lines are adjusted to give the necessary degree of control. A shorter distance, d, or a smaller value

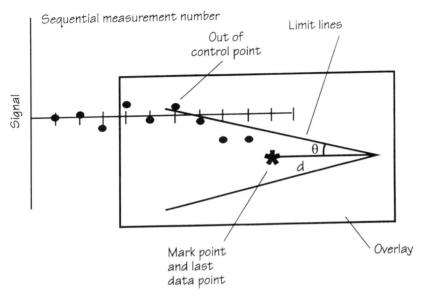

Figure 10.3 Cusum control chart. When the points fall outside the limit lines on the sliding overlay, the process is considered to be out of control and remedial action should be taken.

of θ will trigger an out-of-control situation more quickly when the data varies. Figure 10.3 shows a CUSUM chart.

10.2.2 Control samples

Control samples are of two types, analytical control samples and background samples. Analytical control samples are used for quality control in the laboratory. Background samples are used to determine whether a certain compound is present in the sample under study at a level higher than in the background. Various types of samples are used to determine whether an analytical procedure is under statistical control. Some of the commonly used control standards are given below:

Laboratory control standards (LCS) are certified standards. These are usually obtained from an outside agency and are used to check whether the data being generated is comparable to that produced in other laboratories. These LCS provide a measure of the accuracy of the analysis.

Calibration control standards (CCS) are used to check calibration. The CCS is the first sample analyzed after instrument calibration. It is not necessary to know its concentration, but it is kept for successive comparisons. A CCS may be analyzed after a specified number, perhaps 20, samples have been run. The CCS value may be plotted on a control chart to monitor the statistical control of the analytical procedure.

Matrix control or matrix spike: most environmental samples have a complex matrix, and matrix effects may play an important role in the accuracy and precision of the measurement. **Matrix spiking** is done by adding a known quantity of a component which is similar to the analyte but not originally present in the sample to a portion of the sample. The sample then is analyzed for the presence of the spiked material to evaluate any matrix effects. Matrix spikes are particularly important where the analysis method involves sample extraction. It is important to be certain that the extraction removes most of the analyte. Usually the extraction efficiency, or the matrix spike recovery is required to be at least 70%.

For example, in the analysis of phenols in soil by Soxhlet extraction and GC/MS, deuterated phenol may be used as the matrix spike. The deuterated compound, of course, will not be present in the original sample and it can be easily identified by GC/MS in the same analysis as the target compound. It also has chemical and physical properties which are identical to those of the analyte. This is an ideal case. More often, some compromise must be made in choosing a matrix spike. For example, cleaned soil could be spiked with phenol and extraction efficiency of the method determined. It should be noted that as a matrix, cleaned soil may be significantly different from a real soil.

It should be kept in mind that when a matrix spike is added to soil the spiked compounds may be somewhat loosely adsorbed on the surface. If a PCB solution is added to a soil, it may be fairly easy to extract. In a real contaminated soil, which may have held the PCBs for years, the sorption may be much stronger. Consequently, a matrix spike may be extracted more easily than a pollutant in an actual sample. The extraction spike might produce quantitative recovery, whereas the extraction efficiency for real samples under the same conditions may be quite low. This is especially true for extraction techniques which do not use strong solvents, for example supercritical extraction.

Surrogate spike: surrogate spikes are used in organic analysis. They are compounds that are similar in chemical composition, extraction efficiency and chromatographic behavior to the analytes of interest. When available, a deuterated analog of the analyte is ideal as a surrogate in GC/MS analysis because such a compound behaves exactly as the analyte, and will not be originally present in the sample. The surrogate spike is added to the sample, standard, blanks, and matrix spike. The surrogate spike recovery indicates whether there was a problem with a particular run. For example, consider a set of samples to be analyzed by purge and trap. Deuterated toluene is added as surrogate to all the samples, standards, and the blanks. The recovery of the deuterated toluene in each of the analyses is checked. If it is found that the recovery in a certain sample was unusually low, then that particular analysis must be repeated.

10.2.3 *Background or control site samples*

Many environmental studies are done to prove or disprove a hypothesis. In order to judge whether the concentration of pollutant at a certain site is unusual, it is important to compare it to another area or sample source, referred to as the control site. Often an area considered to be uncontaminated is used as a background site. For example, to determine whether an industrial waste water discharge is significantly increasing the concentration of a particular pollutant in the river, the concentration downstream from the discharge point may be compared to the upstream concentration. The upstream measurement point serves as the control site. This is an example of a **local control site**, which is close to the sample in space and time. An **area control site** is not adjacent to the sampling point but is in the same general area, e.g., same city or county. For example, to determine whether smelters in Dallas, TX, raised the blood Pb level in children living nearby, a control site was chosen in the city that had a similar water supply, ethnic make up, and traffic patterns as the area near the smelter.

10.3 Performance evaluation samples

Evaluation samples are materials which are used to test the measurement process and to be sure that it gives accurate results. These samples are used both in individual laboratories and in collaborative tests between laboratories to test the reliability of a method. The matrix as well as the concentration of an evaluation sample should be similar to the real-world samples for which this method will be used. The certification of an evaluation sample is based on a detailed and careful analysis, usually using several independent analytical methods. In some cases, the materials are specifically prepared to have an accurately known composition. The use of such synthetic evaluation samples should be examined closely to be sure the matrices are really comparable.

Standard reference materials (SRM) are certified standards available from National Institute of Standard and Testing (NIST) in the US. NIST provides a variety of solid, liquid and gaseous SRMs which have been prepared to be stable and homogeneous and which are then analyzed by three independent methods for a number of components. The accurate concentration for each of the analytes is certified. Certified standards are also available from the European Union Community Bureau of Reference (BCR), the Laboratory of the Government Chemist in the UK, from the USEPA, and from various commercial sources which supply standards. These materials are usually expensive, and often samples are prepared in the laboratory, compared to the certified standards, and then used as secondary reference materials for daily use.

10.3.1 Preparation of evaluation samples

In the absence of certified standards, evaluation samples can be made up in the laboratory. These are made by adding the analyte of interest to a suitable matrix which closely resembles the real environmental samples. Spiking may easily be done to liquid or solid samples. However, spikes do not always simulate the actual sample, as we saw in matrix spiking. It may be difficult to obtain quantitative extraction of an analyte strongly bound in a real soil sample, while a soil spiked with a solution of the analyte may give it up easily. It may be necessary to determine experimentally that the spike recovery is similar to that for a real sample.

Uniform mixing is very important in preparing spiked samples. This is relatively simple to ensure in gaseous or liquid samples, but can be difficult in solid samples. Adsorption of trace analyte on the container used to hold the standard can be serious in gaseous samples containing polar compounds and for low concentrations in liquid or gas standards. During the preparation of liquid samples, immiscible liquids may form separate layers or unstable emulsions, which may later separate.

Solids can be spiked either with liquid standards or with other solids. Often a solution of the analyte is added to the solid matrix and the solvent is allowed to evaporate, but this can leave a residue of analyte at a local spot, so proper mixing is necessary to ensure homogeneity. It is better to use a fairly dilute solution for spiking, so that the liquid can be more thoroughly dispersed before it evaporates. If two solids are to be mixed, they should have similar particle size, and the materials should finely ground. It takes hours of mixing in a tumbler or mill to prepare a truly homogeneous solid mixture. If the two solids being mixed have significantly different densities or particle sizes, they will also tend to stratify or unmix when subjected to mild vibration or just everyday handling.

Even gases, especially when they are under pressure in a cylinder, need sufficient time to mix. A newly prepared gas standard may show variable concentrations for several days. Cylinders may be rolled to aid mixing, or the bottoms may be placed in a warm water bath, to induce convection inside the cylinder.

10.3.2 Stability of evaluation samples

The concentration of the analyte in an evaluation sample may change due to physical processes such as adsorption, evaporation, settling, or chemical reaction. So, if a sample is to be stored for any length of time, it is important to know its stability under the storage conditions. Certified standards are given an expiration date by the producer, which should not be ignored.

10.4 Contamination control

Environmental measurement processes are prone to contamination problems. Analytes in many environmental samples are present at trace levels and contamination can be a significant source of error. Contamination can occur at many points in the sampling and analysis. It can occur in the field during sample collection, during transportation, storage, or in the sample workup prior to

Table 10.2 Potential sources of sample contamination

Steps in the analytical process	Contamination sources
Sample collection	Equipment Sample handling steps such as compositing, or filtering Sample preserving additives Sample containers Ambient contamination
Sample transport and storage	Sample containers Cross contamination from reagents or other samples
Sample preparation	Sample handling Dilutions Glassware Ambient contamination
Sample analysis	Instrument memory effects or carry-over Reagents Syringes Glassware, and apparatus

measurement. Some of the more common sources of contamination are listed in Table 10.2.

Sampling devices themselves can be a source of contamination. Contamination may be from the material of construction, from improper cleaning of the sampler, or from carryover from a previous sample. For example, polymer additives can leach out of plastic sample bottles, and organic solvents can dissolve materials from cap liners of sample vials. Adsorbents used in air sampling need to be cleaned prior to sampling because they tend to sorb components from the air during storage. Carryover from previous samples is also possible, if traps are recycled.

In the laboratory, contamination can occur during sample handling at almost any stage of sample preparation and analysis. Contamination can occur from containers and reagents used in the laboratory or from the ambient environment. A metal spatula introduced into a soil sample may leave detectable traces of metal in the sample. Contamination sources are also present during the analytical procedure. A common problem is sample carryover or memory effect, in which the sample components from a previous analysis may be retained in the instrument and show up during the next run.

10.4.1 *Blanks*

Blanks are used to assess the degree of contamination in analytical measurements. They may also be used to correct for relatively constant, unavoidable contamination. A blank is a sample containing a negligible amount of the analyte of interest. Usually blanks are made to simulate the sample matrix as closely as possible. Different types of blanks are used depending upon the procedure and the measurement objective. Some commonly used blanks are listed in Table 10.3. Blanks from the laboratory and field are required to identify all the possible sources of contamination.

Table 10.3 Types of blanks

Blank type	Purpose	Process
System or instrument blank	Establishes the baseline of an analytical instrument, in the absence of sample	Determine the background signal with no sample present
Solvent or calibration blank	To measure the amount of the analytical signal which arises from the dilution solvent. The zero solution in the calibration series	Analytical instrument is run with dilution solvent only
Method blank	To detect contamination from reagents, sample handling, and the entire analytical process	A simulated sample containing no analyte is taken through entire analytical procedure
Matched-matrix blank	To detect contamination from field handling, transportation, or storage	A synthetic sample which matches the basic matrix of the sample is carried to the field and is treated in the same fashion as the samples
Sampling media or trip blank	To detect contamination in sampling media such as filters and sample adsorbent traps	Analyze samples of unused filters or traps to detect contaminated batches
Equipment blank	To determine contamination of equipment and assess the efficiency or equipment clean-up procedures	Samples of final equipment cleaning rinses are analyzed for contaminants

A **system** or **instrument blank** is a measure of instrument background, and is the instrument response in absence of any sample. It is a measure of system contamination or background levels. When the background signal is constant and measurable, the usual practice is to consider that level to be the zero setting. For instance, most spectrometer detectors give a finite electrical signal even when no light is falling on them. This "dark current" is then set to correspond to "no light being transmitted through the sample" essentially subtracting the dark current from whatever readings are taken on samples.

The instrument blank also identifies memory effects or carry over from previous samples. Memory effects may become significant when a low concentration sample is analyzed immediately after a high concentration sample or standard. This is especially true for chromatographic apparatus such as purge and trap or thermal desorber instruments. For example, during the analysis of volatile organics by purge and trap, some components may be carried over in the sorbent trap or on any cold spots in the instrument. It is a common practice to run a deionized water blank between actual samples, to show up such contaminations. These blanks are critical in chromatographic work, where sample components may easily be left in the instrument, to emerge in the next sample. Purge and trap and thermal desorption systems are especially prone to this sort of problem, and running a blank after a particularly concentrated sample is a good practice.

A **solvent blank** checks solvents which are used to dilute the sample in the analysis. Sometimes, a blank correction or zero setting is done based on the reagent measurement. For example, in atomic or molecular spectroscopy, the solvent or reagent blank is often used to provide a zero reading. These blanks measure any absorbance due to the solvent as well as such effects as reflectance from the surfaces of the sample cuvet used in some spectrometers. The sample measurements are then corrected by using this point as the instrument response for 100% transmittance.

A **method blank** is carried through all the steps of sample preparation and analysis as if it were an actual sample. The same solvents and reagents that are used with the actual sample are used here. For example, for the analysis of metals in soil, a method blank may be a clean soil sample which is put through all the extraction, concentration and analysis procedures as if it were a real soil sample. The method blank accounts for contamination which may occur during the sample preparation and analysis procedures, arising from reagents, glassware, or the laboratory environment.

Trip blanks provide information about contamination introduced during sample collection, transport, and storage. Trip blanks are carried through all the steps of collection, handling, and storage in the same way as the actual sample. While it may not be possible to simulate every step, these should be followed as far as possible. For example, during collection of water samples, a sample vial containing deionized water is taken to the field. The cap is opened momentarily and then closed to expose the blank to the ambient air. If a preservative such as acid is added to the samples in the field, this is added to the blank, too. Then the blank is shipped back to the laboratory and stored along with the actual samples. Finally, it is analyzed with the batch of samples it has accompanied. This blank is intended to show up any contamination that may have been absorbed from the ambient air or any cross-contamination that might have occurred during storage and handling.

During an air sampling trip, an unexposed sorbent cartridge similar to the those being used is carried to the field, opened as if it is being installed in the sample collection apparatus, and then repackaged for return to the lab as the samples are. This is then kept with the actual samples all the way through the analysis process. Analysis of this trap will provide information on contamination which happens in transit or in storage before analysis.

A **sampling media blank** is appropriate where filters and sorbent traps are used. A sampling filter or trap is randomly selected before the batch is sent to the field. The blank is analyzed like a sample, in order to find contaminations before the traps or filters are used in the field. This prevents the waste of field samples, which may not be replaceable, by identifying a common source of contamination before the samples are taken. This kind of contamination would be identified in the trip blank, but too late to recover the samples.

Other types of blanks may be employed as the measurement process demands. It should be noted that blanks are effective in identifying contamination, but they do not account for every possible error, such as those due to processes such as volatilization or chemical reaction. Blanks are an important control on the sampling and analytical process. They may be used to correct for contamination if a relatively small, fairly constant contamination is found. More often a blank above a predetermined value is used to reject analytical data, making reanalysis or even resampling necessary. The laboratory SOPs, of course, should spell out what blanks are to be done and how the results are to be treated.

10.5 Quality assurance for the example determinations

In the discussions of specific analyses of air, water, and solid samples, several Example Determinations were given. For some of these examples, it would be instructive to look at the QA and QC procedures which would be associated with these determinations.

10.5.1 Determination of VOC in air by Tenax trapping and thermal desorption

There are several quality assurance procedures which must be done to ensure that reliable data are generated by this method. The traps, first of all, must be checked for contamination before use. Usually they are cleaned on a manifold, with several traps being cleaned at once. One of these traps can be selected at random for checking for cleanliness. This is done by desorbing the trap and analyzing the organics removed. In addition, the pump used to draw air through the trap must be calibrated. The flow through a typical trap is measured using a soap film flowmeter, or a gas test meter, to determine either the flow or the volume drawn through the trap in a given time. Enough measurements must be made to determine not only the flow but the standard deviation of the flow over time and with different traps. The traps also must be checked to ensure that the sample does not break through the trap. A second trap can be installed behind the sample trap. The air sample passes through the two traps before reaching the pump. The back trap is analyzed in the same way as the front one, to determine if any of the analytes have escaped through the front trap. If more than 10% of the analyte is found in the back trap, the sample is considered dubious.

Then the GC must be calibrated. Often a high pressure cylinder containing a gas standard mixture is used for this purpose. The standard is injected using a gas sampling valve. Different volumes or different pressures of standard within the same volume loop can be used for calibration.

A field blank will be performed to ensure that the samples were not contaminated in transit to or from the sampling site. One or more traps are taken to the field, but are not used to take samples. The traps are unpacked, handled as the sample traps, and returned to their sealed containers. These are returned to the laboratory for analysis along with the samples. The blanks in this case are not usually subtracted from the sample analytes, but instead are used to accept or reject a batch of samples. A seriously contaminated blank casts doubt on the entire sampling run.

Finally, the sample peaks must be unequivocally identified. A spiked sample can be helpful for this. A small amount of the standard gas is injected into a duplicate sample, and the peaks which increase their height can be identified.

10.5.2 *Determination of haloacetic acid and trichlorophenol in water*

Chances of error in such an analysis can be quite high because the sample preparation requires several steps. To ensure the quality of the data, it is necessary to be able to show that the analysis is achieving the desired accuracy and precision. Then the purpose of the study must be defined and sufficient samples must be done to ensure that this purpose is achieved. Some of the important QC procedures are the determination of the analyte recovery, the calibration of the method, surrogate or internal standard recovery, and blanks.

Analyte recovery: to vials containing reagent water, appropriate amounts of stock standards containing the different acids are added. Several samples at different concentrations should be tried. The extraction, derivatization and analysis are carried out on each of these. The amount found in the analysis should be greater than 90% of that added, and the relative standard deviation should be less than 5%.

Calibration: quantitation is done using an internal standard to eliminate the effect of variation in injection volume. A measured portion of the internal standard is added to each sample.

The calibration curve is prepared from standard solutions of a minimum of five different concentrations. Calibration standards are analyzed with each set of samples. If the standards show results which differ from those obtained previously by more than 15 or 20%, degradation may have occurred. A new set of standards should be prepared, and the method recalibrated. Moreover, every day, a standard solution of esters corresponding to the haloacetic acids present in the samples is run on the GC to check retention time, and sensitivity.

Surrogate recovery: a measured volume of a solution of a surrogate or internal standard such as 2,3-dibromopropionic acid of appropriate concentration is added to all samples before acidification and derivatization. If the surrogate peak is low or absent, the extraction may have been done at the wrong pH or water in the extract may have interfered with the derivatization. The peak is also used to correct for small volume differences in the GC injection.

Blanks: with each batch of samples, **method blanks** consisting of clean reagent water have to be run. If the blank produces any peaks that interfere with the analytes, the source of contamination should be eliminated. **Instrument** and **reagent blanks** should be analyzed daily. Other blanks may be necessary. **Field** or **shipping blanks** are sample vials containing reagent water and NH_4Cl which are sent to the sampling location with the sample bottles. These are analyzed on return to the lab to identify any contamination during sampling and storage. If this blank shows some contamination, it should be compared to a laboratory reagent blank. If the reagent blank is clean, all the samples are doubtful and resampling will probably be necessary.

10.5.3 *Determination of metals in dry sludge*

The instrument must be properly set up for operation. ICPs differ in their operating procedures and features. The manufacturer's guidelines should be followed to establish detection limits, precision, and background correction settings. For an ICP, alignment of the plasma torch and spectrometer entrance slit is necessary.

Method quality control (or spike recovery): a sludge standard reference material or a similar certified standard from an outside agency may be used. It will be carried through all the preparation steps just as the sample. Good results will be within 5% of the certified value. A lower recovery indicates inefficient extraction or other bias in the system. If a certified material is not available, an analyzed sludge or similar matrix may be spiked with a known quantity of the target metals and the extraction efficiency determined.

Precision: replicate sludge samples are measured to determine the precision.

Blanks: a **method blank** is made by taking all the reagents, but not the sample, through all the steps of sample preparation and analysis. Any contamination during the process will be detected. A method blank is necessary for each batch of samples.

Before analysis, a **calibration blank** and a method blank should be run. The calibration blank is a dilute solution of HNO_3 and HCl. These blanks check for contamination in the instrument or during sample preparation prior to analysis. It is a good practice to analyze the calibration blank between samples. The system should be rinsed with dilute acid for about 60 s between each sample and calibration blank. The analysis of this blank ensures that there is no carryover between samples. If carryover is observed, the system is rinsed with acid until a clean blank is observed. If the sample concentrations are too high, it is diluted to bring it within linear range. Field blanks and other blanks are carried out as necessary.

Spectral interference may be serious in some samples. Instruments usually have computer software that corrects for possible spectral interference. Manual methods are also available that can be used to correct for this type of interference.

Instrument quality control: a standard solution of known composition and concentration is run after every 10 samples. The results of this analysis should be within ±5% of expected value, or corrective action is necessary.

Suggested reading

Csuros, M. (1994) *Environmental Sampling and Analysis for Technicians*, Lewis Publishers, Boca Raton.

Funk, W., Dammann, V. and Donnevert, G. (1995) *Quality Assurance in Analytical Chemistry*, VCH Publishers, New York.

Taylor, J.K. (1988) *Quality Assurance of Chemical Measurements*, Lewis Publishers, Boca Raton.

Study questions

1. What are the requirements for a standard operating procedure (SOP)?
2. What are some of the considerations used in establishing control limits on a control chart?
3. Discuss the importance of selecting proper background or control sites for sampling.
4. Discuss how the use of one type of blank may eliminate the need to do some other type of blank.
5. What purposes do spiked matrix samples serve?
6. What are some of the QC steps necessary for the following measurements?
   a. Purge and trap analysis of VOCs in water.
   b. Determination of heavy metals in drinking water
   c. Measurement of dissolved Pb in drinking water.
   d. Sampling and analysis of solvent vapors in your laboratory using sorbent cartridges and thermal desorber.
   e. Sampling and analysis of anions in air particulates by ion chromatography.
   f. Collection and analysis of air samples in canisters for VOC determination.
   g. Determination of pesticides in soil.

Appendix 1 Percentiles of the *t*-distribution

Confidence level of 2-sided interval *df*	20 *t* 0.60	40 *t* 0.70	60 *t* 0.80	80 *t* 0.90	90 *t* 0.95	95 *t* 0.975	98 *t* 0.99	99 *t* 0.995
1	0.325	0.727	1.376	3.078	6.314	12.706	31.821	63.657
2	0.289	0.617	1.061	1.886	2.920	4.303	6.965	9.925
3	0.277	0.584	0.978	1.638	2.353	3.182	4.541	5.841
4	0.271	0.569	0.941	1.533	2.132	2.776	3.747	4.604
5	0.267	0.559	0.920	1.476	2.015	2.571	3.365	4.032
6	0.265	0.553	0.906	1.440	1.943	2.447	3.143	3.707
7	0.263	0.549	0.896	1.415	1.895	2.365	2.998	3.499
8	0.262	0.546	0.889	1.397	1.860	2.306	2.896	3.355
9	0.261	0.543	0.883	1.383	1.833	2.262	2.821	3.250
10	0.260	0.542	0.879	1.372	1.812	2.228	2.764	3.169
11	0.260	0.540	0.876	1.363	1.796	2.201	2.718	3.106
12	0.259	0.539	0.873	1.356	1.782	2.179	2.681	3.055
13	0.259	0.538	0.870	1.350	1.771	2.160	2.650	3.012
14	0.258	0.537	0.868	1.345	1.761	2.145	2.624	2.977
15	0.258	0.536	0.866	1.341	1.753	2.131	2.602	2.947
16	0.258	0.535	0.865	1.337	1.746	2.120	2.583	2.921
17	0.257	0.534	0.863	1.333	1.740	2.110	2.567	2.898
18	0.257	0.534	0.862	1.330	1.734	2.101	2.552	2.878
19	0.257	0.533	0.861	1.328	1.729	2.093	2.539	2.861
20	0.257	0.533	0.860	1.325	1.725	2.086	2.528	2.845
21	0.257	0.532	0.859	1.323	1.721	2.080	2.518	2.831
22	0.256	0.532	0.858	1.321	1.717	2.074	2.508	2.819
23	0.256	0.532	0.858	1.319	1.714	2.069	2.500	2.807
24	0.256	0.531	0.857	1.318	1.711	2.064	2.492	2.797
25	0.256	0.531	0.856	1.316	1.708	2.060	2.485	2.787
26	0.256	0.531	0.856	1.315	1.706	2.056	2.479	2.779
27	0.256	0.531	0.855	1.314	1.703	2.052	2.473	2.771
28	0.256	0.530	0.855	1.313	1.701	2.048	2.467	2.763
29	0.256	0.530	0.854	1.311	1.699	2.045	2.462	2.756
30	0.256	0.530	0.854	1.310	1.697	2.042	2.457	2.750
40	0.255	0.529	0.851	1.303	1.684	2.021	2.423	2.704
60	0.254	0.527	0.848	1.296	1.671	2.000	2.390	2.660
120	0.254	0.526	0.845	1.289	1.658	1.980	2.358	2.617
∞	0.253	0.524	0.842	1.282	1.645	1.960	2.326	2.576

Appendix 2 Critical values for the f-test at 5% level of significance

n_1 / n_2	1	2	3	4	5	6	7	8	9	10	12	15	20	24	30	40	60	120	∞
1	647.8	799.5	864.2	899.6	921.8	937.1	948.2	956.7	963.3	968.6	976.7	984.9	993.1	997.2	1001.0	1006.0	1010.0	1014.0	1018.0
2	38.51	39.00	39.17	39.25	39.30	39.33	39.36	39.37	39.39	39.40	39.41	39.43	39.45	39.46	39.46	39.47	39.48	39.49	39.50
3	17.44	16.04	15.44	15.10	14.88	14.73	14.62	14.54	14.47	14.42	14.34	14.25	14.17	14.12	14.08	14.04	13.99	13.95	13.90
4	12.22	10.65	9.98	9.60	9.36	9.20	9.07	8.98	8.90	8.84	8.75	8.66	8.56	8.51	8.46	8.41	8.36	8.31	8.26
5	10.01	8.43	7.76	7.39	7.15	6.98	6.85	6.76	6.68	6.62	6.52	6.43	6.33	6.28	6.23	6.18	6.12	6.07	6.02
6	8.81	7.26	6.60	6.23	5.99	5.82	5.70	5.60	5.52	5.46	5.37	5.27	5.17	5.12	5.07	5.01	4.96	4.90	4.85
7	8.07	6.54	5.89	5.52	5.29	5.12	4.99	4.90	4.82	4.76	4.67	4.57	4.47	4.42	4.36	4.31	4.25	4.20	4.14
8	7.57	6.06	5.42	5.05	4.82	4.65	4.53	4.43	4.36	4.30	4.20	4.10	4.00	3.95	3.89	3.84	3.78	3.73	3.67
9	7.21	5.71	5.08	4.72	4.48	4.32	4.20	4.10	4.03	3.96	3.87	3.77	3.67	3.61	3.56	3.51	3.45	3.39	3.33
10	6.94	5.46	4.83	4.47	4.24	4.07	3.95	3.85	3.78	3.72	3.62	3.52	3.42	3.37	3.31	3.26	3.20	3.14	3.08
11	6.72	5.26	4.63	4.28	4.04	3.88	3.76	3.66	3.59	3.53	3.43	3.33	3.23	3.17	3.12	3.06	3.00	2.94	2.88
12	6.55	5.10	4.47	4.12	3.89	3.73	3.61	3.51	3.44	3.37	3.28	3.18	3.07	3.02	2.96	2.91	2.85	2.79	2.72
13	6.41	4.97	4.35	4.00	3.77	3.60	3.48	3.39	3.31	3.25	3.15	3.05	2.95	2.89	2.84	2.78	2.72	2.66	2.60
14	6.30	4.86	4.24	3.89	3.66	3.50	3.38	3.29	3.21	3.15	3.05	2.95	2.84	2.79	2.73	2.67	2.61	2.55	2.49
15	6.20	4.77	4.15	3.80	3.58	3.41	3.29	3.20	3.12	3.06	2.96	2.86	2.76	2.70	2.64	2.59	2.52	2.46	2.40
16	6.12	4.69	4.08	3.73	3.50	3.34	3.22	3.12	3.05	2.99	2.89	2.79	2.68	2.63	2.57	2.51	2.45	2.38	2.32
17	6.04	4.62	4.01	3.66	3.44	3.28	3.16	3.06	2.98	2.92	2.82	2.72	2.62	2.56	2.50	2.44	2.38	2.32	2.25
18	5.98	4.56	3.95	3.61	3.38	3.22	3.10	3.01	2.93	2.87	2.77	2.67	2.56	2.50	2.44	2.38	2.32	2.26	2.19
19	5.92	4.51	3.90	3.56	3.33	3.17	3.05	2.96	2.88	2.82	2.72	2.62	2.51	2.45	2.39	2.33	2.27	2.20	2.13
20	5.87	4.46	3.86	3.51	3.29	3.13	3.01	2.91	2.84	2.77	2.68	2.57	2.46	2.41	2.35	2.29	2.22	2.16	2.09
21	5.83	4.42	3.82	3.48	3.25	3.09	2.97	2.87	2.80	2.73	2.64	2.53	2.42	2.37	2.31	2.25	2.18	2.11	2.04
22	5.79	4.38	3.78	3.44	3.22	3.05	2.93	2.84	2.76	2.70	2.60	2.50	2.39	2.33	2.27	2.21	2.14	2.08	2.00
23	5.75	4.35	3.75	3.41	3.18	3.02	2.90	2.81	2.73	2.67	2.57	2.47	2.36	2.30	2.24	2.18	2.11	2.04	1.97
24	5.72	4.32	3.72	3.38	3.15	2.99	2.87	2.78	2.70	2.64	2.54	2.44	2.33	2.27	2.21	2.15	2.08	2.01	1.94
25	5.69	4.29	3.69	3.35	3.13	2.97	2.85	2.75	2.68	2.61	2.51	2.41	2.30	2.24	2.18	2.12	2.05	1.98	1.91
26	5.66	4.27	3.67	3.33	3.10	2.94	2.82	2.73	2.65	2.59	2.49	2.39	2.28	2.22	2.16	2.09	2.03	1.95	1.88
27	5.63	4.24	3.65	3.31	3.08	2.92	2.80	2.71	2.63	2.57	2.47	2.36	2.25	2.19	2.13	2.07	2.00	1.93	1.85
28	5.61	4.22	3.63	3.29	3.06	2.90	2.78	2.69	2.61	2.55	2.45	2.34	2.23	2.17	2.11	2.05	1.98	1.91	1.83
29	5.59	4.20	3.61	3.27	3.04	2.88	2.76	2.67	2.59	2.53	2.43	2.32	2.21	2.15	2.09	2.03	1.96	1.89	1.81
30	5.57	4.18	3.59	3.25	3.03	2.87	2.75	2.65	2.57	2.51	2.41	2.31	2.20	2.14	2.07	2.01	1.94	1.87	1.79
40	5.42	4.05	3.46	3.13	2.90	2.74	2.62	2.53	2.45	2.39	2.29	2.18	2.07	2.01	1.94	1.88	1.80	1.72	1.64
60	5.29	3.93	3.34	3.01	2.79	2.63	2.51	2.41	2.33	2.27	2.17	2.06	1.94	1.88	1.82	1.74	1.67	1.58	1.48
120	5.15	3.80	3.23	2.89	2.67	2.52	2.39	2.30	2.22	2.16	2.05	1.94	1.82	1.76	1.69	1.61	1.53	1.43	1.31
∞	5.02	3.69	3.12	2.79	2.57	2.41	2.29	2.19	2.11	2.05	1.94	1.83	1.71	1.64	1.57	1.48	1.39	1.27	1.00

n_1 and n_2 are the degrees of freedom of the numerator and denominator respectively.

Index